Productivity Machines

History of Computing
William Aspray and Thomas J. Misa, editors

For a complete list of books in the series, please see the back of the book.

Productivity Machines

German Appropriations of American Technology from Mass Production to Computer Automation

Corinna Schlombs

The MIT Press
Cambridge, Massachusetts
London, England

This book was set in Stone Serif and Stone Sans by Jen Jackowitz.

Library of Congress Cataloging-in-Publication Data
Names: Schlombs, Corinna, author.
Title: Productivity machines : German appropriations of American technology
    from mass production to computer automation / Corinna Schlombs.
Description: Cambridge, MA : MIT Press, [2019] | Series: History of computing
    | Includes bibliographical references and index.
Identifiers: LCCN 2018047659 | ISBN 9780262537391 (pbk. : alk. paper)
Subjects: LCSH: Production engineering--Germany--History. | Technology
    transfer--United States--History. | Industrial
    productivity--Germany--History. | Technical assistance, American.
Classification: LCC TS73 .S35 2019 | DDC 670--dc23 LC record available at https://
    lccn.loc.gov/2018047659

To my parents, Clara and Paul Schlombs

Contents

Acknowledgments   ix

Introduction   1

1   **Measuring Productivity at the US Bureau of Labor Statistics**   13

2   **Weimar Rationalization Appropriates American Productivity**   43

3   **The Marshall Plan's Productivity Revolution**   77

4   **US Management and Labor Debate Political and Economic Values of Productivity**   107

5   **German Perceptions of US Productivity**   135

6   **Codetermining German Labor Relations**   169

7   **IBM: An American Corporation in Germany**   195

8   **Computing Technology: Productivity Promise or Automation Threat?**   219

Conclusion   249

Notes   259
Bibliography   313
Index   341

Acknowledgments

Fortunately, books are not measured by the productivity of their authors in words per hour, for I have spent more time researching, contemplating, writing, and revising the words in this book than I care to admit. I would like to thank my advisers, who guided my initial forays into the transatlantic history of computing at the University of Pennsylvania: Ruth Schwartz Cowan, who nudged me into choosing a topic that took advantage of my German background, always encouraged me to do my best, and generously shared professional advice as well as personal concerns; Nathan Ensmenger, who introduced me to the history of computing and was a steadfast supporter with levelheaded advice; Tom Misa, who readily supported me even before he joined my committee as an external adviser, and who has continued to back my work through helpful comments and advice; and Ron Granieri, who joined my committee late to lend his expertise in German history, and who continues to cheer me on. I would also like to thank the faculty and grad students in the History and Sociology of Science Department who created an always intellectually engaging community in a cooperative environment with teatimes (thanks, Jeremy!), journal clubs, and office conversations in which new ideas could be freely tried, critiqued, improved, and if necessary, discarded. I would particularly like to mention my writing groups with Andi Johnson, Emily Pawley, Divya Roy, Hilary Smith, and Dominique Tobbell. Since then, the Rochester Institute of Technology (RIT) has become the institutional home for my work, and my thanks go to dear colleagues in the history department and across our college for their unwavering personal and professional support as well as to our college's deans for accommodations to balance work and life.

Funding from National Science Foundation Scholars Award #15020127, research awards from the Hagley Library and Museum as well as the Truman Library Institute, and a Paul A. and Francena L. Miller Faculty Fellowship Award and several faculty research grants from my college enabled me to conduct additional archival research, and expand the scope and time frame of my research. I have also received support through a National Science Foundation Dissertation Improvement Grant, a scholar-in-residence fellowship at the Deutsches Museum, the Adelle and Erwin Tomash Fellowship in the History of Information Processing from the Charles Babbage Institute, an IEEE Fellowship in Electrical History, and a dissertation research fellowship from the University of Pennsylvania.

The Tensions of Europe network provided an initial intellectual home for my work, thanks to Gerard Alberts. Many companions in SHOT's special interest groups in computing and gender studies, SIGCIS and WITH, have become colleagues, mentors, and friends over the years, particularly Amy Bix, Nina Lerman, Arwen Mohun, and Erik Rau. Countless commentators and copanelists have provided helpful insights and feedback, and shaped and improved my thoughts on quantification, codetermination, labor, gender, and corporate relations; most prominently among them are Janet Abbate, Keith Breckenridge, Will Chou, Patrick Chung, Jim Cortada, Helena Durnova, Larry Frohman, Shennette Garrett-Scott, Ellen Hartigan-O'Connor, Mar Hicks, John Krige, Jan Logemann, Joris Mercelis, Tom Mullaney, Mary Nolan, Jason Parker, Ben Peters, Joy Rohde, Andy Russell, Perrin Selcer, Edie Sparks, Ying Jia Tan, Ksenia Tatarchenko, Steve Usselman, Lee Vinsel, Heidi Voskuhl, Audra Wolf, and Jeff Yost. Eden Medina shared early advice on the book revision process, and Dan Raff compassionately accompanied my professional pursuits. Tamar Carroll, Nathan Ensmenger, Christine Keiner, Emily Pawley, and Rebecca Scales read chapter drafts, and Rena Searle patiently commented on numerous chapter drafts in addition to sharing (busy family) life in Rochester.

My research would not have been possible without the support of many librarians and museum professionals and volunteers. First and foremost, my thanks go to Morna Hilderbrand at RIT's interlibrary loan department who obtained countless books from libraries close and far, and who always worked magic when I needed to keep books longer. I'd also like to mention the late Fritz Kistermann, a volunteer at the Haus zur Geschichte der

IBM Datenverarbeitung, then in Sindelfingen, Germany, who spent three full days with me so that I could investigate IBM employee magazines in the museum's holdings; and the late Karl Ganzhorn and two former IBM employees in Germany who shared their time and expertise in oral history interviews. Nelson Lees and his staff at the Center for Economic Initiatives, and many other archivists, responded to inquiries for oral history transcripts, obscure yearbook entries, image permissions, and other requests. The indefatigable Susan Matheson helped clarify my thoughts, and Katie Helke, Virginia Crossman, and the staff at the MIT Press shepherded my project through the publication process, always with timely and invaluable advice.

Productivity Machines has a companion website at https://productivity machines.com. The website makes available core primary sources, such as US government studies on productivity and automation, reports by German visitors to the United States, documents on management and labor relations in the United States and Germany, and sources on the Marshall Plan's Productivity Program. Originating from the National Science Foundation's requirement for data management, the site assembles archival sources from the United States and Germany for the benefit of readers as well as teachers and students, in one single, easily accessible place. I would like to thank Mason Lezette and Nick Stanek—both museum studies students at RIT—for their excellent work in selecting, organizing, and presenting the website content as well as programming and designing the site (and many fun conversations along the way!), and Frances Andreu, Rebekah Walker, and the staff at RIT's Digitization Lab for their professional support.

Finally, the love and support of my family and friends carried me in my personal and intellectual journeys across the Atlantic and into the history of productivity. My parents nourished my curiosity, and unwaveringly supported me personally and professionally even when I moved away from their lives and values; their steadfast sense of what is socially appropriate has inspired my investigations into class relations across the Atlantic. Gunnar, Sanja, Jana, Nico, Michaela, Ansgar, Erik, and Theresa made visits to Germany (with and without research) into family fun. Brittany made every minute Anna and Ronan spent with her enjoyable, freeing many hours for me to write. Anna, thank you for letting me share your fun in reading, writing, and learning, and for slowly embracing this book even if it often meant spending time at my desk. Ronan, thank you for being your inquisitive and sometimes-sleepy self when you came into my office for a hug after your nap. And Shashi, I'm grateful to you for bringing me across the Atlantic, for being with me through my professional journey, and for reminding me of the many important things in life.

Introduction

The drive for ever-greater productivity infuses our daily lives. To achieve this elusive goal, we are bolstered by a growing array of technologies, such as productivity apps that manage our tasks and schedules to increase our daily accomplishments, magazine articles and self-help books that advise us how to improve our productivity, and colleagues and friends talking about how productive their days have been. This push to do more is ever present. From Stone Age hoes to the steam engines of the first Industrial Revolution, technologies have allowed humans to get more done in the same amount of time. Also, the objective to grow and expand has been inscribed in the capitalist rationale to use resources in the expectation of future gains. But it was not until the late nineteenth century that analysts in the United States began to measure how new technologies—and particularly the mass production technologies of the second Industrial Revolution—increased workers' ability to create more without spending more time; in short, to be more productive.

Germans observed closely as Americans debated how the ability to make more in the same time would affect economic relations. Would fewer workers make more things? Or would workers work fewer hours? How much would workers be paid? Would they be paid more for making more things? Or would they be paid the same for working the same hours? What was to happen with all the things produced? Who would buy the many things that could be made? Who would afford and use them? How much would and should people spend for the many things available to them? Many Americans concluded that workers were to be paid more for making more things so that they could afford more of the things produced, and at lower prices. Thus, they imbued the new system of mass production with a set of economic and political values that were to guide the relationships between workers, manufacturers, and consumers.

The soft power of America's new economic productivity helped establish the country's place among other nations, even when it retreated in isolationist policies. In 1941, the editor of *Life* magazine, Henry Luce, claimed the twentieth century to be "the American Century." He argued that American music and films as well as American machines and patented products had already established the United States as a world leader in culture, technology, and economics, if not politics.[1] In the decades before 1941, people around the globe studied American mass production technologies, and Americans actively transported their new economic and political convictions to other places around the globe. After 1941, in the emerging Cold War competition, officers for the Marshall Plan, the US aid program formally known as the European Recovery Program, sought to instill productivity-mindedness in Western Europeans in an effort to rebuild European economies, win European workers to the ideals of the US economic model, and strengthen the coalition of the capitalist West against the Soviet Union. My analysis shines a fresh light on the role of America's technology and economic system in the country's ascendency to world power by investigating American and German thought on productivity and technology, and how that affects economies.[2]

A variety of technologies were instrumental in the long American ascent to power. Rifles, axes, and hoes, along with many other technologies, had enabled white pioneers to conquer the American West and its native peoples. Once the Western frontier closed, attention shifted beyond the North American continent to East Asia and South America. Now, large civil engineering projects—such as roads and canals—as well as books and vaccines aided US government and missionary activities from the Philippines and China to Panama. Technological mastery shaped the perception that different peoples had of themselves and each other and founded hierarchical relations between those commanding technologies and those being subjugated to new technological regimes. Real or perceived technological supremacy was often couched in racial terms, legitimating missions to uplift foreign peoples and "civilize" them by incorporating them into an industrialized, modern way of life.[3] Productivity technologies from mass production to electronic computers formed the pinnacle in this technological ascent, embodying the essence of the American economic system and social and labor relations and establishing American leadership with regard to European countries.

Productivity Machines examines transatlantic exchanges of productivity technology and culture between 1920 and 1960, with World War II as a turning point. It unravels the historical roots of productivity—a widely popularized economic concept that originated as a statistical measure of economic output per worker or per work hour, calculated by the US Bureau of Labor Statistics (BLS). The book traces the development of productivity measures and the emergence of productivity culture in the United States in the 1920s, explores the machines that became emblematic technologies for increasing productivity, and investigates transatlantic exchanges of productivity ideas and technologies before and after World War II. A seemingly objective measure, productivity came to encapsulate the American economic system abroad.

I analyze the transatlantic circulation and appropriation of productivity culture and technology to better understand the US capitalist system in international relations.[4] To do so, I study the interpretative flexibility of productivity in the transatlantic context—that is, how different groups have seen productivity at different times, and how their views have changed over time.[5] The first two chapters set the stage, looking at the period before World War II. In the 1920s, BLS officers developed productivity indexes to analyze economic and labor changes resulting from technological improvement in the late nineteenth and early twentieth centuries. Immediately, as chapter 1 shows, questions arose as to who should benefit from productivity increases, and if and how much workers should share in these benefits. At the same time, Germans and other Europeans traveled to the United States to study its technological improvements, encapsulated in Henry Ford's production line, as chapter 2 discusses.

After World War II, the US administration sought to reshape European economies based on the US model, and the Marshall Plan's Productivity Program—the focus of chapters 3 and 4—became a vehicle to do so. A programmatic core of the Marshall Plan, the Productivity Program provided technological assistance to Europeans for longer-term economic reconstruction while the majority of the Marshall funding was spent on shipments for immediate relief from food and fuel shortages. In the emerging Cold War, the promise of higher standards of living for all due to higher productivity was used to counter communist promises of a better life from a communal economy. US officers now also promised that labor, management, and consumers were to equally share the benefits from productivity. Productivity came to mean more than increased worker output through technological improvement. In the eyes of US officers, executives, and labor unionists, productivity offered the chance of a peaceful social revolution that would reshape labor relations, eliminate social differences based on class, and open opportunities for a higher standard of living for all. They tied productivity to the political values of free enterprise—free competition between privately owned companies without government or union interference—and collaborative labor relations based on plant-level collective bargaining between a local union and corporate management.

But West German unionists feared that relinquishing their say in managerial decisions would expose workers to further exploitation rather than allow them to participate in the benefits of productivity. Therefore, German labor unions fought for codetermination—that is, the right of worker representatives to participate in corporate decisions through works councils and labor representatives on boards of trustees. Continuing labor relations traditions from the Weimar Republic during the interwar period, codetermination also rejected the American model of collaborative labor relations. Chapters 5 and 6 cover German responses to American productivity, and chapter 7 provides an in-depth look at the labor relations of one US company—the International Business Machines Corporation (IBM)—in Germany. Chapter 8, finally, turns to questions about productivity technologies that culminated in debates over the new iconic productivity technology—electronic computers—and the prospect of automated factory and office work.

Productivity History: The Literature

Productivity Machines builds on the work of historians and philosophers of technology who have shown that technologies are not objective, neutral artifacts. Philosopher Andrew Feenberg developed the most theoretically elaborate version of this argument in his critical theory of technology, while historians of technology have shown how social and political values shaped technologies, and technologies in turn shaped social and political relations. Underlying these works is a constructionist understanding of technology: technological progress is not inherent in technologies, and technologies don't determine their uses. Unpacking the "black box" of technology shows that technologies are always contextual, shaped by their social contexts and in turn shaping their social contexts. Thus, Feenberg argued that technologies carry with them the values of those involved in their creation and implementation; in our society, these values are often those of industrialists, managers, or others in power. Computers, Feenberg suggested in an example, could be used for surveillance or communication, thereby giving rise to different uses and social relations.[6] Providing a fine-grained study of individually pursued technology transfers as well as governmentally mediated technology transfers, I reveal that American and German officials, executives, engineers, and workers continued over decades to negotiate the values of productivity technologies, such as assembly lines and, later, electronic computers. Thus, values were imbued in technologies during their design and implementation, as Feenberg and others have argued, as well as during their transfers from one locale to another.[7]

Feenberg developed a socialist utopian theory, arguing that involving a wider range of social groups in the creation of technology would help shape technologies for more socially desirable outcomes.[8] In Feenberg's Marxist analysis, it is financial, capitalist, or technological elites that control technologies. By contrast, I highlight that labor unions were also at the table—sometimes disregarding the views of rank-and-file union members, sometimes even to the latter's detriment. As such, chapter 8 demonstrates that US labor unions welcomed technological change, but sought to protect their members—and only their members—from the immediate consequences through retraining and relocation programs. German labor unions likewise embraced technological change and productivity technologies, from Ford's assembly lines to electronic computers. This book thus provides a fuller cast of those involved in the construction of the values of technology and complicates the picture. Rather than just elites on the one side and everyone else on the other, it shows that a diverse set of historical actors had a seat at the table, including government officials and labor unions. Still, the views of those not at the table—for example, individual workers fearing the loss of their jobs from computer automation—remained excluded from the process.

With this book, I advance the internationalizing of the history of computing. The history of computing has long been dominated by a US focus, which has led to the overgeneralization that because computing technology developed in a particular way in the United States, it would develop the same way elsewhere. But in recent years, excellent studies on computing in the Soviet Union, Chile, and elsewhere have begun to reveal that computing was shaped by local conditions and needs, which in turn shaped local technology, business, and politics.[9] In Chile, for example, computers and cybernetics were used to control a socialist economy. Like these studies, I show that computers where imbued with new meaning locally. By doing so, this book unpacks economic factors that were taken for granted in James Cortada's *Digital Flood*, to date the only book-length treatment of transfers of computing technology between countries. Identifying patterns of dissemination, Cortada argues that characteristics of computing technology as well as local social, economic, and legal conditions, among them the standard of living, affected dissemination. Cortada contends that a higher standard of living, measured as GDP, promoted the adoption of computing technology.[10] *Productivity Machines*, by contrast, closely examines the economic debates surrounding computer productivity. It shows that computers were shaped by local economic models and in turn shaped these models. The standard of living did not merely serve as a conduit for the dissemination of computing technology; rather, the introduction of computing technology raised questions of social justice about how income was to be distributed within a country.

Almost two decades ago, Walter LaFeber called for integrating technology into the history of American foreign relations. Based on Thomas Hughes's work, LaFeber proposed a constructionist approach, arguing that technology allowed for the discussion of political, economic, and social choices, and he encouraged diplomatic historians to analyze how policy makers related their views of technology to their foreign policy goals.[11] Historical research on technology in international relations so far has often focused on military technologies that advance US interests through hard power, such as torpedoes and nuclear bombs.[12] *Productivity Machines* adds a focus on a new set of technologies: the mass production methods of the second Industrial Revolution as explored in chapter 2, and the electronic computers of the third Industrial Revolution examined in later chapters. Analyzing the history of the Marshall Plan's Productivity Program, this book goes beyond histories of postwar economic reconstruction. Part of an emerging history of civilian technologies in international relations, from aviation to aerospace and hydroelectric dams, this book explores how technologies have shaped US foreign political and economic relations and how, in turn, technologies were shaped by them.[13] Highlighting the constitutive nature of foreign relations, *Productivity Machines* provides deep insights into their complexity, with its vast cast of characters that interwove official and public diplomacy, including diplomats, corporate executives, union officers, engineers, social workers, and students. More importantly, it features the mass production and electronic automation technologies that manufactured consumer products, from cars and refrigerators to ballpoint pens, that facilitated new standards of living and economic and social relations, and that undergirded the ascendancy of the United States as an international market empire.

Political economists have identified differences between liberal market economies, like the United States and Great Britain, and coordinated market economies, like Germany and other continental European countries. These varieties of capitalism are distinguished by different forms of labor relations, training, and mobility, and different social systems of production as well as the generation and transfers of technical knowledge.[14] Similar to these political economists, chapters 5 and 6 identify the US economic model based on a dynamically growing economy, free enterprise, and collaborative labor relations, and the West German model of a social market economy with corporative elements based on the state protection of competition and codetermination. But rather than conceiving of these varieties as ideal types—that is, abstract and hypothetical concepts—they investigate how these varieties emerged in the intricate and complex transatlantic interactions after World War II.[15] Doing so, they reveal inconsistencies and motivations in each economic model. In the United States, unsettled questions about the role of labor and economic planning—resulting in the free enterprise campaign—led to heated debates among those involved in the Marshall Plan's Productivity Program and uncertainties about exposing European visitors to US realities. In West Germany, the ideal of the social market economy called for the state to protect free competition, seeing the state's social role in opening economic opportunities for everyone. Free competition confronted calls for nationalization, and attempts by labor unions and business organizations to establish their roles in economic decisions, resulting in a complex amalgam of free competition and corporative influence.

Productivity Machines investigates transatlantic productivity transfers with a focus on the German reception of American productivity. Americans often differentiated little between European countries. For example, the Marshall Plan administration created the Organization for European and Economic Cooperation (OEEC) to encourage European cooperation within the aid program instead of creating independent national programs. This book thus has a European dimension insofar as Americans addressed Europe as a whole. The European reception of American productivity, however, frequently differed from country to country. While American foreign policy makers may have generally imagined Europe as a tabula rasa on which they could imprint their economic model after World War II, West Germany promised to be the best place to do so. Even before World War II, Germans as well as Italians had eagerly studied American productivity while other Europeans showed less interest. After World War II, Germany's industries were razed by wartime destruction, and its economic and political elites were delegitimated through their cooperation with the Nazi regime, presenting a clean slate in American eyes.[16]

In addition, Americans thought that Nazi state corporatism was responsible for the destruction of democracy in Germany, and in the late 1920s, the growing corporatist influence of interest groups like labor unions and business associations as well as the state intervention in economic relations had sidelined political decision making. US occupation forces therefore banned cartels and interest groups in an effort to strengthen parliamentary democracy; a free competitive economy, they reasoned, would bolster democratic institutions.[17] West Germany not only seemed to promise the best chances of implementing an American productivity program and its economic values; it also presented the most urgent need to reshape political and economic relations to create political stability in Central Europe, and prevent another authoritarian regime and military conflict. The German reception of American productivity thus is not representative of responses in other European countries. Rather, my study shows that even under the seemingly conducive and exceptionally important circumstances in West Germany, American productivity was critically debated and was appropriated only with modifications that suited local conditions in Germany.

Productivity overlaps with efficiency because both measure output per (labor) input. Efficiency, the older of the two terms, became closely associated with Frederick W. Taylor and scientific management, and scientific management engineers played important roles in the transatlantic history of productivity, even if they are not always identified as such. For example, they promoted efficiency technologies in Europe, and served as technical consultants and advisers to the Marshall Plan's Productivity Program. By focusing on productivity and not efficiency, I sidestep debates among contemporaries, historians, and social scientists about the relation between scientific management and Taylorism as well as the change of Taylorism from rigid methodological principles to encompassing labor questions such as the level of wage rates.[18] Instead, I follow the lead of historical actors who rallied around the newer concept of productivity with which they could associate broader—albeit shifting—ideas about management and workers, wages, consumers, and prices. Concentrating on productivity also allows me to draw on a broader cast of historical actors including engineers, labor and management representatives, and government officers.

Productivity ideas, from the mass production technologies of the second Industrial Revolution to the information technologies of the third Industrial Revolution, flowed mostly eastward across the Atlantic.[19] Yet, not all American technologies and ideas were appropriated, and there were domestic developments in Germany. The strong German machine tool industry, for example, continued to pursue a general-purpose approach and did not follow the increasingly inflexible Detroit automation—an approach that came to be seen as advantageous by the 1970s. Or in the computing field, German pioneers like Konrad Zuse and Heinz Nixdorf laid the basis for a homegrown computer industry, and large electric companies like Siemens and AEG-Telefunken entered the field.[20] But despite political protection— for example, only German manufacturers (not IBM) were eligible for an early federal purchasing program for scientific computers—the German computer industry failed to be internationally competitive. By the late 1960s, IBM dominated 70 percent of the computing markets in West Germany as well as other continental European countries.

Key Questions

Productivity Machines addresses three key questions. First, how did electronic computers replace the assembly line as the icon of American productivity? During the 1920s, individual Americans, such as Ford, promoted their economic ideals at home and abroad; at the same time, Europeans of different social and national backgrounds, including Germans seeking to stabilize the German economy after World War I, eagerly studied American productivity.

Following World War II, US Marshall Plan officers encouraged the model of American productivity in European countries. Helping companies recover from wartime destruction and promoting steps toward a larger European-wide market, they created suitable conditions for the adoption of electronic computers. By focusing on productivity, *Productivity Machines* links the assembly line to electronic computers and provides insights into different modes of technology transfer, from the individual endeavors of the inter-war years to the government-mediated programs after World War II.

Second, what characterized capitalist countries in the "West" during the Cold War? Purportedly, the United States and Western European countries formed an alliance of capitalist countries during the Cold War. Yet debates about productivity reveal differences between these Western economies. For example, Americans conceived of computers as productivity machines that promised increases in living standards through economic growth, while Europeans saw computers as automation machines that caused technological unemployment and increased economic disparity. Consequently, American unions sought to implement what they called collaborative labor relations, while West German unions fought for codetermination. Taking a fresh look at the Marshall Plan and its related productivity debates, this book contributes to a better understanding of the continuing variety of capitalist models in the Cold War "West."

And third, in what ways did the meaning of productivity change from the decades before World War II to the decades after? In the 1920s, Ford and other Americans promoted their notions of productivity, promising higher standards of living through higher wages and economic growth, supported by mass production and consumption. Germans enthusiastically adopted American productivity technologies even as they were divided about the technologies' emerging social implications, such as changing gender relations. After World War II, Marshall Plan officers devised an ambiguous notion of productivity that promised the contradictory goals of cooperative labor relations and free enterprise. While Americans promoted the idea of increasing economic profits, which were to be distributed, Europeans saw profits as a fixed amount that had to be distributed, which pitted social classes against each other. This book thus provides insights into the evolution of class relations on both sides of the Atlantic.

Despite its focus on the Productivity Program, *Productivity Machines* does not assess the impact of the Marshall Plan; others have done so before.[21]

Also, although the Marshall Plan included agricultural productivity programs, and Europeans traveling to the United States paid attention to American productivity on farms and in offices, these topics are not covered in the book. Instead, I look at industrial productivity because it shaped the labor relations that characterize a modern capitalist society. Moreover, I do not evaluate the social and political implications of productivity. My close study of productivity culture and technology in transatlantic relations shows that Americans replaced an economy of scarcity—in which static economic gains had to be redistributed—with an economy of abundance—in which everyone could partake in ever-growing economic gains. It is beyond the scope of the book to analyze how much different social groups partook in the gains from higher productivity, and how widely productivity affected social disparities and questions of social justice, because that would require consulting different archival sources. By promising higher standards of living, productivity culture also appears to encourage the use of ever-increasing natural resources instead of conserving the same. These larger environmental history questions of how an economy based on dynamic growth affects the use of natural resources are beyond the scope of the book too. By studying productivity in transatlantic relations, however, *Productivity Machines* hopes to raise awareness of these essential as well as pressing social and environmental questions.

Productivity Machines provides a history of capitalism in transatlantic relations from the 1920s, when the United States became an economic model, to the post–World War II period, when countries formed the supposedly homogeneous Western capitalist bloc during the Cold War.[22] The German economist and sociologist Werner Sombart had asked in 1906 why there was no socialism in the United States—a question based on the observation that the United States was the only fully formed capitalist society that had not developed a socialist working class, with a working-class consciousness and a political labor party.[23] Driven by this sense of US exceptionalism, US entrepreneurs, engineers, union officers, workers, and government officials brought US productivity technologies and economic values abroad. But Europeans and Germans appropriated the US model to local political and economic values, and different varieties of capitalism persisted on both sides of the Atlantic. Investigating these transatlantic exchanges, *Productivity Machines* disentangles assumptions about technology, labor, markets, competition, and the role of the state as well as how they were debated

and changed over time. These assumptions often did not form a coherent whole. American productivity—even if it failed at reshaping Western
economies after the US model—helped establish US technological prowess,
and facilitated the United States' ascent to global economic and political
leadership.

1 Measuring Productivity at the US Bureau of Labor Statistics

In March 1927, Ewan Clague wrote that there was "no immediate and exact connection between high productivity of labor in an industry and the wages paid in that industry."[1] Clague was later referred to "as the nation's chief philosopher of figures" and acclaimed for his "integrity and penchant for accuracy," which "made him acceptable to both business and labor." In 1927, however, he was fresh out of graduate school in economics at the University of Wisconsin and had been hired to conduct research on productivity for the US Bureau of Labor Statistics (BLS).[2] He had just published a series of articles describing the productivity of eleven industries and had introduced a new measure of productivity, called the "productivity index," that laid the groundwork for the various productivity indexes that the BLS still publishes today, most prominently the quarterly Major Sector Productivity index (figure 1.1).

Clague's controversial statement concluding that productivity and wages were not correlated was based on the data that he had collected across industries as diverse as automobile manufacturing, petroleum refining, and meatpacking. These studies were undertaken by the BLS because of concerns over how new technology caused the loss of skilled jobs, and had introduced women and children to the workforce to tend these new machines. A few years earlier, BLS officers had suggested that technological change improved productivity rates and that technologically advanced companies should be emulated. But in 1927, Clague helped refine and expand the BLS's calculations of labor productivity. He was paid through the Department of Labor's Conciliation Service, the agency that mediates collective bargaining negotiations, so he was aware that labor unions, after two prosperous business years with stable prices, were expected to demand

Figure 1.1
Ewan Clague in 1946, when he was named commissioner of the Bureau of Labor Statistics.
Source: In author's possession.

"sharing in the gains in production" in upcoming contract negotiations.[3] Clague's extensive research showed that there was no connection between high productivity and high wages.

While the BLS may at first sight appear like a technocratic and dull agency, it nowadays governs the financial well-being of most American citizens. Aside from productivity indexes, the BLS also calculated a cost-of-living index, later named the consumer price index, which measures the change in the price of a representative group of consumer goods and services, and has been a highly influential statistical estimate for decades. Productivity and the cost-of-living index presented two sides of the same coin: both contributed to measuring how much workers needed to work to afford a certain standard of living. While productivity measured a worker's output in a given time, presumably based on the assumption that workers producing more would be remunerated better, the cost-of-living index tracked prices to determine how much money workers needed to achieve a certain standard of living.[4] Notably, the BLS focused on cost-of-living calculations in times with more labor union friendly policies or more labor union influence—World War I, the New Deal, and World War II—and it concentrated on productivity calculations in times of more industry friendly policies or less labor union influence—the 1920s and after World War II.[5] By the 1950s, collective bargaining agreements in the automotive and other industries tied automatic wage adjustments to changes in the consumer price index, and in the 1970s and 1980s, poverty thresholds, social security benefits, and income tax brackets became tied to the index, affecting federal programs such as food stamps, school lunches, and veterans' pensions. Tiny changes in the consumer price index thus affected thousands of dollars of corporate salary payments, and by the 1980s, 50 percent of the federal budget was indirectly or directly tied to the index, putting the BLS under intense public scrutiny. Not surprisingly, the BLS's cost-of-living indexes have received more historical attention than its work on labor productivity.[6]

Productivity measurements, calculated as output per worker or man-hour, had their roots in a late nineteenth-century BLS study of technological change that compared methods of hand production with machine production. Over the 1920s, the BLS significantly changed its methodological approach to productivity measurements. In 1923–1924, the BLS still conducted a study of productivity in selected consumer industries that, following the model of the nineteenth-century study, provided narrative descriptions of the production methods that were based on information from a limited number of voluntarily reporting companies. By 1926–1927, Clague calculated and published productivity indexes for different industries, based on statistical data originally raised for other purposes, such as the biannual Census of Manufactures. During the 1920s, BLS officers also began to use the term "productivity" more regularly. BLS officers still used alternative terms in the early 1920s such as "production rate" or "productive capacity," and devised tables of "relative productivity" that compared the output for different production steps for hand and machine production methods.[7] By 1927, Clague explained and popularized the term "productivity" in his above-quoted article in the *American Federationist*.

BLS officers, employers, and workers often had different views of these statistical measurements. For example, during World War II, labor unions began to publish their own competing cost-of-living index because they argued that under the conditions of wartime wage and price controls, BLS officers undercalculated the price increases since they did not sufficiently take into account the disappearance of low-cost goods from store shelves and the decreasing quality of the products offered.[8] Examining how relevant social groups viewed the productivity measurements—in other words, the interpretative flexibility of productivity—will thus help us better understand social relations between government experts, workers, and employers.

BLS officers sought to establish productivity as an objective measurement of work output. Certainly they were aware of the limitations of the measurement. Productivity was neither a measure of industry efficiency nor worker efficiency—that is, it did not measure how much a given industry or a worker produced with a fixed input of time and material. For example, workers' output was affected by a variety of things such as production methods, equipment, efficiency of management, capacity utilization, and supply arrangements, and output per production worker did not take into account things like the work of other groups of employees, vacation time, waiting time, call-in time, and other periods that were paid but not actually worked. In addition, changes in the quantity of materials, electric energy, or the amount of capital consumed per unit may have important effects on the costs per unit of output, too.[9] Theodore Porter has shown that scientific quantifications, such as statistical numbers, create the *appearance* of impartiality and fairness, which lends authority to decisions. In Porter's words, numbers are "a way of making decisions without seeming to decide."[10]

My goal for this chapter is to trace the historical roots of productivity measurements in the BLS to provide a baseline for tracking the changing meanings and interpretations of productivity by different social groups over time. While Clague was later praised for the objectivity and impartiality of his work for the BLS, *Productivity Machines* unpacks the interpretative flexibility of productivity—that is, the different meaning that productivity has had over time for different social groups such as BLS officers, employers, trade unions, and workers—domestically as well as in transatlantic relations.[11] As later chapters in the book will demonstrate, productivity took center stage after World War II in the Marshall Plan, which provided a guideline for the economic reconstruction of postwar Europe and shaped the US economic and political relationship with Western European Allies in the emerging Cold War conflict. US Marshall Plan officers used BLS productivity measurements, yet they no longer assumed that there was no connection between productivity and wages, and demanded that gains from higher productivity be shared equally between owners, workers, and consumers.

This chapter focuses on productivity, how it was initially measured and interpreted, and how that changed during the 1920s. It first looks at a BLS study of technological change from 1898 that shaped the bureau's later work on productivity. Second, the chapter discusses field studies of productivity in the early 1920s that show that BLS officers sought to promote technological change because they believed it increased productivity. Third, the chapter explores how new productivity measures were created in the late 1920s that moved from narrative reports using information from a handful of voluntarily reporting companies to calculating a numerical index of productivity based on statistical information reported in census data and other statistical reports. Finally, the chapter examines Clague's argument that productivity was not proportionally related to wages, and his conclusion that labor ought to let management set productivity and wage levels.

BLS's 1898 Study of Technological Change

The BLS, founded in 1884, is the oldest division within the US labor administration. It was created in response to pressure from the labor movement that charged that the federal government did not occupy itself with the concerns of working people.[12] The labor movement had pushed for a government agency that would lobby for working people and would help pass legislation. When President Chester A. Arthur appointed Carroll Wright—a gifted administrator but politically moderate and statistically untrained—as the first BLS commissioner, however, he created a toothless agency that limited itself for its first three decades to discrete, topical studies, partly because a lack of funding forbade more regular data collection.[13] Yet with the United States undergoing rapid changes during the second Industrial Revolution and the widespread introduction of machine production methods, technological change was one of the areas in which BLS officers conducted research.

The BLS is today best known for its cost-of-living index, which has also drawn the most public scrutiny of the office's work. Issued regularly since the 1910s, the cost-of-living index required the BLS for the first time to regularly collect data, perform calculations, and publish the results. Calculated in relation to base prices from 1913—the first year that the BLS gathered data—the cost-of-living index served to set wage levels during World War I. With war production agencies guaranteeing repeated wage hikes linked to the rising cost of living, the wartime arbitration of labor negotiations also put labor unions on an equal footing with employers, requiring rational arguments for why wages were just.[14] For the first time, the cost-of-living index gave labor an important statistical tool.

It was not the cost-of-living index, though, that initially influenced the BLS's productivity research in the 1920s; rather, BLS officers modeled their first research questions and methods after a study of hand and machine labor from Commissioner Wright's era in the previous century. The study was an elaborate example of the BLS's early topical studies, requiring four years of fieldwork. In 1894, Congress charged the BLS with "investigat[ing] and mak[ing] report upon the effect of the use of machinery upon labor and the cost of production, the relative productive power of hand and machine labor, the cost of manual and machine power as they are used in the productive industries, and the effect of wages as they are used by women and children."[15] This requirement was motivated by the 1893 recession and growing concerns over how changes in production technologies impacted labor.

In November 1894, BLS officers and hired agents set out to compare "the operations necessary in producing an article by the old-fashioned hand process and by the most modern machine methods, showing the time consumed by the workmen and the cost of their labor for each operation under the two systems."[16] The notions of hand and machine production were simplifications: "hand" labor included the use of often-simple tools such as saws, hammers, chisels, picks, shovels, knitting needles, and so on, and "machine" labor allowed for some operations to be conducted manually. In the words of the report, hand labor was to be understood as a "primitive method of production which was in vogue before the general use of automatic or power machines, and which still exists to some extent in remote rural sections, or occasionally even in towns, while the machine method is the one generally in use at the present day."[17]

The voluminous report, almost sixteen hundred pages and two volumes, titled *Hand and Machine Labor* and published in 1898, described the production methods for over 670 products in manufacturing, agriculture, mining, and transportation in an impressive amount of detail, down to listing the age and pay of workers for every single production step.[18] Yet it drew only limited conclusions on the overall labor effects from technological change. It generally stated that wages had increased since the introduction of power machinery along with the employment of women and children in its operation. But the report cautioned that the research provided no basis to decide whether this increase was caused by the use of machinery, a higher standard of living, increased productivity, all these causes combined, or other causes. The report also did not identify a clear trend regarding employment numbers from the introduction of machinery. It indicated that in some cases, there was surplus of labor, and in other cases a demand for labor. Assuaging labor unions and others concerned about technological employment, the report stated that there had been an overall increase in industrial production to meet increasing demands, which resulted in an overall increase in the number of people employed in the machine production system rather than the hand production system.[19] In other words, this was a consumerist argument for economic growth through consumption. In addition, the report stated that in the transition from hand to machine production, the savings from decreasing the time to make a product were greater than increases in labor costs—possibly assuaging management concerns over increasing salaries. The report explains this gap with the "wonderful inventiveness of the age in which we live" and larger rewards for labor.[20]

An Example: Manufacture of Red Bricks

A closer look at the manufacture of red bricks, one of the industries that BLS officers would again study in the 1920s, will help us better understand the 1898 model. BLS officers and agents visited brick factories across the United States in what they conceived to be the centers of the industry. The officers and agents conducted interviews to learn about the production methods, and where available, they consulted payrolls for information on salaries and labor costs. Once they had data for two or more factories, they compared them for accuracy and selected the "better" set—that is, the set for the factory that was more complete—for inclusion in the final report.

The report included tables representing these raw data on hand and machine production for comparison. Officers identified fifteen steps each for both the hand and machine production of standard red bricks. While the operations for burning the bricks were largely identical between hand and machine production, the preparation of the bricks was different. Some of these differences were related to the quality of the clay and the processes applied. The machine method used a stiff-mud process where the clay was directly formed into bricks, using its original consistency from the pit, but it required furnishing power and firing the boilers. The hand method, by contrast, applied a dry-clay process that required drying and crushing the clay before it was molded, with two additional operations.[21] While the machine method used steam power to dig, move, grind, and mold the clay, and otherwise used manual power, the hand method used horsepower to move, crush, and grind the clay, and otherwise used manual labor. In both hand- and machine-powered establishments, only men were employed— indicating that women and children did not tend the new machinery in all industries.

It does not appear that salary ranges for workers changed significantly, at least for the brick manufacturing industry, with the introduction of machinery, with the exception of slightly higher pay for foremen and engineers.[22] Salary ranges between the two methods were comparable, but the top pay for the machine method was greater than the top pay for the hand method, and there were more salary categories under the machine method. Under the machine method, salaries were $35 to $45 a month for workers, $60 to $90 a month for foremen and engineers, and $5.50 a day for the man overseeing the burning. Under the hand method, most workers were paid $39 per month, the molders and brick setters received $62 to $82 a month, and the man overseeing the burner got $4 a day. There were also a few day laborers working at $1.25 a day.[23] It is unclear how many days per month the workers were required to work.

In addition, the report tabulated and discussed these raw data in a separate analysis and summary. Here we learn that it took a total of 7 hours, 29.3 minutes to produce 1,000 bricks with the machine method and almost three times as long, or 20 hours and 36.7 minutes, with the hand method. Most of the time savings stemmed from the excavation and preparation operations where steam shovels dug the clay, and then cars transported it to the machine house, while in the hand method, workers with picks, shovels, and horse carts performed the work. It took 31.5 minutes with the machine method to excavate the clay for 1,000 bricks and transport it to the molding machine, and nine times as long at 4 hours and 40 minutes to do so with the hand method. Also, the brick-molding machine grinded and molded the clay in one step, taking 13.5 minutes for 1,000 bricks, while the hand methods required two separate steps, taking twelve times as long, or 2 hours and 40 minutes.

The machine facility employed many more workers than the hand facility, with 119 workmen compared to 21. The labor cost amounted to $1.1678 for 1,000 bricks for the machine method, and almost three times as much, or $3.0005, for the hand method.[24] Focusing on the production methods, the report rendered the workers exchangeable; they only appear with their age, sex, occupation, and pay, indicating that the introduction of machinery had made it easy to switch out workers. Although technological change resulted in the de-skilling of workers, BLS officers at least embraced it as a factor in reducing production time.

Field Studies of Productivity: The Early 1920s

More than twenty years later, BLS officers returned to studying the effects of technological changes. In 1923, they undertook a study of productivity in about a dozen consumer industries. This section analyzes one of these reports—for the red brick manufacturing industry—to show how BLS officers now sought to promote technological change to increase industrial productivity. Methodologically, BLS officers followed the model of the 1898 study on hand and machine production in an effort to determine the effects of technological change, comparing the time required for each production step in the hand and machine production methods of the 1890s with their own data from the 1920s. The final results of these studies were *narrative* reports that described the latest machine production methods—a form of report that while not uncommon for the statistical methods of the day, even if surprising from our expectations today, was devoid of numerical data, and based on a handful of voluntarily reporting companies and other sources. The 1923 study—following the example of the 1890s' study—squarely positioned research on productivity as concerned with technological change, and BLS officers promoted technological change by claiming it increased productivity, although their data did not necessarily bear out such claims.

In the first few decades of the twentieth century, labor relations in the United States were characterized by conflict between management and workers over the control of production and the productivity level. This was the time of union organization and attempts at collective control over work conditions, such as hours and pay as well as work processes—that is, by whom the work was to be done, and how. Skilled workers had traditionally controlled their own work and pace, supervised by skilled foremen. For example, a skilled worker could go about making a wooden chair by first assembling the pieces and then sanding them, or by sanding the pieces first and then assembling them. He was able to make many decisions on his own about how he went about his work, and management often had limited knowledge of production methods. Scientific management, however, sought to change that by studying work steps, planning the overall production flow, and determining the individual work steps—frequently with resistance from workers. The managerial control of production processes often went hand in hand with the introduction of machinery, which led to de-skilling: the reduction of the level of skill required to carry out a job. Workers were now interchangeable, and they were sometimes women and children. They tended the new machines, and were trained and supervised by management. Some skilled workers became tool and die makers, responsible for the construction and maintenance of the new machinery. Intense conflict and strikes arose over questions related to workers' skills, such as work classification and pay grades.[25] At the same time, early twentieth-century factories faced high turnover rates, particularly among unskilled machine tenders—over 400 percent at Ford in 1913—who usually stayed with a job only for a few months and, particularly in economically strong times, worried little about finding a new job.[26] The rapid expansion of industrial production meant that unemployment from new machinery was not a concern; de-skilling was.

At the same time, workers fought for what they called the "living wage." For most of the nineteenth century, Americans had upheld the idea of small, independent producers—such as farmers or cabinetmakers—who formed the basis of a democratic republic founded on independent citizens. They viewed wage labor—working by the hour for someone else—as something acceptable only on a temporary basis; wage labor was seen as "slave labor" or "prostitution." After the Civil War, advancing industrialization and its ensuing technological changes rapidly increased the demand for wage labor, and it became clear that many workers—and not only immigrants and African Americans but also white Americans—would work as wage laborers for their entire lives. Trade union rhetoric now began to slowly change from rejecting wage labor to determining what were fair and unfair wages, encoded in the notion of the living wage. The living wage was to afford workers an "American standard of living." It was different from a mere minimum wage, although in the early twentieth century, progressive reformers co-opted the notion of the living wage and then often used it synonymously with the minimum wage.[27] The BLS's work on productivity in the 1920s marked a shift toward studying work—an issue pursued by management—and away from the bureau's cost-of-living studies tied to wage demands in the previous decade.

Despite these drastic changes in work conditions and labor relations, BLS officers used the methodological model of the 1898 study in 1923. Also, they sought to understand technological change and its effects on labor, and wrote their reports in ways that promoted further mechanization. This time, BLS officers chose to study about two dozen consumer industries, compared to the over 670 product units in the 1898 study; they included red brick manufacturing, men's and women's clothing, slaughtering and meatpacking, baking, the manufacture of agricultural implements, and newspaper and magazine printing.[28] For each industry, the bureau collected detailed data but only from a handful of sources, including articles in the trade press, information provided by trade associations, information from voluntarily reporting companies across the country, and information from on-site company visits in the larger Washington, DC, area. Such reliance on a small number of voluntary sources was common statistical practice in the 1920s. While sampling methods and ideas of representativeness had been developed in the late nineteenth and early twentieth centuries, they were not widely used until national market surveys and election polls in the 1930s were shown to be better than reporting on a voluntary basis.[29]

Letters from the bureau to solicit information from companies reveal a continued interest in technological change. For example, the BLS informed companies that it conducted a study of industries "in regard to the change from handicraft to a factory stage." To raise more, and more comparable, responses, the officers eventually requested specific information such as the number and kind of operations performed in the companies, which operations were performed by hand or machines, the names of the machines used, and the speed of production per hour for each operation per individual worker.[30] The BLS thus honed in on information on production per worker per hour—in other words, productivity.

For most industries, BLS officers at least created a table of "relative productivity" that compared the production per hour per operator for each operation under machine production methods in 1898 and in the early 1920s with earlier hand production methods.[31] For those industries for which they were able to raise sufficient information, BLS officers proceeded to produce an industry report.[32] Considering that these industry reports were composed by an office of "labor statistics," they were surprisingly narrative. Most reports were about five typewritten pages and described the current technology in use as well as, in some cases, an overview of the historical development of technology. Numerical data were scarce. Often, the reports provided only some information on the overall employment in a given industry—information that could be gleaned from the biannual Census of Manufactures without conducting any further calculations. Aside from such descriptive numbers, no further statistical calculations were undertaken.[33]

A Field Report: Common Brick Manufacturing

Among these industry reports, the one on the common brick manufacturing industry stood out in completion, length, and informational detail; it will therefore serve as an example to explain the BLS method. A closer analysis of the report also shows how BLS officers used the study to promote technological change. Archival records reveal that this report relied on detailed information from only four production facilities—a surprisingly small number compared to our statistical practices today.[34] In addition, it included information on 79 companies from across the country on their production rates, manufacturing costs, average full-time hours, and earnings; all these data were most likely taken from the Census of Manufactures.[35]

Like the other industry reports, the brick manufacturing report included a lengthy description of production methods such as explanations of the three ways of producing bricks: the stiff-mud process, soft-mud process, and dry-clay process. The consistency of mud differed in these processes, requiring different ways of forming and handling the bricks and involving different types of labor and time requirements.[36] For example, the stiff-mud process allowed for the easy handling and stacking of the bricks, while in the soft-mud process, the clay was so soft that the brick could not be handled, and the bricks were automatically dumped from their molds onto metal pallets for drying, without being touched by hand. The dry-clay process, finally, required months-long "weathering" or drying of the clay. About 57 percent of the companies in the report used the stiff-mud process, 33 percent used the soft-mud process, and only 10 percent used the dry-clay process.[37] The stiff-mud process achieved the highest productivity measurements, and the dry-clay process the lowest.

The report provided the description of the production methods in what it initially called a "typical" plant. Later on, the authors of the BLS report admitted that the described plant may be an "extreme case," but emphasized that it "serves to illustrate what is possible in the mechanical hauling of brick."[38] In other words, the described plant may have been a model of mechanization that the BLS authors wanted the industry to aspire to rather than a typical plant. At this stiff-mud plant, a "steam shovel with an extra long boom" at once scooped up two cubic yards of clay, sufficient to make 780 bricks, and loaded the clay into a car; a small gasoline engine pulled the car to the foot of an incline, where a cable then hauled the car to the machine house. In the machine house, the clay was formed into bricks in a fully mechanized process. First, the clay cars were tilted, and the clay was dumped onto the hopper of a granulator that cut up the rough clay. The clay then went through a pair of conical corrugated rolls that threw out any stones. Next, an elevator carried the clay to the second floor, where it fell into a "pug mill" and was mixed. The clay then dropped into a brick-forming machine that created and cut 40,000 to 50,000 standard-size bricks an hour. Workers then hand set 840 bricks onto each dryer car in such a way that the bricks could later be lifted by a fork crane. A transfer car drew the dryer car into a steam dryer for twenty-four to thirty hours, and then brought the bricks to the kiln, where an overhead crane lifted the entire car to the right position in the kiln. The kiln used a steam- or oil-run "blowing process" to burn the bricks. Afterward, the crane lifted the bricks out of the kiln and loaded them onto trucks. The report emphasized that during the whole manufacturing process, the bricks were touched by hand only once, when workers removed them from the belt and set or "hacked" them onto the dryer car.[39]

This described model plant was located in the Chicago district where the clay was unusually plastic, and the carbonaceous matter in the clay reduced the time of burning to only fourteen hours, compared to forty-eight hours or up to fourteen days in other districts with different clay quality.[40] As a result, Chicago plants had a much higher output than plants elsewhere, with plants in the Chicago area producing between 20,000 and over 42,000 bricks per hour. In comparison, production among the stiff-mud plants elsewhere in the country was significantly lower: at 10,000 to 15,000 bricks per hour, the three highest-producing plants manufactured less than the least productive Chicago plant, and five plants produced less than 2,500 bricks per hour, or a mere 6 percent of the production of the model Chicago plant. While the high degree of mechanization set the model plant apart from the other Chicago plants, the quality of the Chicago clay set apart all Chicago plants from plants in other parts of the country.

Based on the report, it is impossible to determine how much the different factors, such as the degree of mechanization or quality of the local clay, affected a plant's productivity. While the BLS authors tended to emphasize the high degree of mechanization as a factor increasing productivity, Ralph P. Stoddard, the secretary manager of the Common Brick Manufacturers' Association, pointed to other factors affecting the productivity of a plant such as the distance of the clay supply from the plant, the character of the labor supply—with some operating on piecework, and others on daily tasks—and the length of the workday because plants in the south generally worked longer than plants in the north. Expressing doubts with regard to the BLS's attempts at calculating productivity rates, Stoddard stated that it "would be rather complicated to attempt to equalize these varying conditions."[41] Comparing the Chicago plants to plants elsewhere, it appears that the different clay quality explains the gap of 20,000 bricks per hour between the lower-producing plants and the gap of 20,000 to 30,000 bricks per hour for the higher-producing plants. In contrast, among the Chicago plants, higher mechanization led to an increased productivity of 10,000 to 20,000 bricks per hour, while among plants across the country, higher mechanization—as well as other factors such as distance from the clay supply, character of labor, and length of the workday—only led to productivity increases of up to 10,000 bricks per hour. In other words, the *quality of the clay* appears to have had a larger overall effect on a plant's productivity than the *degree of mechanization*. In addition, it is questionable whether the required investment for machinery would have been profitable for smaller plants that had a much lower output and therefore lower margins.[42]

Notably, it appears that none of the other industry reports for the 1923 study were ever finalized or even published; some of the reports remained in the stage of mere tables, and others in various shorter or longer drafts of narrative reports. The archival records reveal no indication of why BLS officers appear to have aborted their study. Were they disillusioned by the low response rates to their inquiries? Did the studies prove too expensive to cover a sufficiently large set of industries? Did BLS officers see the need to switch their methodology to include newer industries for which no comparative productivity data from previous hand and machine methods were available? Did narrative reports no longer appear as sufficient evidence in the eyes of possible audiences? Regardless of the reasons, the study makes it clear that the BLS's work on productivity in the 1920s was situated within the larger context of concern about technological change, and it was in this context that Clague and others picked up the BLS productivity research a few years later, now with new methodology.

Quantitative Productivity Measurements: The Late 1920s

In 1926, BLS commissioner Ethelbert Stewart asked John R. Commons, a leading economist and labor historian, to recommend a graduate student to conduct productivity research. Commons, as well as many of his colleagues at the University of Wisconsin, encouraged his students to serve the state and nation; Commons had his students attend committee hearings at the Madison capitol, study current political issues, and become involved in the legislative process. When Commons recommended Clague for the position, Clague—who had finished his coursework but not his dissertation—agreed, resisting the temptation of an immediate academic career and following the advice of Commons, who said that ten years of practical experience would turn him into a better teacher. Clague moved to Washington, DC, while his wife finished her coursework back in Madison. He initially roomed at the Brookings Institution—which was then in its first decade, and was still a school and research institute—and engulfed himself in his new work and the intellectual program at Brookings.[43]

Clague was part of a growing number of academically trained statisticians joining the BLS during this time, contributing to the bureau's professionalization.[44] Although Stewart directed Clague to use the 1898 study of hand and machine labor as a model, Clague instead decided to calculate a productivity index because he considered it to be "useful" for labor unions and management.[45] His research thus marked a methodological shift in the BLS's productivity work: the BLS moved from narrative reports relying on field studies to a numerical index based on census and other statistical data. No longer did a few companies that voluntarily chose to report stand in for a whole industry; the BLS now relied on comprehensive data for all companies in that industry. No longer did descriptive narrative provide the overview on technological status and historical development; the BLS now provided numbers that assessed an industry's productivity. And no longer did in-depth information guide the search for what was going on; BLS officers now qualified the accuracy of the indexes because they needed to adjust gaps and uncertainties in the statistical data for their calculations.

The BLS's new work on productivity was similar to its work on the cost of living from the 1910s in that it was published as an index—that is, a measure of the change from a baseline. Practically, both indexes—the productivity and cost-of-living indexes—posed challenges that the officers needed to solve: for the cost-of-living index, they needed to decide which goods to include in the calculation; for the productivity index, they needed to decide how to calculate the total production, such as when an industry produced more than one good. The shift from a narrative report to a numerical index can be seen as a shift from a more holistic tradition of statistical research that aimed at describing a human community by facts such as climate, natural resources, and population, and even discussed representations in cross tables as too reductionist, to a more arithmetic tradition of statistical research that described a human community based on written records such as baptisms, marriages, and burials in the form of predetermined tables with numbers, weights, and measures.[46]

While Clague helped usher in a methodological shift in the BLS's productivity research, he continued his predecessors' concerns about technological change. In 1926, he opened a series of articles in the *Monthly Labor Review*, the house magazine of the US Department of Labor, contending that the United States was undergoing "a new industrial revolution which may far exceed in economic importance that older industrial revolution" through mechanization in late eighteenth-century England. But while many were aware of the "great improvements in machinery, process management, and output," the technological advances were not being described in comprehensive terms, although they might constitute "the most remarkable advance in productive efficiency in the history of the modern industrial system."[47] Clague's productivity index was to provide this comprehensive description of productive efficiency increases from technological change.

The most important challenge for Clague was to derive sufficiently reliable and suitable information from data sets that were not intended to provide the kind of information that he was looking for. Clague selected eleven industries for his studies: rubber tires, automobiles, petroleum refining, cement manufacturing, iron and steel, flour milling, paper and pulp, cane sugar refining, slaughtering and meatpacking, leather tanning, and boots and shoes.[48] Chosen based on the availability of statistical data on production and employment, this selection included newly emerging industries such as rubber tires and automobiles that had been excluded from the field studies of the early 1920s because of a lack of comparative information. Clague used large data sets such as the biennial Census of Manufactures—which included the numbers of wage earners and salaried employees, standard hours of labor per week, and often a detailed summary of production—monthly employment data gathered by the BLS, and data collected by the US Chamber of Commerce in cooperation with trade associations.

Clague qualified the indexes because of the limitations of the data that they were based on. "It is not pretended that these indexes are perfect," Clague wrote, "or that they measure with absolute accuracy the productive efficiency in the industries over the period in question. There are many gaps and uncertainties in the figures, as will be evident from the discussion below on the methods of constructing the indexes; but when all due allowances are made for inaccuracies the results are still striking enough to leave little room for doubt as to the meaning and importance of the recent developments in industrial production."[49] Since manufacturers did not regularly keep precise records of the total man-hours and other important information, however, the necessary data would likely never be available.[50] Clague therefore concluded, "In the absence of the statistical data necessary to work out any index of productivity of this nature, the best that can be done is to try to approximate this result from the figures available."[51] Of course, the seemingly objective index numbers eradicated these judgments and uncertainties.

Rather than providing absolute numbers, Clague developed indexes of the total production and total man-hours, calculated in relation to a base year, usually 1914 or earlier for industries for which data were available.[52] Theoretically, it would have been possible to measure output in products per man-hour, such as five hundred cigarettes per hour. Yet many industries produced multiple products, and it was impossible to express the output in actual products. The paper industry, for example, produced different classes of paper from newsprint and boxboard to wrapping paper and fine writing paper that all required different amounts of time and labor in the production. An index based on the total tonnage of paper produced would not reflect these differences and would be problematic because the production of different kinds of paper did not fluctuate with each other, meaning that the industry might one year produce more of a more labor-intensive type of paper, and the next year more of a less labor-intensive kind of paper. In addition, it was possible to measure the percentage of change more accurately than it was to measure a phenomenon itself. In other words, a concrete expression of the actual level of productivity could be misinterpreted because it conveyed a level of accuracy and definiteness that the statistical data did not warrant; an index number, by contrast, showed a relative increase or decrease without any implication about the actual level of productivity. Index numbers thus avoided overstating the statistical knowledge and were also applicable to industries with multiple products.[53]

Calculating the Productivity Index

To calculate productivity, Clague needed two pieces of information: the total production in an industry in a given year, and the total number of man-hours worked in the industry that year. The total production was then divided by the total man-hours to calculate the industry's productivity for a given year.

For the first figure, total production, Clague needed to adjust the statistical basis. He relied on data from the Census of Manufactures and the Chamber of Commerce, which tended to report data originally gathered by trade associations. This data posed several problems. First, for some industries such as the chemical industry, the products were so varied and dissimilar that it was impossible to construct any index based on output, even if there was a common unit of measurement in the industry. Second, for various industries, there was no common measurement for different products; in the cotton industry, for example, cloth was measured in square yards, printed cloth was measured in linear yards, yarn was measured in pounds, and fine goods was measured in pieces—and all these measurements were inadequate to reflect the amount of work or quality of the product, such as the thickness and fineness of the weave. And third, for other industries, no production figures were available at all.[54] To overcome these difficulties, Clague relied on one of four methods. For one, he could calculate the amount of raw materials consumed, particularly in industries where a single raw material was part of virtually every product, such as the blast furnace industry where the output could be measured in tons of pig iron as well as iron ore consumed. Clague could also focus on one or two key products that adequately reflected the state of the industry, such as in the iron and steel industry where pig iron and steel ingots were only two out of fifty to one hundred products, yet they were intermediate products and thus virtually all finished products contained these two. Third, he could develop separate indexes for products within an industry and assign weights to each to calculate the overall production in the industry, as in the cigar and cigarette industry. And finally, he could use an index distinct from products or raw materials such as the total of spindle hours—that is, the machine run time in the cotton industry.[55]

For the second figure, the index of the total man-hours, Clague also adjusted the statistical basis, calculating the index of the total man-hours in a two-step process. In the first step, he determined the number of employees in a given year, using census and month-to-month employment data from the BLS. In the second step, he determined the average hours worked in a given week and year. He multiplied the number of employees by the average hours worked per year to get the number of total man-hours.[56] These employment figures required adjustment because of six issues. First, the standard daily or weekly hours were significantly reduced from 1914 to 1925. Second, the standard hours didn't take into account overtime during prosperity or part-time work during depression. Third, entire plants may have closed for two or three days a week during downturns, without the closure being reflected in employment data. Fourth, office employees may have worked different hours than salaried workers in the factory. Fifth, the amount of turnover in the workforce varied with prosperity and depression. And sixth, voluntary or involuntary absences, which increased or decreased with business conditions, were not accounted for. To adjust for the reduction of working hours, Clague used Census of Manufactures data on the average full-time hours per week per wage earner. The effects of the business cycle on employment (issues two through five) could not properly be taken into account except for cases where special studies on wages and hours were available. And the question about the hours for office employees proved not to be problematic for the period under investigation because it only affected a productivity index if the proportionate number of office employees in an industry changed significantly.[57]

It should be noted that like in the studies from the early 1920s, productivity appeared to be affected by factors other than technological improvements: fluctuations in the business cycle, demand for specialty products, and industrial growth. As Clague emphasized, productivity indexes revealed little about production *methods* in an industry. A low productivity index could simply mean that an industry was already established and had developed highly productive production methods that left little room for improvement—and thus resulted in a flat productivity index, as in the boot and shoes industry. By contrast, new industries with rapidly growing demand and output as well as rapidly improving production methods showed rapidly rising productivity indexes, such as the rubber tire and automobile industries. Therefore, it was impossible to make direct comparisons between industries based only on productivity indexes.[58]

All eleven industries under investigation experienced increases in productivity over the years 1914 to 1925, yet these increases varied widely from over 200 percent in the rubber tire industry to a mere 6 percent in the boot and shoe industry. The rubber tires and automobile industries saw the largest increases with indexes of 311 and 272, indicating 211 percent and 172 percent productivity increases, respectively, over 1914.[59] Both were new industries that experienced vast expansions of production between 1914 and 1925; the output—not the productivity—of rubber tires even increased sevenfold.[60] Both industries also likely benefited from improvements in production techniques as they grew.[61] The next-largest increases in productivity were in petroleum refining, cement manufacturing, and the iron and steel industry. Petroleum refining was a comparatively new industry, and the cement industry had gone through a period of rapid expansion since 1919, allowing existing factories to be used to their capacity and increasing the productivity of the industry. The iron and steel industry showed quick responses to the overall economic situation, and was also subject to labor unrest, both of which affected the industry's productivity.[62] Four industries

Table 1.1
Labor Productivity Index in Eleven US Industries, 1914 = 100

Year	Rubber tires	Automobiles	Petroleum	Cement	Iron and steel	Flour	Paper and pulp	Cane sugar	Slaughtering	Leather	Boots and shoes
1914	100	100	100	100	100	100	100	100	100	100	100
1915					120						
1916		120			124						
1917		133			109		101				
1918		90			103		101			98	
1919	130	136	92	103	100	96	194	79	93	101	105
1920		150			115		102			99	
1921	190	193	111	124	94	118	94	82	119	126	115
1922		249	126		136		118		125	130	116
1923	266	270	135	132	139	128	116	102	128	134	107
1924	301	262	163	143	137		128	114	129	131	107
1925	311	272	183	161	159	140	134	128	127	126	106

Source: Adapted from Anonymous, "Productivity of Labor in Eleven Industries," 37.

showed more moderate increases in productivity: flour milling, paper and pulp, cane sugar refining, and slaughtering and meatpacking. In industries supplying basic food products—such as flour milling as well as slaughtering and meatpacking—more constant productivity indexes were to be expected because of a relatively stable demand regardless of the economic cycle.[63] The paper and pulp industries benefited from the introduction of new machinery in the 1920s, leading to increases in the productivity index, yet the overall productivity increases in standard products such as newsprint and wrapping paper were counterbalanced by lower productivity in specialty products. The lowest productivity increases were in the leather tanning as well as boots and shoes industries. Productivity in these industries was closely tied together because most leather was used for the production of shoes and boots. Productivity in both industries was negatively affected by the rise of demand for specialty and novelty shoes that allowed only for short production runs and defied mass production methods (table 1.1).

In addition to productivity differences *between* industries, there could be vast disparities in productivity *within* an industry, between highly productive plants and less efficient ones. In the boot and shoe industry, for example, less productive plants remained competitive because they focused on the production of specialty rather than mass-produced products, while in the flour milling as well as slaughtering and meatpacking industries, less productive but local companies benefited from low transportation costs. Similarly, in cement manufacturing, the existence of easily accessible raw materials might outweigh productivity advantages. Together, these factors made it "possible that small firms will continue in such industries for years, not able to develop any great increase in productivity or reduction in costs, but nevertheless able to maintain their position in the industry."[64]

Clague's work marked a shift in the BLS's productivity measurements from descriptive narratives based on limited data to reports based on quantitative information. Published in the *Monthly Labor Review*—the BLS's journal of information, research, and analysis on the labor force, the economy, employment, wages, prices, and more—the indexes reached an audience of economic experts in local, state, and federal government administrations as well as corporations, labor unions, and academia. Addressing an expert audience, Clague laid open the methodological considerations behind the productivity indexes: the problems of measuring total production and the limits of calculating the total work time. While the index was not a perfect measure, it was the best approximation. Soon, however, the index would take on a life of its own, devoid of such methodological considerations. The seeming objectivity of the productivity index measured the amount of work necessary to achieve a certain standard of living, while the cost-of-living index provided a complementary measure of the price levels affecting the standard of living. Employers may have been more concerned with how much work they got for a certain wage, and workers may have been more concerned with how much they could afford from their wages.

Productivity and Wages

With the productivity indexes published in 1926 and early 1927, Clague addressed the relation of productivity and wage in a 1927 article on "Productivity and Wages in the United States" in the American Federation of Labor (AFL) magazine, the *American Federationist*. He argued that there was no immediate connection between productivity and wages.[65] Clague's modest family background as well as his studies predisposed this self-described progressive to a labor-friendly position. His father was an immigrant farmer from Ireland, and Clague was a first-generation college student who supported himself by working on farms during the summer, and delivering papers and waiting tables during the academic year. He expected to eventually become a teacher, but found mentors who made it possible for him to study with Commons in Wisconsin. Clague thought of himself as being "very sympathetic and friendly to organized labor" because of his graduate work.[66] He described the atmosphere at the BLS as one of independence from organized labor, which he explained was presumably due to the need to obtain voluntary figures from employers. Given his own labor-friendly disposition, he was surprised to find antagonism toward labor among some BLS officers, who, for example, criticized a labor strike.

 Driven by the BLS's tradition of concern over technological change, Clague admitted that the effect of a plant's arrangement could not be "measured in any numerical way," and that it was even difficult to discuss such arrangements so as to keep individual plants from being identified. At the same time, he held that "no one who has seen the C.F.&I. plant could fail to be impressed with the poor arrangement, the extra transportation required, etc.," while other plants "have a straight continuous flow of materials thru to the finished steel."[67] While Clague planned a number of publications—studies addressed to a larger, public audience and numerical measures for an expert audience—he and his collaborators were laid off in 1928, and he only published a single article from the study.[68] Clague insisted that there was no proportionate relation between productivity and wages. This could be seen as an attempt to tame possible uses by labor unions of the index in future contract negotiations in order to demand wage increases based on a seemingly objective number.

Before Clague, other BLS officers had raised the question of productivity and wages during the 1920s. In a 1921 address to the Annual Convention of the Association of Governmental Labor Officials of the United States and Canada, BLS commissioner Stewart argued that while "for years the slogan of both capital and labor has been 'A fair day's wage for a fair day's work,'" and while cost-of-living statistics have allowed fixing a fair day's wage on the basis of living costs, "there is nothing tangible in any large industry that can be used to determine a fair day's work."[69] At the same time, a proposed study to improve performance in railway shops was not undertaken because of its implications for wage negotiations. As the proposal noted, the outlined plan rested "primarily on confidence between the organized workers and the railway management"; productivity improvements as a whole would become "one of the guiding principles in wage determination" and establish a "premium on efficient production."[70] Unable to find a "railway management that is intelligent enough to realize its opportunity and inaugurate an experiment along the lines suggested herein," the BLS kept the aborted study proposal confidential.[71]

Similarly, Clague had argued before that three corporate groups could be involved in productivity improvements: workers, investors, and managers. Productivity could be increased based on an invention by a worker, research conducted at the expense of investors, or new processes installed by managers. It was therefore difficult to assign credit for increases in productivity. In addition, those helping increase productivity may not be the ones benefiting from it. As Clague recognized, "the benefits of increased output may not go at all to the class largely responsible for it." For example, in monopolistic markets, "capitalists" were often able to "reap most of the benefits." In other cases, such as the linotype machine, a powerful union enabled workers "to get first chance at learning the new machine and secured for them a fairly good share of the benefits." And in competitive markets, "nearly everything" might "go to the public in reduced prices."[72]

By late 1927, Clague discussed the relation between productivity and wages in more detail, again only to dismiss a proportionate relationship. He began by emphasizing the importance of real wages—that is, the purchasing power of the money received by workers—rather than nominal wages, because under conditions of inflation, an increase in the nominal wage could still mean a decrease in the purchasing power if price increases were larger than wage increases. Clague then distinguished wages from earnings, with wages meaning the rate of pay per hour, while earnings expressed the amount received per week, per month, or per year. Although wages allowed for easy comparison, mere consideration of wages was meaningless to workers who were also concerned about their earnings in order to pay for their living. For example, workers may in principle desire a reduction in hours, but such a decrease may not be practical if it results in a reduction of overall earnings.[73] Therefore, Clague stressed, there was "no immediate and exact connection between high productivity of labor in an industry and the wages paid in that industry."[74]

The reason for this disconnect between productivity and wages, Clague explained, lay in the distinction between the so-called quantity principle and value principle. The quantity principle demanded that the more the persons in an economy collectively produced, the more goods (and services) they had to distribute among themselves, and the richer everyone was. It suggested that higher labor productivity led to greater wealth. The value principle, by contrast, held that the more goods produced, the less each of them was worth, and the lower their price would be. It called for higher protective tariffs, for instance, to ward off an influx of additional goods and competition. In the second half of the nineteenth century—called a "period of abundance" by Clague's dissertation adviser, Commons—the quantity and value principle were in conflict. Individuals and small groups were interested in a restriction of output to get more wealth for themselves by doing as little as possible, while the nation was interested in the highest-possible production because the higher production of wheat or cotton meant cheaper bread or clothes for consumers. British trade unions thus sought to restrict output and were attacked for this policy as selfish, although the unlimited production led to periodic crises of depressions, overproduction, and unemployment.[75] In the United States, businessmen in the 1880s and 1890s recognized that guided or regulated production would prevent such crises. Forming combinations and trusts, they inaugurated what Commons called the "period of stabilization."[76]

By the 1920s, Clague recognized business as the regulating economic factor. As he wrote almost with clairvoyance shortly before the Great Depression hit, "the business man is the one who is responsible for adjusting the amount of production to the needs of the country." A businessman who stopped "producing in time to prevent a glut on the market" served the national interest better than one who kept producing at highest capacity until a situation of overproduction led to plant closures, bankruptcy, and economic crisis.[77]

Under these conditions of a period of stabilization, Clague argued, trade unions could cooperate with unrestricted productivity and "safely guarantee complete cooperation in increasing productivity." Following the same view, the AFL had already announced that it would pursue a "policy of determining wages on the basis of productivity, and, wherever employer cooperation can be secured, the unions will guarantee unrestricted production on this basis." In conclusion, Clague underscored that the correlation between wages and productivity was not high within individual industries:

> Where productivity is highest, the wages are likely to be high, too, but the wages will not be by any means in proportion to productivity; and where productivity is lowest, the wages will be higher than productivity would justify. The workers in industries where productivity is high ought not to permit themselves to be misled into expecting wage rates so high that the industry will not pay them, and, on the other hand, workers in industries of low productivity must continue to rely upon bargaining power and an appeal for a decent standard of living.[78]

Clague recommended that workers not restrict their output if they wanted steady work at good wages, and that productivity benefits be equally divided between higher wages and shorter hours for all workers.

As a correlation of earnings and productivity from the 1910s to the 1930s showed, both followed similar overall trajectories but occasionally showed reverse trends, such as when earnings dipped while productivity continued to increase in the mid-1920s (figure 1.2). These numbers appeared to confirm—and possibly even motivated—Clague's cautionary position.

Clague's article may have anticipated union demands that did not come to pass—at least not in the form of a major union campaign. In the early

OUTPUT PER MAN-HOUR AND "REAL" HOURLY EARNINGS

MANUFACTURING 1914–1939

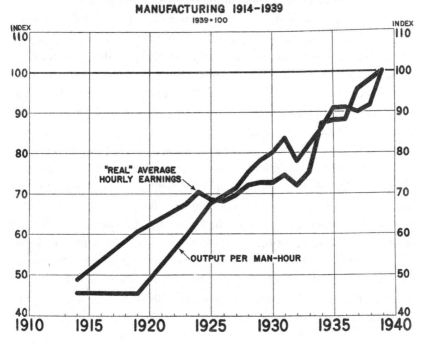

Figure 1.2

Productivity and earnings in the United States, 1910–1940.

Source: Ewan Clague, "The Consumer Price Index and Productivity Statistics," speech delivered at the University of Wisconsin at Madison, April 22, 1959, box 13, RG 200, NARA.

twentieth century, workers had come to think of themselves as consumers, demanding a living wage that would allow them an American standard of living. At the same time, scientific management had promoted a tendency to individually assess a worker's work and set the wage accordingly. In response to such individually assessed wages, often realized in task-and-bonus or differential piece rates, unions had fought for a graduated scale of standard rates in the 1910s that recognized a variety of skill levels.[79] In other words, unions had fought for standard rates rather than individually assessed premium rates, and had tied these standard rates to the roles of skilled workers and consumers. Union demands in the United States had typically not been tied to productivity.

Before he was laid off, Clague himself started a new productivity study, with his productivity index gaining attention within the Department of Labor as well as from companies such as US Steel. With two colleagues, he embarked on an in-depth study of productivity in the iron and steel industry, surveying blast furnaces, open-hearth steelworks, and Bessemer steelworks across the United States for six months.[80] Combining statistical analysis and fieldwork, this study methodologically coupled the BLS's older work with Clague's newer work on indexes. Yet when US Steel eventually withheld its data, the study fell apart, and only one publication resulted from it: an article in the bulletin on independent merchant blast furnaces.[81] When the BLS could no longer finance his position, Clague moved on to positions as a researcher at Metropolitan Life Insurance, Yale University's Institute of Human Relations, and the University of Pennsylvania's School of Social Work before returning to federal service in 1935 for the Social Security Board and the Unemployment Insurance System. In 1946, he was named BLS commissioner—a position he held for almost two decades; in this role, he would bring the bureau's productivity measurements to the Marshall Plan.

Conclusion

The BLS's productivity index went on to a distinguished domestic and international career, although the attention of BLS officers in the 1930s first shifted back to the cost-of-living index. During the union-friendly New Deal years, work on the cost-of-living index aligned the BLS with the liberal analysis of underconsumption that sought to protect and expand (hitherto-weak) labor organizations to build and maintain mass purchasing power. During World War II, the cost-of-living index again served mobilization efforts, like in World War I, albeit this time not without controversy. In the postwar years, however, productivity would be in the limelight again. Domestically, General Motors and the United Auto Workers, in a surprise collective bargaining agreement in 1948, tied future wage increases to pro-ductivity measures as well as to the cost-of-living index, by now renamed the consumer price index. In the succeeding years, many large corpora-tions followed the example of this so-called elevator clause. Internationally, productivity became a conceptual core of the Marshall Plan, a vehicle to bring the US model of capitalism to participating Western European countries. BLS officers supported the Marshall Plan administration with productivity calculations and assessments of productivity levels in Western European economies.

For this later work, BLS officers built on the methodological sophistication of their productivity calculations from the 1920s. During this decade, they moved from calculations based on a small number of voluntarily reporting companies to calculations based on encompassing census data. They moved from narrative descriptions of production methods to a numerical index calculating relative changes in production output per input. They shifted from occasional studies to regularly published indexes. And they moved from discussions within an expert community to at least considerations on how to educate a larger public audience on productivity. Before World War II, aside from the index and a small number of articles, productivity did not win wide recognition and remained largely an internal affair to the BLS. As later chapters in this book will show, this was to change greatly after World War II, when productivity gained a great deal of publicity in industry periodicals and the general press.

This chapter also demonstrates that the origins of the BLS's productivity research were situated in the context of concerns over technological change. Already the first BLS study of production methods in 1898 had been motivated by worries over changes brought on by industrialization as well as the effects of women and children tending to the newly introduced machines. With employment changing seasonally, and with the economic cycle and high turnover rates, unemployment from technological change was not a primary concern; de-skilling was the issue. By the 1920s, BLS officers suggested that technological change improved productivity rates, although their studies did not necessarily bear this out. Still, their reports presented technologically advanced companies as models to be emulated. In the work of the BLS, productivity was thus intimately tied to technological change.

The BLS's periodic shifts between cost-of-living research and productivity research correlated with larger public attitudes toward labor unions. During the 1910s, the New Deal years, and World War II—times of strong union organizing—the office focused on its cost-of-living index; in the less union-friendly periods of the 1920s and after World War II, officers worked on productivity. This chapter thus began to reveal the interpretative flexibility of productivity, this seemingly objective statistical measurement, which meant different things to different social groups. In the remainder of this book, we see how this interpretative flexibility played out with regard to labor and class relations in the transatlantic context. Yet after World War II, when European countries asked the Marshall Plan administration for information on the economic and social effects of the latest technological change, automation, US administrators refused to give a response. In Europe post–World War II, technological advancement would be viewed as a threat to workers' employment altogether, not merely as a threat to lowering the skill levels of workers. Also, Clague and Stewart's decoupling of productivity from wages during the 1920s stood in stark contrast to later claims by the Marshall Plan administration that higher productivity would lead to higher standards of living in Western European countries, promising at least a trickle-down economy, if not outright salary increases from productivity. But before turning to the postwar period, we will first look at transatlantic productivity exchanges in the 1920s, and the social and economic values that American and German entrepreneurs and trade unionists gave to the iconic productivity technology of the second Industrial Revolution, Ford's mass production lines.

2 Weimar Rationalization Appropriates American Productivity

Beginning in the early 1920s, a wave of Weimar Germans traveled across the Atlantic to learn more about the United States. They were interested in its economy, culture, and geographic and climatic conditions with the goal of improving their own country's economy. One of the first Weimar Germans to visit was Carl Köttgen, an electrical engineer and member of the Siemens board of directors. In his widely read 1925 travel report, he stated that "only when every single person recognizes that progress depends on working more, will we heal again."[1] Köttgen looked to the United States for a model for rebuilding the German economy after the First World War and became a leading figure in the German rationalization movement, where industrialists and engineers, sometimes in collaboration with labor unionists, sought to "rationalize" German industry—that is, increase industrial efficiency through technological and organizational improvements such as mechanization, division of labor, work organization, and in some prominent cases, continuous flow production.[2] While most German visitors embraced such technological and organizational improvements, the most powerful American voices—including industrialists—linked these production methods to high wages and low prices—a connection rejected by most Weimar Germans, particularly industrialists. Thus German technical rationalization appropriated American productivity.

After returning home, the Weimar Germans did not agree on many things. Köttgen's report stood out from the cacophony of voices, and many subsequent travelers emphasized his call for longer and more intensive work rather than costly technological improvements. One of the strongest criticisms came from the German socialist trade union association, which felt that if workers wanted to have a say in the ongoing economic debate and shape economic policy, they needed firsthand knowledge of the American economy. Setting out to assess the intensity of work—the pace and output—in the United States, the group returned with the realization that it was not high intensity but "high salaries [that] caused the blossoming of the economy." The trade union group concluded that in Germany, the "general recipe against the disease of the economy" was "to produce a lot and consume little"; therefore, Germany had "low salaries and long work hours." But in the United States, workers had high salaries, worked fewer hours a week, and could purchase goods at lower prices. Business profits were based on small margins and large volumes. These conditions allowed new production organization and technologies to flourish.[3]

Many American entrepreneurs welcomed the Germans with open arms and unlocked their factory gates to show off their new production methods and economic success. Businessmen who were owners of small or medium-size companies, and who were ethnically German, proudly displayed their successes, lambasting German industrialists for their narrow-minded thinking, lack of ambitions, and social preconceptions. In larger companies, workers who spoke German were asked to guide visitors through the facilities. Going beyond merely accepting visitors from abroad, many leading American entrepreneurs tried to describe what they thought was the cause of their success. Henry Ford's autobiography, for example, was translated into German, French, Spanish, and Japanese, among other languages, carrying his ideas about successful business practices abroad. The department store magnate Edward Filene traveled to Europe annually, promoting similar economic views. Promising a higher standard of living through new production technologies, mass production, and mass consumption, they promoted their form of capitalism, which has been called welfare capitalism.

Welfare capitalism was a system of labor relations, emerging in the United States in the early decades of the twentieth century, in which firms took over responsibilities for the welfare of their employees from weak unions and a weak state.[4] These firms often paid comparatively high wages, provided insurance programs and medical services, and offered amenities, such as clubs and vacation homes, in return for their employees not joining trade unions; at the most, these firms accepted a company union. The Ford Motor Company was one of the most prominent examples of a welfare capitalist company, albeit a somewhat idiosyncratic one that paid an unusually high wage of $5 a day and closely monitored workers' personal lives through the company's Sociological Department. Other leading examples of welfare capitalist companies in the United States include the National Cash Register Company, Eastman Kodak, Sears Roebuck, and Thompson Products.[5]

In contrast to the officers of the Bureau of Labor Statistics (BLS), which we discussed in chapter 1, welfare capitalist entrepreneurs frequently imbued technological change with ideas about labor relations. Ford, for instance, saw continuous flow production as part and parcel of high wages and low prices. This chapter focuses on Ford and his ideas because Germans primarily looked to Detroit and took the Ford Motor Company to be representative of the US economic system. Of course, Germans also became ardent adherents of the time-motion studies associated with the efficiency improvements and the scientific management movement building on Frederick Taylor's work.[6] But Germans formed their own term, "Rationalisierung," referring to industrial restructuring and technological modernization, new workers shaped by scientific management methods, and new consumer products. The German rationalization movement revolved around the adoption of technical science and systematic organization for the production process and other areas of productive activity such as housework, from rationalized worker seating that reduced unnecessary movements to rationalized kitchen designs.[7] While the rationalization movement in Germany evolved through close encounter with the American model, the two are not congruent. In this chapter, I use the term "rationalization" for the technology-focused approach to industrial improvement in Germany.[8]

In the 1920s, transatlantic exchanges were based on individual initiative, not government programs. It was up to individuals like Ford and Filene to propagate their ideas of US capitalist relations abroad. And it was up to individuals such as Köttgen or union officers—traveling on their own or, more often, their organization's dime—to look at what the United States had to offer. After World War II, such individual exchanges would continue, but as we will see in chapter 3, the US Marshall Plan's Productivity Program would take over, pay for, and plan the trips for scores of European visitors to expose them to the program's idea of higher standards of living through productivity, free enterprise, and collaborative labor relations. But in the 1920s, German visitors planned their own itineraries—although there were some almost-obligatory stops, such as Ford's factories and the Chicago slaughterhouses. German visitors varied widely in their perception of US companies, production methods, and economic relations.

While a few writers focused only on Ford and his company, most visitors sought to capture more broadly the US economy, culture, and geographic and climatic conditions. Some, like Köttgen, dismissed US production technology as unrealistic for German conditions because the machinery would require large investments, and it was more profitable to continue paying low salaries. Others, like engineer Franz Westermann and the union delegates, saw new technologies and organization as essential for transforming the German economy. Westermann lauded American workers for their motivation, and accused German workers and labor unions of opposition to technological change, while the German union delegates as well as bourgeois writers such as the newspaper editor Arthur Feiler sympathized with the cause of US labor unions. Almost all Weimar observers, however, sought to wrap their minds around American prosperity by comparing nominal wages and purchasing power, price levels, and consumption patterns with those in Germany. The wide variety of views and arguments cut across social, political, and ideological lines, and defy easy categorization of German perceptions of the US model; it is impossible to distinguish Weimar writers along binaries such as bourgeois and working class, left and right, technocratic and technophobe.[9]

The goal of this chapter is to discuss transatlantic exchanges concerning ideas about labor, technology, and social relations in the 1920s. After a brief overview on foreign activities by US entrepreneurs in the 1920s, the chapter examines Ford's economic ideas as well as his view of the close relation between the continuous flow of production, high wages, and low prices. It concentrates on Ford not necessarily because his ideas were idiomatic of the US economy, but rather because many German visitors perceived them to be so. Then, the chapter takes a closer look at the wide-ranging German perceptions of the US economic model. After a brief prosopography of German visitors and a description of the rationalization movement in Germany, it discusses Köttgen's call for workers to work more intensely and for longer hours, and then turns to the response from union representatives. They treated production methods with brevity in their report, while they perceptively understood the social and economic relations, such as the link between welfare capitalist measures and the divisions between skilled and unskilled workers. This chapter also points to transatlantic differences, with Americans—including industrialists—increasingly linking productivity with high wages, while Weimar Germans—particularly industrialists—embraced technological rationalization and rejected demands for high wages. The chapter ends with a brief discussion of rationalization during the Nazi period.

Spreading the American Vision

With the United States wearing the mantle of rescuer of Western civilization after its victorious intervention in the First World War, Europeans increasingly looked across the Atlantic, seeking a vision of modernity in US mass production methods. In the centuries before, transatlantic exchanges had been lively, but Europe held the center stage. In the age of exploration, American explorers maintained correspondence networks, sending specimens of seeds, plants, and bones to learned European centers in Paris and elsewhere, where these specimens were incorporated into contemporary systems of knowledge. By the early nineteenth century, the young American republic began to seek European technologies and the technical knowledge embodied in craftsmen and engineers. Thus, mechanized weaving looms made their way across the Atlantic (despite a British export ban), accompanied by craftsmen who knew how to set up and operate the cutting-edge machinery. Well-to-do Americans like Moncure Robinson traveled to Europe and brought back new engineering wonders such as railroad technology, adapting them to local environmental and financial circumstances.[10] Throughout the nineteenth century, young Americans continued to tour Europe for education and inspiration, and they often reshaped European ideas to fit American conditions; for example, German research labs as places of theoretical learning and scientific practice turned into American scientific graduate schools.[11] But it wasn't until the turn of the twentieth century that Americans began to liberate themselves from their peripheral position.

Increasingly, Americans felt that they had something to offer the rest of the world, and they shifted from expanding their western frontier to expanding their influence and interests globally. In particular, nationally operating firms started to expand abroad in the late nineteenth and early twentieth centuries, such as Singer, Standard Oil, NCR, and International Harvester, as well as telegraph and telephone companies, insurance companies, and drug, film, and explosives companies like DuPont and Kodak. Initially they used independent sales agents, but then hired salaried export managers, opened sales and distribution branches, and eventually built finishing, assembly, or manufacturing plants abroad.[12] In addition, missionaries such as the student volunteers of the YMCA carried Christian values abroad, American philanthropists began to engage across their country's borders, and journalism and Wild West shows—horse-riding and shooting performances that reenacted historical events of America's western frontier—carried emerging mass culture abroad. The US government assisted these international activities by working to reduce foreign restrictions, negotiating beneficial tariffs, and building a naval program to provide protection at sea; but it refrained from policies that influenced global markets and politics.[13]

Although the US government retreated into political isolationism after World War I, the 1920s were characterized by increasing US international engagement, frequently spearheaded by business entrepreneurs and professionals who continued to carry their social, economic, and technical ideas abroad. The YMCA and the Rotary International founded chapters in Europe that promoted the ideas of service and an American standard of living that clashed with the strict European divisions between upper bourgeois and working classes.[14] American companies continued to expand their international operations, sourcing raw materials, such as oil and rubber, from around the world as well as increasingly marketing their mass-produced wares abroad. Historians of international business have argued that the two decades following World War I were a time of lower international engagement—and the war certainly interrupted the colonial flow of money and people between the centers of the British and other European empires and their colonies.[15] But many US companies expanded their international operations in European and South American countries during this time. To bypass tariff barriers, increasing numbers of US companies set up local assembly and even manufacturing plants in foreign countries.[16]

For instance, all three predecessor companies of IBM had engaged in international operations since the early twentieth century: the International Time Recording Company had sales representatives in foreign countries, the Computing Scale Company opened a factory in Toronto in 1902, and the Tabulating Machine Company entered into licensing agreements with British and German companies.[17] Similarly, the Ford Motor Company exported its first vehicles in the company's founding year, 1903. A year later it established a manufacturing plant in Canada, just across the river from

Detroit, and one in Manchester, England, in 1909. International operations expanded quickly with the establishment of continuous flow production in 1912–1913. Ford opened an Irish manufacturing plant immediately after the war and incorporated companies in France, Denmark, Spain, and other European countries. Through the Canadian company, it expanded into areas of the British Empire such as South Africa and Australia, and established operations in the Middle East and Japan.[18]

The International Chamber of Commerce (ICC), founded in 1919, was an important vehicle to promote free trade and help the economic reconstruction after the war.[19] US businessmen—and most of them were men—joined the ICC seeking international economic cooperation at a time when the US Congress retreated from international political cooperation and refused to ratify US membership in the League of Nations, the international body that president Woodrow Wilson had created during the Paris peace negotiations in an effort to help postwar reconstruction and avoid future military conflicts. Trade barriers in particular proved to divide the US business community and politicians. While American ICC members, mostly large international companies, fought for the reduction of tariffs, smaller companies operating only in the United States sought higher tariffs to protect their markets. To the chagrin of American ICC members, the US government solved this conflict by engaging in "most-favored-nation" policies—bilateral agreements that granted individual countries favorable access to US markets in return for US access to their markets—and establishing high tariffs for all other countries. Also, the US government refrained from canceling war debts, which meant that those debtor nations were unable to purchase US goods. Instead, the US government encouraged businesses to extend private loans under government supervision to other countries, as in the case of the $100 million loan by J. P. Morgan as part of the 1924 Dawes Plan to address the German hyperinflation crisis. By contrast, ICC members preferred trade agreements for tax-free zones; some members in fact envisioned a "United States of Europe" with a tariff integration that would eliminate internal European tariffs.[20]

Among the American founding members of the ICC was the retail magnate Filene. An eccentric bachelor, Filene was best known for his Bargain Basement in Boston, a store where consumers could get wares for cheap, thereby helping to accelerate American consumerism. Filene pursued progressive employee relations, founding the Filene Cooperative Association, which ran its own credit union and food cooperative. Employees also elected four of the nine directors on the company's board, and among other things, voted to close the store on December 26, one of the busiest shopping days of the year. A leading spokesperson for industrial unionism, in 1919 Filene also created the Twentieth Century Fund, which in its initial years funded union initiatives. Ousted from his family's business in 1928 because of his liberal management ideas, Filene continued to travel to Europe almost every summer, visiting spas and hobnobbing with local business and political elites, although he did not expand his own retail operations into Europe, where department stores continued to offer their wares to mostly bourgeois customers at more elevated prices, and only a few nickel-and-dime-store chains existed on suburban peripheries. While it is not clear whether Filene talked about his progressive employment policies in Europe, he did seek to disseminate the idea of distribution—mass distribution. Convinced that nondistribution, not overproduction, was the obstacle to being profitable, he advocated marketing measures such as product design, packaging, and salesmanship in addition to low pricing.[21]

IBM's Thomas J. Watson also became active in the ICC and was elected president in 1937. In his keynote address to the biennial ICC congress, held that year in Berlin, he promoted his slogan of "World Peace through World Trade," expressing his view of the important role that private business was to play in international relations (figure 2.1).[22] His conviction, however, that he had persuaded Adolf Hitler to refrain from military aggression for the sake of international trade, as well as his ill-fated decision to accept the German Cross of Merit award from the Nazi government—the second-highest civilian award for a foreign national—reveals a certain naivete in international political relations.[23]

Watson's IBM was then still an emerging office machine company; it had only changed its original name, the Calculating-Tabulating-Recording Company, to the International Business Machine Company in 1924. During the 1920s, the company had also expanded into South American markets as well as growing its operations in European markets, creating manufacturing facilities in Sindelfingen, Germany, in 1924, in Vincennes near Paris in 1925, in Berlin in 1934, and in Milan in 1935. Watson, the company's visionary chairman, had traveled to both South America and Europe to establish his business, often with his wife and sometimes even

Figure 2.1

In 1938, Thomas J. Watson's slogan "World Peace through World Trade" was inscribed on this building plaque that greeted visitors to the IBM headquarters building at 590 Madison Avenue in New York.

Source: IBM, "The Creation of the World Trade Corporation," Icons of Progress, IBM 100, https://www.ibm.com/ibm/history/ibm100/us/en/icons/ibmworldtrade/. Courtesy of International Business Machines Corporation, © International Business Machines Corporation.

his four young children in tow.[24] But the real incarnation of the American businessman, at least in the eyes of most Europeans, was Ford.

Ford and the American Economic System

In the early 1920s, Ford was at the height of his national and international power, with his company producing six out of ten cars in the United States, and audiences abroad devouring his autobiography, *My Life and Work*, even more so than his contemporaries at home. The German translation in 1923, for example, caused a sensation, going through thirteen printings in its first year, and selling over two hundred thousand copies.[25] It also motivated German businessmen, economists, political scientists, newspaper editors, union representatives, and others to travel to the United States in search of a model for German reconstruction after the war—trips that would include an almost-obligatory stop in Detroit. German visitors often (mis)took the Ford Motor Company as representative of American business, technology, and productivity. The company was in some respects exceptional compared to typical American companies in the 1920s. The productivity of the Ford Motor Company increased more than threefold during the 1920s, surpassing other industries as well as other automobile companies. Also, Ford had embraced welfare capitalist measures comparatively early, although in a unique and especially paternalistic fashion with the infamous $5-a-day wage.[26] And in contrast to BLS officers who warned that there was no proportionate relation between productivity and wage increases, Ford felt strongly that a worker's higher productivity ought to be remunerated through higher wages. It was to the Ford Motor Company that Germans primarily looked in the 1920s.

Ford idealized his company's labor and customer relations in his auto-biography, even as they had already begun to unravel. The company had initially assembled cars one by one, with skilled craftsmen responsible for multiple steps, and with each part brought to the stationary car. This changed as the company increased production: now the cars moved on the assembly line from worker to worker, and the production process was even-tually divided in thirteen-second increments, with every worker respon-sible for only a tiny step. For example, tightening a screw could involve three workers and three steps: the first worker placed the screw, the second worker screwed it in, and the third worker tightened it. While Ford had at first purchased parts from third-party vendors, the company began to pro-duce all the parts, except for rubber components, in-house. As production was divided into smaller and smaller steps, unskilled workers, trained on the job, replaced skilled craftsmen, some of whom found new positions in production planning as well as the making and maintaining of machine tools. Ford freely conceded that he did not invent flow production—he was inspired by the Chicago meatpacking industry—but his company did advance flow production to an unprecedented degree.[27]

Ford engineers analyzed the company's production process to eliminate waste in both time and materials, as Ford emphasized. For example, they conducted time-motion studies to determine the best position of worksta-tions and tools to eliminate workers' need for movement; in particular, they reduced the need for workers to carry pieces from station to station, which had previously taken up to a third of a worker's time. Instead, the assembly line or, wherever possible, gravity-driven slides moved parts from station to station. Workers were positioned so that the least amount of movement was required of them. Likewise, the company eliminated material waste wherever possible such as by cutting parts from sheets of metal to minimize leftovers, or melting and reusing metal shavings. This focus on maximiz-ing the efficiency of his company's production also led Ford to sever the connections with his initial investors, including the Dodge brothers in a divisive lawsuit, so that he could reinvest profits as well as prioritize long-term corporate expansion and production improvements over short-term dividend payments.[28]

Ford also had strong convictions when it came to labor relations and salaries. In January 1914, his company introduced the $5-a-day wage (which by the early 1920s, was increased to $6 a day). Technically, this was not a guaranteed salary but rather a bonus for which workers had to qualify; Ford himself called it a "profit-sharing plan."[29] Based on approval by the company's Sociological Department—which monitored workers' private lives, from their family and housing situation to drinking and other habits—workers who had been with Ford for six months or longer qualified for bonus payments if they met certain efficiency thresholds. The bonus was calculated so that each worker would make at least $5 a day, giving those with lower salaries a larger bonus. For instance, a worker with an hourly rate of 34¢ received a profit of 28.5¢ per hour, which resulted in $5 for eight hours, while a worker with an hourly rate of 54¢ received a profit of 21¢ per hour, resulting in $6 a day.[30]

These high wages achieved multiple goals in Ford's view. First, they addressed labor turnover. In 1913, the company had to hire fifty-four thousand workers to fill thirteen thousand workstations. The need to take in and train new workers plummeted because workers stayed at least a few months to qualify for the higher wages, and they tended to stay on longer because of the high wages. Second, the wages aimed to increase efficiency. Different from a dividend paid at the end of the year, the bonus payments were immediately tied to production goals. More generally, Ford argued that higher wages allowed workers to meet their family needs without worries, and, being worry free, they could focus their energy more fully on their work. Third, the wages created demand for the ever-increasing number of cars. Not only would workers be able to purchase cars themselves; workers would spend their wages on other products, bringing money into circulation, making the community and other enterprises richer overall, and thus creating more demand from other people as well.[31]

This leads us to Ford's ideas about prices and customer relations. Ford was convinced that there would be enough customers for his cars as long as the price point was right and argued that with every price cut, he reached a new stratum of potential customers. As a consequence, he rejected the idea of market saturation, foreseeing instead an unlimited demand for automobiles. Farmers often owned two cars rather than just one by the 1920s, and many owned a truck, too.[32] The price of the Model T rose occasionally to finance the company's expansion, such as in 1909–1910, when the company built the Highland Park plant. More typically, the company *cut* prices to reach more customers. Ford claimed that the company cut prices regardless of production costs; that is, it accepted lower profits in the years following price cuts. According to Ford, production costs usually caught up with price cuts, as the lower price exerted pressure on further increasing the efficiency of production.[33]

As is well known, Ford was an autocrat and a man of strong convictions.[34] His $5-a-day wage was criticized for the company's patriarchal interventions into workers' private lives. By the early 1920s, the Sociological Department had been largely replaced by corporate police, who at one time had about 10 percent of the workers enlisted as spies to prevent organizing efforts. This led to workers being dismissed for minor infractions such as misplacing their badge or staying too long in the restroom.[35] Despite these idiosyncrasies, Ford's was a welfare capitalist company. Sure, the $5-a-day program was unique, but good wages were a typical feature of welfare capitalist benefits. In addition, Ford incorporated other welfare capitalist features, such as clean and well-aired workplaces, extensive insurance and pension systems, an employee suggestion program, and job security—all in exchange for workers not organizing through a union, be it a company or national union. Until the eve of the Second World War, Ford refused to negotiate with any union in the United States or abroad, claiming that the company already offered higher wages and shorter hours than union representation could promise.[36] And Ford's frequent price cuts and reinvestment of profits into the company's growth were not typical for other welfare capitalist companies. Yet it was to Ford's Motor Company that Germans looked as a model.

It is important to keep in mind that Ford's ideas about production, wages, prices, and customers were part and parcel of a larger, indivisible system. Without high wages and low prices, there would not be enough customers to warrant the incredible expansion of Ford's company. Without high wages and low prices, the company would have overproduced and collapsed. Many German visitors, however, looked only at Ford's production technologies, and disregarded his ideas about wages and prices—or, at the best, considered them bizarre.

Weimar Visitors and Rationalization in Germany

Countless German industrialists, engineers, union representatives, social workers, members of parliament, academics, journalists, and writers traveled to the United States during the 1920s to study American production methods. Whether it was to renew business contacts after a disruptive decade, open new markets, or simply satisfy curiosity motivated by Ford's autobiography, most Germans included a stop in Detroit on their itineraries, as mentioned earlier, and many published travel reports on their return.[37] In certain German circles it was considered rude to ask someone whether he—or in a few cases, she—had visited the United States; a more appropriate question was whether the person had had an individual tour of the Ford factory or had even talked to Ford himself.[38] Alice Salomon, a social worker, was asked on her arrival in New York whether she had already finished her memoir—a question that certainly indicated there were a large number of German visitors and that they had a tendency to write about their American journeys.[39] The question may also indicate a suspicion that Germans arrived with preconceived notions of what they would find on the new continent.

German visitors formed a cacophony of voices, some reporting on their visits from the analytic viewpoint of an academic, others from the personal experience of engineers amazed by what they saw, and yet others with the intent of uncovering the downsides for workers. They usually arrived via a major Eastern seaport, traveled by train, and spent between a few weeks and several months in the United States. Some remained in the rust belt states, others made it as far as Chicago, and a few traveled all the way to the West Coast. While some only visited major sites of American productivity—such as the large cities on the Eastern Seaboard, Ford's factories in Detroit, and the Chicago meatpacking district—others made a point of veering off the beaten path and discovering more unusual sites of US production. While many commented, at least in passing, on larger social and cultural aspects, such as the role of women, class relations, race, religion, and prohibition, they all wrote about US production methods—that is, Ford's continuous flow production. German visitors, however, differed in how they perceived Ford's production methods as well as his ideas about labor and class relations.

Some, like Köttgen and Westermann, largely ignored Ford's social and economic ideas, and instead focused on the technical aspects of mass production, such as mechanization and the division of labor. In doing so, they implied that rationalization—increased industrial efficiency—was a technological phenomenon, without social or economic implications. Others, like the political economist Friedrich von Gottl-Ottlilienfeld and the group of union delegates, primarily concentrated on Ford's larger social and economic ideas, seeing them as indivisible from the production methods.[40] Positive perceptions of Ford's production methods cut across class lines; most liberals and labor-friendly individuals of all classes generally had a more positive perception of Ford than conservatives did. For example, German engineering professor Paul Riebensahm called Ford's ideas "unrealistic," and full of "youthful enthusiasm" and "factual incoherence"— statements that could also be read as a bourgeois rejection of Ford's lack of formal education.[41] Weimar Germans also addressed the question of how to organize their own social and economic relations. How were employers and employees to relate to each other? How were classes to relate to each other? What was the appropriate role of the government in the economic sphere? What were to be adequate levels of wages, prices, and the standard of living? How long and under what conditions were employees to work?

These discussions unfolded in the context of German recovery from the Great War. Weimar Germany emerged from the war with two needs: the need to meet the reparation demands of the Versailles Treaty, and the need to redefine social relations after the demise of the authoritarian German Reich, accompanied initially by violent street fighting. Social Democratic, social liberal, and Christian Democratic politicians swept into the seats of power, replacing conservative bourgeois parties. Soldier and worker uprisings in the last war days and early peace years led to the formation of works councils that participated in managerial decisions, achieving, for example, an economy-wide eight-hour workday. Pressures from paying reparations contributed to the hyperinflation of 1921–1924, and the Dawes Plan led to stabilization by extending US credit to Germany, delaying reparations payments, and ending the French and Belgium occupation of the mining and steel industries in the Ruhr area.

In this environment, large companies in various German industries— such as coal mining, machine toolmaking, and electrical machine manufacturing—pursued efficiency measures to cut production costs. The Ruhr coal mines mechanized their production in the first half of the 1920s, replacing manual work and explosives with pneumatic hammers, undercutting machines, and more efficient transportation systems under- and aboveground. Large electric companies, such as Siemens, introduced work organization and planning methods in the first half of the 1920s, to be followed by new electric carts for materials transportation and continuous flow production in the second half of the 1920s.[42] Economists at the time, and historians later, were concerned that rationalization measures increased unemployment before the economic crisis of 1929 and therefore contributed to the rise of the Nazi Party.[43]

A rationalization movement emerged in Weimar Germany that grappled with relations between the state, entrepreneurs, and workers. One of the theoretical forerunners of the movement, von Gottl-Ottlilienfeld "coined the German term "Rationalisierung." Von Gottl-Ottlilienfeld sharply distinguished Fordism from Taylorism—Taylor's scientific management methods to increase labor productivity. He criticized Taylorism for its highly normed and therefore inhumane work conditions, massive costs for work organization that devoured the cost savings from higher efficiency, and failure to pass the savings from higher efficiency onto society in the form of lower prices. Von Gottl-Ottlilienfeld argued that Fordism allowed for the development of the personality of all employees, including workers, through high salaries and opportunities for promotions; that the continually lowered prices avoided overproduction; and that the reinvestment of profits instead of paying dividends would eventually mean the end of the financier class.[44] He called this Fordist system "white socialism," "leader socialism," or a form of "private property without capitalism"—in other words, a corporatist way to overcome capitalism without a Bolshevik revolution.[45]

It was, however, the industrial captain Carl Friedrich von Siemens who institutionalized the rationalization movement in Germany by founding the Weimar Reichskuratorium für Wirtschaftlichkeit (Weimar RKW) in 1921, originally charged by the German economic ministry to extend the standardization measures adopted during the First World War.[46] The Weimar RKW was to "raise the general level of prosperity by cheaper, more plentiful and better quality products," as outlined by its bylaws.[47] To do so, the Weimar RKW acted as an umbrella organization that organized and coordinated the efforts of existing associations, such as the Deutsches Institut für technische Arbeitsschulung (DINTA, German Institute for Technical Vocational Training), which organized systematic apprenticeships and management training, and the Reichsausschuss für Arbeitszeitermittlung

(REFA, Imperial Committee for the Determination of Working Hours), which designed, measured, controlled, and evaluated work processes and promoted incentive pay systems. Initially, only entrepreneurs and politicians were members of the Weimar RKW. In 1925, when the Weimar RKW first pursued government funding, it added Social Democrats to its membership, although they amounted to only 5 percent of the total membership. The Weimar RKW thus presented a model of cooperation between industry and the Weimar state.[48]

In addition, there were other homegrown approaches to German recovery and economic reorganization. One of them was the planning approach of Walther Rathenau and Wichard von Moellendorf. Based on experience with wartime materials management, they proposed organizing councils consisting of entrepreneurs, labor representatives, and state officials who would set prices, allocate raw materials and market shares, and determine economic policy. They envisioned class cooperation in the public interest.[49] Another domestic approach was a version of industrial democracy (*Wirtschaftsdemokratie*), developed by Fritz Naphtali out of the free union movement; it aimed for reform of the economic system with the eventual goal of achieving a socialist society. This system envisioned larger political measures, such as the control of cartels and monopolies, rather than measures at the company level.[50] Because all societal groups—including free unions—supported technical improvement, technical measures became the focus of the German rationalization movement.

The open question was, in what kind of larger societal relations should rationalization technologies be embedded? This is where the disagreement arose. Entrepreneurs such as von Siemens saw rationalization as a means for social uplift through increased production and lower prices but rejected the call for higher wages. Labor unions appreciated immediate benefits from new technologies, such as less physically strenuous work and possibly higher purchasing power, and continued to aim for larger social reform. Scholars have argued that there was a distinct German rationalization movement, and that rationalization was a German-grown concept. It differed from US productivity in that the level of wages remained an open question.[51] In view of this open question, Weimar Germans looked across the Atlantic for solutions—although again, different people came back with different answers.

It was unclear what to take away from the American model. A number of Germans professed that it was impossible to understand the US system without having experienced it. Westermann, for example, emphasized that Germany and the United States had fundamentally different economies, production methods, and ways of thinking. He felt that it was impossible for a German who had not traveled to the United States to understand the country and for an American who had not traveled to Germany to understand Germany. As exasperated as he was about Americans berating Germans for their failure to properly make use of technical ability, he was just as hopeless about the possibility of explaining the American circumstances to Germans on his return.[52] Other writers like political economist Moritz Bonn pointed to continuing diversity within the US economy, where half of the firms had not modernized.[53] And the engineering professor Paul Riebensahm, after six weeks of traveling in the United States, stressed that its complexity and diversity defied any description. He berated Germans for focusing on Ford's flow production and ignoring the many small specialty manufacturers that didn't use those production methods, such as Yale and Towne Manufacturing Company, a producer of locks with small keys. In Riebensahm's view, straightforward depictions based on numbers, measures, and drawings could help increase the competitive advantage of individual companies. A real understanding of the US system, by contrast, called for a communal assessment that integrated a broad range of industries and different social viewpoints, from corporate directors to production engineers and foremen.[54] A closer reading of the travel reports shows how Köttgen and the trade union group appropriated American production methods for German conditions.

Köttgen: Work Harder and Longer

One of the earliest German visitors to the United States was Köttgen, a Siemens deputy director of the Weimar RKW who would later shape the rationalization movement in Germany. He visited the United States in fall 1924, together with the Weimar RKW's secretary, Professor A. Schilling. Many people disagreed strongly with Köttgen's travel report, but his views affected German thoughts on production methods for the next twenty-five years. As a young engineer, Köttgen joined Siemens and Halske in Berlin in

1894 and rose through the ranks until he became the company's chairman in 1921. Köttgen had pursued his career in the management of a paternalist company that had introduced, in the late nineteenth and early twentieth centuries, typical welfare capitalist measures such as medical care, a canteen, paid vacation days, nurses, company housing, employee saving schemes, and eventually profit sharing, albeit without direct interference in workers' private affairs. When Köttgen left for the United States, he was responsible for rationalization efforts at Siemens, and had already introduced the standardization of parts and products, new cost accounting systems, work bureaus for work planning, the standardization of warehouses and office work, and a punch card system, and the implementation of a conveyor belt system—one of the first in Germany—was underway.[55]

While Köttgen introduced rationalization methods, including continuous flow production, his travel report stood in a tradition of critical German voices that had questioned the low quality of mass-produced goods and insisted on labor-intensive quality production before the First World War, and now argued that the lack of capital rendered technological modernization unaffordable.[56] Similarly, Köttgen asserted that Ford's mass production was not suitable for Germany because of environmental and economic differences between the two countries. (Two decades later, following World War II, Germans would point to these same conditions.) Instead of rationalized production methods helping increase the output per worker (*Produktion per Mann*), Köttgen called for German workers to work longer hours to rebuild the German economy from wartime destruction. He thus used the US model to demand a repeal of the eight-hour workday won by the soldier and worker uprisings in the period immediately following World War I.[57]

The most important differences between the United States and Germany outlined in Köttgen's report were the United States' beneficial climate, resources, and size. US agriculture was less labor intensive due to sunny southern locations—New York City is at the same latitude as Rome—as well as regular rain, new soils that required less fertilizing, and large fields that allowed for the use of teams of horses or tractors. A lower percentage of the US population was employed in the agricultural sector: only 23 percent, compared to 43 percent in Germany.[58] Abundant natural resources—such as coal, iron, and waterpower—and agricultural products—such as cotton—allowed for cheaper industrial production. Although wages in the United States were higher, Köttgen argued that the standard of living in both countries was comparable because housing and finished items, such as clothing, cost about twice as much in the United States as in Germany.[59]

In his study, Köttgen focused on a comparison of economic conditions in the United States and Germany, and added a rich, ninety-page appendix filled with tables and statistics to support his arguments. Compared to other travel reports, he devoted a mere three pages of his seventy-page essay to his description of Ford's company; in these three pages, he discussed the history of the company and its manufacturing system, the price of Ford automobiles, road construction, and gasoline prices. Nowhere did Köttgen mention Ford's larger social and economic convictions, even just to refute them.

In Köttgen's view, the large domestic market was the only reason that allowed Ford's mass production methods to emerge, which meant that such production methods were unsuitable for the smaller markets of Germany and possibly even Europe. Indeed, Köttgen thought that even the large US market could bear only a single automobile factory like Ford's, and he contended that the US automobile market would soon be saturated, rendering the production methods obsolete.[60] Finally, he claimed that the need for a large, up-front investment in machinery rendered Ford's mass production inflexible, and he felt that the lower labor costs in Germany made expensive investment in machinery less profitable.[61] Under German conditions, therefore, Köttgen called for an immediate extension of weekly work hours to expand the overall production, in addition to higher productivity per hour in the longer term.

Köttgen explicitly rejected one major characteristic of American welfare capitalism: high wages. He stated that employee representatives often asked that wages be increased, arguing that higher salaries led to increased purchasing power, which in turn led to higher demand and economic growth. He complained that "even persons called to the representation of people"—a not-uncommon snipe against members of parliament in Weimar Germany—insisted on increased wages. As we have seen, Köttgen's opinion about wages was diametrically opposed to that of Ford, who saw high wages as an intricate part of his company's production system, helping reduce turnover and create demand for his vehicles. Köttgen also noticed other typical features of the American welfare capitalist system, albeit without recognizing them as part of an economic approach. For example, he said that there were no public health, accident, disability, and pension insurance systems in the United States. In his view, US companies took care of their employees' needs in an effort to keep their staff healthy and content, providing medical consultations, hospitals, and dental treatment on-site and often during work hours, and also offering corporate savings banks as well as accident, pension, and life insurance.

To summarize, in the 1920s, Köttgen was an example of a German visitor who did not recognize some of the features of American welfare capitalism and explicitly rejected others, molding a critical German reception of the US economic system after World War II. Like Köttgen, other German visitors selectively adopted American rationalization technologies, such as the assembly line at Siemens, but dismissed the values of the American production system, such as higher wages. While the technologies were imported by German companies, the economic values imbued in them were not.

German Trade Union Group Embraces Ford's Productivity

Köttgen's report, together with a number of similar ones, was the primary motivation for the Allgemeiner Deutscher Gewerkschaftsbund (General German Union Association), the German socialist trade union organization, to send its own delegation to the United States. The union delegates quoted from Köttgen's book in their report, and were particularly enraged by Köttgen's opinion that the output of the German economy could be increased by 20 percent through longer hours and more intensive work.[62] Others, like Westermann, disagreed with Köttgen's pessimistic view of mass production but still took swipes against unions, complaining that plant directors in Germany always had to introduce improvements "against the will of the workers" and generally accusing workers of "soldiering"—that is, working more slowly than their capacity.[63] Because of contradictions between travel reports, union representatives felt that they needed firsthand knowledge of American economic and social conditions in order to continue shaping German economic policy. A fifteen-member union group visited the United States from September to November 1925 to "recognize and explain the underbelly" (*Schattenseiten*) of the US economic and social systems.[64] The group returned convinced that Germans needed to improve their production methods and work organization to increase their economic output and recover from wartime destruction.[65] It was particularly enamored by the technical and social organization of the Ford Motor Company, although the group criticized Ford's unwillingness to recognize unions.

The union delegates agreed with Köttgen that the German economy was not to become an exact copy of the US one. The essence of the German economy could not be changed through thoughtless duplication, imitation of individual parts of the US production system, or the "inoculation" of the American spirit; in their words, they could not transfuse their economy with a different blood type. Like Köttgen, they emphasized the differences in geographic size, population, and markets between the United States and Germany with the help of detailed graphs and tables. They also recognized a difference in social attitude, which they attributed to stronger entrepreneurial forces and a shorter history. As an example, they noted that to eliminate custom borders in Germany in the nineteenth century took a full generation's time, while no such hardened economic traditions existed in the United States.[66] Yet they considered the United States' shorter history—or rather, smaller historical burden—to be even more important regarding class relations—an important aspect that we will return to.

Of course, the praise of American productivity was the backdrop to the union trip, and work organization along with the intensity and speed of work were among the first topics in the union report. The union delegates provided a few poignant illustrations that depicted the differences in work organization and supported their argument that work organization led to seemingly higher productivity in the United States. For example, in Germany, masons worked from fixed scaffolding, and assistants put bricks in random piles at their feet; masons needed to mix the mortar themselves, and every day they needed to bend down thousands of times to pick up bricks and mortar, and then lay bricks up to head height. By contrast, masons at construction sites of similar size in the United States worked from hanging scaffolding that could be moved up and down; there was no bending or working overhead. Bricks were supplied to them in a handy way, and they were not expected to do any side work. Similar differences in foundation work were noted: in both Germany and the United States, foundations were usually made of concrete, but large US construction sites typically poured concrete rather than tamped concrete as in Germany, thus avoiding the cumbersome tamping. Also, US companies used concrete mixers with pouring towers in contrast to the arduous process of manually transporting concrete as in Germany. In a strident conclusion, the union delegates accused those authors who compared the foundation work in Germany and the United States in terms of cubic meters per man per day, without consideration of the different conditions, of being "demagogue[s]," and those who compared the productivity of US and German masons according to the simple formula "bricks per manhour"—the BLS definition—of not understanding the conditions of the job.[67]

With regard to production speed, the report stated that the United States and Germany were comparable in many other sectors, such as the textile industry (highly productive in both countries), furniture making, and machine making, and that the speed of conveyor belts varied from company to company in both countries, preventing easy comparisons. Therefore, the union delegates concluded that the higher productivity of the US economy and US workers was not *"caused by higher physical intensity of work."* They emphasized that *"the difference in production methods"* caused the differences in productivity, putting the ball back into the court of employers.[68] (This laid the basis for strong union support for rationalization and rationalization technology that the German union movement maintained and confirmed after World War II, as we will see in chapter 8.) To counter reports by many other visitors—mostly engineers—who had lauded the willingness of American workers to work, they pointed to workers' discontent in the United States, particularly among unskilled workers, who were no more willing to work fast and hard than were German employees.[69]

The union delegates also suggested that German workers desired better work organization. They stated that in the United States, foremen were responsible for providing technical drawings and other devices to workers, while special assistants replaced a worker's broken tools within minutes; they asserted that German workers would be happy if "an appropriate work organization would allow them to turn such annoying dawdling into a rhythmic work pace" that relieved them from "pilgrimages" from one department to another in search of a technical drawing, or waiting in "mass assemblies" in front of a company's supply of jigs and tools.[70] In this conclusion, the union delegates were united with Riebensahm, who criticized Köttgen's call for two additional hours of work per day; this would not make up for the three to four times higher productivity in the United States.[71] While Köttgen dismissed US rationalization methods as not applicable because of their cost and the need for a larger market, the union delegation and others demanded more rationalization methods in Germany.

Labor, Class, and Race Relations

In contrast to the surprisingly cursory treatment of production technologies—particularly compared to other Weimar travel reports—the union group paid close attention to labor and class relations in the United States. In this way, their report was likely informed by the leading socialist analyst Werner Sombart, who had argued that a higher standard of living ("reefs of roast beef and mountains of apple pie"), favorable attitude toward capitalism, political inclusion through universal male suffrage, the dominant two-party system, and higher social and geographic mobility (the western frontier) had prevented American workers from embracing socialism.[72] Without using the term "welfare capitalism," the unionists described welfare capitalist measures and realized that these measures served to avoid unionism—a factor they resented. Yet they made an exception for Ford, due to what they perceived as his goodwill toward his employees. For example, they noted that other nonunion companies occasionally used spies to detect and deflect early union organization efforts, but that Ford did not do so (actually, he did). As in other travel reports, Ford took on a special position in the union report, although here not for Ford's production methods but instead for Ford's labor relations, and the fact that Ford organized his company and wages in ways that allowed workers, even unskilled ones, to comfortably make ends meet. They called this "Fordist" socialism or, in von Gottl-Ottlilienfeld's words, "white" socialism, in a bow to Ford's corporatist promise of overcoming class conflict through higher wages and lower prices—two elements that were integral to Ford's philosophy but were not present in Germany.

The union report included a list of all the social services provided by many US corporations: free consultations by physicians and dentists, legal counseling, loan assistance, health insurance, schools, playgrounds, and, in some cases, workers councils—internally organized company unions, which the German union delegates considered a parody of the idea of union organization. They emphasized that these benefits by far surpassed accident insurance and surviving dependent insurance—the only benefits that other visitors usually mentioned in their travel reports, and ones that the union delegation would have liked to see organized by the state, rather than by insurance companies or the companies themselves, to avoid speculation and better protect workers' interests. The union delegates accused other German visitors of being "entirely incapable" of recognizing the diverse manifestations of corporate benefits—or worse, they were being deliberately dishonest by withholding such information, and "cheerfully and sanguinely" reporting the opposite.[73]

The union delegates clearly made a connection between these forms of corporate benefits and the lack of union organization, thus recognizing this key feature of "welfare capitalism" without using the term. They saw corporate benefits as a cost that entrepreneurs paid to keep away union organizations. In the view of the union delegates, welfare capitalist entrepreneurs had learned from modern industrial science that investing in corporate benefits produced "excellent financial yields": these benefits helped increase productivity as well as provide the company with a good social reputation. US entrepreneurs even presented themselves as friends of industrial democracy by including workers in corporate matters, as long as workers did not commit themselves to national labor unions. Workers were to cooperate democratically with entrepreneurs within a corporation to promote corporate success and thus promote the interests of both sides.[74]

The union delegates noted that wages—nominal wages as well as spending power—were comparatively higher in the United States than they were in Germany. They thought that the reason for this was union organization rather than welfare capitalism. In terms of nominal wages, they found wages in general to be about four times as high as nominal wages were in Germany, including for unskilled workers. They found the highest wages in organized trades, such as printers and masons, and the next highest in metalworking trades. Hourly wages for the metalworking trades were between $0.90 and $1.10 in major northeast cities. With a currency exchange rate of $1 to 4.20 reichsmark, this was about four times the hourly wage of German metal workers at 0.90 reichsmark.[75] The spending power of wages in the United States and Germany was more difficult to compare because of different consumption patterns. For example, German staple products, such as potatoes and pumpernickel bread, were significantly more expensive in the United States, while the food typically consumed by US working families, such as meat and dairy, would be comparable to what is usually offered in "bourgeois restaurants" in Germany and therefore unaffordable on German working-class wages. While food and clothing were cheaper in the United States, accommodations were more expensive, often requiring workers to spend about a quarter of their wages on rent. Homeownership, however, was more common in the United States than in Germany, even among the working class.[76]

They argued that at Ford—as well as at many other nonunion companies—union contracts helped raise wages. Wages were not set by "free competition"; that is, wages were not based on an individual worker's capability and productivity, or the supply and demand for jobs. While Ford rejected any negotiation with labor unions and set wages of his own free will, an unnamed US union officer observed that it was "curious" that Ford's wages were always just 10¢ over the hourly contract wages. For example, the maximum wage for mechanics at Ford was $1.375 per hour or up to $11 a day (for an eight-hour day), compared to the $1.25 per hour maximum wage for mechanics in Detroit on average.[77] This, however, applied only to skilled workers.

Unskilled workers at Ford earned significantly more than the average wage for unskilled workers in the United States—up to 30 or 40 percent higher. The average hourly wage for unskilled workers in the US metalworking industry was $0.469; at Ford it was at least $0.625—34 percent more. Even sweepers at Ford earned the $5 daily wage, or 42 percent more than the $3.50 daily wage typically earned by street sweepers in Detroit. Yet unskilled workers on Ford's production line were 30 to 40 percent more productive than unskilled workers at other car manufacturers. Paying high wages to avoid worker turnover made sense in the eyes of the union delegates. After all, workers at Ford did not achieve their high productivity until after they had received several months of training, when they had internalized the pace and movements of the jobs that they were responsible for. While the physical effort was not significantly higher than the efforts of workers at other plants, Ford workers needed to learn specific physical movements and could no longer be considered unskilled workers. Therefore, Ford's wages for "unskilled" workers were equivalent to the average wages of the most highly paid unskilled workers in the construction industry; workers on Ford's production line earned up to $0.77 an hour, and assistant construction workers in Detroit earned $0.75 an hour.[78] With black miners in the South earning about half the wages of organized miners in the North, the German union delegates concluded that the dividing line was not between negotiated (union) and nonnegotiated (nonunion) wages but rather between wages that were directly or indirectly influenced by union negotiations, such as Ford's wages in the North, and wages that were set under conditions of "free exploitation," usually in the South.[79]

Many Weimar visitors also commented on class relations in the United States. Allianz insurance executive Rudolf Hensel, for example, reported that there were no "classes and castes, rank, standing and occupation" in the United States; free persons encountered each other as equals. Hensel considered the lack of different class compartments in trains as well as the lack of fences around homes as expressions of a classless society.[80] Similarly, editor Arthur Feiler commented that the US working class participated in the growing prosperity, and that "instead of class antagonism the numerous prosperous, hopeful workers in America feel a frank, honest business relationship toward their employer" that could "at times mean heated struggles for every cent of wages," but was free of the "deeply rooted, centuries-old bitterness, always present in countries where class differences can never be bridged."[81] Westermann noted the lack of offices (*Würde*) and titles, and that respect was based on competence and achievements instead. He also pointed out that workers seemed to believe that higher wages through technical improvements would overcome class conflict.[82] Social worker Alice Salomon expressed her hope that "if any country ever solved the capital-labor problem, it would be done" in the United States because of the country's education system, in which children "grow up less fettered by class distinctions and class-consciousness" than in Europe.[83]

Likewise, the union delegates mentioned repeatedly in their report that relations between workers and entrepreneurs in the United States were freer and more convivial than in Germany with its centuries-old class distinctions handed down from medieval feudal societies, as Sombart had already noted two decades earlier. They commented on the fact that in many US companies, neither the tone of interactions nor the simplicity of the interior design of the work space and office buildings conveyed a sense of distance or prejudice between workers and management. For example, offices resembled factories with exposed red brick walls that were not even whitewashed, only a few charts for decoration, and technical draftsmen seated in rows in front of iron easels. This architectural design was backed up by a more familiar tone of interaction within companies, with employers and employees addressing each other by first name. While the delegates acknowledged that this tone might be chosen to create "loyal employees"—who are more beneficial to companies than unwilling or hostile ones—they insisted that the familiar tone still expressed more respect. Also, US workers were much more assertive than German ones and lacked a servile and menial disposition, and this attitude prevented US workers from developing a sense of class consciousness.[84]

By contrast, in the young Weimar Republic, large social groups did not look beyond the working-class background of Friedrich Ebert, the head of the Social Democratic Party (Sozialdemokratische Partei Deutschlands, or SPD) and elected German president, and considered it a "visible expression of the deterioration of morals" that a "saddler's assistant" had claimed the highest political office. But in the United States, many politicians and entrepreneurs, including Secretary of Labor James Davis, proudly emphasized their working-class origins, sometimes in hopes of garnering positive public opinion. Instead of class divisions between the bourgeoisie and the working class, the union delegation diagnosed a division within the working class itself: the clear division between skilled, "white" workers—mostly of northern or central European origins—and unskilled workers, often African Americans, or more recent immigrants from eastern or southern Europe; in other words, race replaced class in the United States.[85] Earning lower wages that did not provide them with access to the proper "American standard of living," unskilled workers of color frequently lived in squalid quarters and had trouble making ends meet. Also, children from lower-class families lacked access to proper schooling and education that could supply them with better chances for social mobility.

In contrast to Köttgen, the union delegates concluded that the vast geographic size and abundant resources of the United States is not what caused a higher standard of living, since these conditions also required a greater investment in transportation networks, automobiles, and other infrastructure; rather, the technologies and work organization, together with the larger role of consumption, is what caused the higher standard of living. In the United States, labor unions and even manufacturers recognized the larger economic role of higher wages. In their view, the opening, building, and exploitation of the domestic market through the creation of purchasing power by the common masses was the secret of the US economy. "High wages and low prices, larger turnover for smaller profits: from this practice, the technological wonders and work organization grew by themselves (automatically)."[86]

Rationalization under Nazism

With the Great Depression, the attractiveness of the American economic
model waned. While the American system of mass production with high
salaries and low prices had enabled unprecedented economic growth, it
seemed unable to control the volatility of the business cycle. Among European countries, Weimar Germany was affected more quickly and severely by
the economic decline because the private financing of the Dawes and Young
Plans had tied the German economy to American financial markets. Massive
withdrawals of international credits resulted in a banking crisis and unprecedented unemployment, rising from 1.4 million unemployed in September
1929 to 6.1 million in February 1932. The rising Nazi regime exploited the
economic and social insecurity, mobilizing support from those who feared
socioeconomic decline, such as the lower middle class. Like Weimar Germans before them—or even more so—Nazis embraced American technologies and created programs for mass-produced consumer items, but they
appropriated American technology for their own racial and social ideology.[87]

Hitler personally admired Henry Ford's technological innovations and
agreed with Ford's anti-Semitic critique of capitalism. In the early 1920s,
before gaining his leadership role in the Nazi movement, Hitler already
recommended Ford's publications for reading. Once in power, the Nazi
regime continued to embrace American mass products; for example, Coca-
Cola—nowadays idiomatic to Americanization—increased its number of
bottling plants in Germany from five in 1934 to fifty in 1939. Also, Germans continued to travel to the United States to study US mass production
technologies, albeit in declining numbers. For instance, Hermann Göring's
organization for German rearmament advertised traveling to the United
States; the German travel agency Lloyd organized forty trips to the United
States in 1937; and in the same year, the German employer and employee
organization Deutsche Arbeitsfront (German Labor Front) conducted three
tours to the United States. As in the Weimar period, these trips typically
included visits to the Ford Motor Company.[88] But the Nazi regime usurped
the social and economic meaning of American mass products, embedding
them in their ideology of a national community, the *Volksgemeinschaft*.

The Volksgemeinschaft promised all Germans equal opportunity based
on their individual performance and regardless of their social background; at
the same time, it excluded all those that were deemed inferior or unwanted

by the Nazis: Jewish persons, political opponents such as Communists and Social Democrats, homosexuals, disabled persons, and the Sinti and Roma. Thus, the Volksgemeinschaft overcame class distinctions but secured and deepened the racial segmentation of the German population.[89] At workplaces, the Volksgemeinschaft was realized through a focus on performance that promised individual advancement as well as increased productivity. For example, the German Labor Front—which usurped labor unions and employer associations in an attempt to transform workplaces from sites of class conflict to well-ordered racial communities—promoted pride in work through slogans such as "Work Ennobles" (*Arbeit Adelt*), and conducted performance competitions between apprentices and plants. It also included a program called "Beauty of Labor" (*Schönheit der Arbeit*) that merged paternalistic business practices and scientific management methods to beautify workplaces as well as increase productivity through programs to reduce noise, create green on-site spaces, provide new washing and changing facilities, install better ventilation and lighting, and incorporate proper nutrition in company canteens.[90]

The German Labor Front also conducted a mass consumption program of so-called people's products (*Volksprodukte*), promising that all members of the German Volksgemeinschaft would participate in a higher standard of living and overcome class-based consumption patterns. State-sponsored mass consumer products included a radio receiver—a highly popular product and the only successful program that also served propaganda purposes for the Nazi regime—a television, refrigerators, housing construction, seaside resorts for vacationing, and Volkswagen cars. The consumer product program was in an ambiguous relation to other Nazi policies, such as rearmament and economic autarky. For instance, while refrigerators generally competed for resources with military products, they also promised to extend the life of perishable foods, thus reducing the need to import foodstuffs.[91] Economists have argued that Nazi economic planning prioritized producer goods over consumer goods, even before active military mobilization started in 1936, and that planning did not reduce income disparities. The programs, though—by introducing new products and exploiting mass production—attempted to address the conundrum of how to make consumer items available to the larger population, including working-class households, without increasing wages.[92] The now-infamous Volkswagen project exemplified these dilemmas.

Spurred by the successes of the people's radio receiver, the people's car was prematurely announced in 1934. Hitler, a self-proclaimed car fanatic who regularly spoke at the annual automobile exhibition in Berlin, asserted the title of father of the Volkswagen. With 80 percent of cars in Germany in the early 1930s used in trade and industry for transportation and deliveries, the Volkswagen was to turn cars from an investment product into a consumer good; Hitler also hoped that car manufacture would help with economic recovery and overcome unemployment. The German automobile association Reichsverband der Deutschen Automobilindustrie initially took charge of the project, and the technical design of the car was given to Ferdinand Porsche, who traveled several times to the United States to study automobile and production technologies at Ford and other car manufacturers.[93] The Volkswagen was eventually designed as a compact car with four wheels (instead of three, when it was first designed as a microcar) and enough space for four adults, or two adults and three kids, at a price tag of 990 reichsmark. By 1937, the first prototypes were completed and received broad public acclaim, yet the project also faced serious budget and schedule overruns, and the German Labor Front took over.

The price of 990 reichsmark—set when the car was designed as a microcar—exhibited conflicts between consumer, autarky, and rearmament policies. Three factors determined who would be able to afford a Volkswagen: the price, the purchasing power, and the maintenance costs. Since the Nazi regime sought to avoid salary increases, which would have expanded purchasing power, only the price and maintenance costs could be influenced. The price became the main focus of propaganda. Although experts knew that it was unrealistic and would require major subventions, the price was regarded as Hitler's promise to the people and was not adjusted until the end of the war. But the onetime purchasing price was less significant than the ongoing maintenance costs for gas, oil, a garage for protection, tires, repairs, and insurance. The prices of gas, oil, and tires, however, even increased in the second half of the 1930s due to the Nazi autarky policy and its restriction on imports; the autarky policy thus unintentionally limited the number who could afford a Volkswagen. Realistically, the main target group were households with an annual income of 7,000 to 8,000 reichsmarks—that is, the bourgeois households of self-employed persons, white-collar employees, and civil servants, not industrial workers. In other words, the Volkswagen reinforced rather than erased consumption differences based on class. In 1938, the German Labor Front created a savings program for the Volkswagen, and by the beginning of the war, 270,000 households participated in the program, with mostly higher-income households and only 10 percent salaried employees and only 5 percent workers, indicating the persistence of consumption patterns.[94]

In the same year, construction began on the manufacturing site near Fallersleben—today called Wolfsburg. The factory's flow production methods were based on Porsche's observations in the United States and help from German engineers at Ford whom Porsche wooed back to Germany. The factory was to reach its full capacity in three construction phases. The predictions of production volume diverged widely: from 40,000 cars per year—comparable to Opel manufacturing numbers—to the Nazi propaganda numbers of 1.5 million cars annually—more than Ford. The Volkswagen factory was hailed as the "largest plant in the world."[95] While such a production volume might have helped reduce the price of the car, the German market would not have been able to absorb this number of cars; a much larger market would have been needed. This shows that in the Nazi mind, mass production was tied to the policy of living space (*Lebensraum*), that is, the idea that Germany needed more territory to fulfill its potential. Indeed, Hitler believed that the large US domestic market was the main reason for its high productivity and therefore that obtaining more territory would increase productivity in Germany. Technology, in his view, was secondary.[96] With the beginning of the war, however, material shortages prevented mass production scales, and the factory instead made airplane parts, army vehicles based on the Volkswagen design, and bombs—using forced labor.[97] While the Nazis failed to produce a consumer car, they proudly claimed German dominance in car racing. In anticipation of the car, the regime built up the highway infrastructure—a public works project that reduced unemployment and was meant to serve as a mode of transportation for the leisure activities of a classless Volksgemeinschaft, but eventually fulfilled military purposes.[98]

Ford Motor Company not only served as a technological and organizational model for the Nazi Volkswagen project but also pursued its own operations in Germany. Like other US companies, such as General Motors (which had acquired the German car manufacturer Opel in 1929) and IBM, Ford sought to protect its foreign investment as long as possible. But its operations in Germany were eventually put under Nazi custodianship, making communication and control as well as the return of proceeds impossible, and leading to a reinvestment of profits. For example, Ford had opened a subsidiary in Berlin in 1925, and started assembling cars and trucks in 1926, because tariffs on *parts* were lower than those on fully assembled *cars*. In 1931, the company opened a large assembly plant in Cologne. While the Nazis looked to Ford as a model, they also perceived of the company's operations in Germany as competition to the Volkswagen project and treated the company as foreign, although IG Farben held a 35 percent capital share. By contrast, Opel was able to foster its image as a German company; as the largest car exporter in Germany, it also helped the Nazi government earn foreign currency that was needed for rearmament, and the company was initially invited to participate in the Volkswagen project.[99] Both Ford and Opel ended up supporting the Nazi military effort through the production of vehicles and weapons systems, as did many other US companies.[100]

While the Nazi regime rejected some forms of Americanization, such as jazz and Hollywood films, overall it remained open to American mass production technologies and products, but the new construction of the Volkswagen plant may have presented an exception. Many established companies, such as Siemens and Daimler-Benz, appear to have introduced new production technologies and flow production in the late 1920s. In the 1930s, they mainly changed their organization and personnel policy with the help of organizations from the Weimar rationalization movement, such as the Imperial Committee for the Determination of Working Hours (REFA) and the German Institute for Technical Vocational Training (DINTA), which conducted time-motion studies and developed training programs.[101] With the Nazis claiming to have eliminated class conflict in the workplace, though, they presented Nazi Germany itself as an alternative to socialism and capitalism. From the perspective of the economic disaster of the Great Depression, there was no reason to prefer capitalism and liberal democracy over a new system that promised an egalitarian community under the watchful eyes of a strong leader.[102]

Conclusion

While the perceptions of Germans traveling to the United States varied widely, all the visitors were fascinated with American production methods, particularly Ford's continuous flow production, and their reports focused on rationalization and rationalization technologies. Some, such as Kött-gen, considered American production methods unsuited for Germany because of its smaller market size and the need for capital investments, but many engineers and academics were more positive; the trade union group even called for the introduction of rational production methods in Germany, thus putting the onus on German industrialists. With regard to welfare capitalism, however, their perceptions varied more widely. While no visitor used the term "welfare capitalism," the union group was the most perceptive, describing welfare capitalist measures in detail but criticizing welfare capitalism for its rejection of unionism. Other visitors paid no attention to the question of labor relations, wages, and consumption, and still others dismissed welfare capitalist measures as unsuitable in the German economic situation. The union delegates were also perceptive with regard to class relations in the United States, and identified divisions within the working class between unskilled workers—often African Americans and more recent immigrants—and skilled workers—craftsmen and union members, frequently of northern European origin. Although the perceptions cut across class lines, conservatives were more critical, if not outright dismissive, of Fordism while progressives more openly embraced various aspects of it.

While the US government retreated from international engagement after World War I, US companies continued to be involved in international operations in the decades following the First World War and may have even expanded their operations. In these international operations and contacts, US professionals and businessmen conveyed their ideas about technology, labor relations—often welfare capitalist ideas—markets, and business models. Filene, Watson, Ford, and many others did so not in a systematically organized way but rather in a multitude of individual interactions, some of which were moderated by associations such as the International Chamber of Commerce. Together, they promised that productivity would lead to a higher—American—standard of living.

German fascination with American technology continued into the Nazi period. Despite the autarky policy that walled the German economy off from foreign imports, the regime continued to promote the adoption of American technology. At the same time, however, these technologies gained new meanings as they were embedded in the ideology of the Volksgemein-schaft. These transatlantic transfers of productivity ideas during the 1930s certainly deserve more study; they are not discussed here in more detail because West Germans after World War II referred back to the model of Weimar labor relations, rendering the Nazi period a temporary divergence.

German visitors generally agreed that they did not want to turn Germany into a direct copy of the United States, but they disagreed in terms of how much the United States could act as a model for European development. Köttgen's perception of US companies and their production methods prefigured post–World War II German perceptions of American productivity: endowed with enormous natural resources and large markets, American companies could achieve unprecedented productivity levels, but those conditions didn't apply to Germany. Although many Weimar visitors disagreed with Köttgen's call for longer workdays and a higher work pace, many of his ideas survived into the next two decades and formed the basis of the German response to the American economic model after the Second World War. To convey his view, Köttgen used statistical measures as objective conveyors of the American system. In the previous chapter, I traced the origins of one such statistical measure, productivity. In the remainder of the book, I look at how productivity came to stand for the American economic system, and how it was used to convey this system across the Atlantic. The idea of productivity eventually came to represent contradictory features, such as free enterprise and collaborative labor relations, that had their roots in the transition from welfare capitalism to industrial democracy in the interwar period.

3 The Marshall Plan's Productivity Revolution

In spring 1952, the US administrator Everett H. Bellows declared at a Marshall Plan conference, "I hope to shock you when I tell you that our main purpose is to create a benign social revolution."[1] Bellows worked for the Marshall Plan's Productivity Program, which promised higher standards of living through higher productivity. Its goal was to lure European workers away from Communism and shape European economies after the model of US capitalism. To do this, the program tied political values to productivity, even though some of those political values—such as free enterprise and collaborative labor relations—contradicted each other. Officers of the Productivity Program wanted to create integrated markets with mass production and distribution; they wanted to dynamically expand European economies through higher wages and lower prices, and ultimately change social relations throughout Europe.

The approach of Bellows and his colleagues in the Productivity Program differed from that of Ewan Clague at the Bureau of Labor Statistics (BLS), discussed in chapter 1, in one important detail: Clague claimed that there was no "immediate and direct" connection between productivity and the wages paid in an industry, but two decades later, officers of the Productivity Program demanded that gains from productivity be shared equally between industrialists, workers, and consumers. This was a major shift. Indeed, Productivity Program officers made the sharing of productivity benefits a major criterion in their selection of plants to showcase the benefits of American-style productivity in Europe—the so-called demonstration projects. They chose plants that shared the benefits of increased production with their employees and customers through higher wages and lower prices, and refrained from cartel agreements, competing for raw materials and markets instead of allocating them in agreement with other firms. These plants did not necessarily display high technological efficiency. In fact, some of the plants considered for demonstration projects appear to have operated in rather-unproductive ways, lacking proper cost accounting systems, organizational structures, or delegation of managerial responsibilities. Plants could be considered for the program even if, after productivity improvements, they would still be far from ideal regarding high technical efficiency. Instead, plants were chosen to demonstrate the workings of the American system of free enterprise and collaborative labor relations.

The Productivity Program also marked a shift away from individuals—such as Henry Ford, Edward Filene, or countless Weimar visitors—transferring productivity ideas and technologies. Now, the transfer was mediated by government agencies. These agencies relied on significant support from business and management associations as well as national trade unions, which served the government in advisory functions, contracted with the program in consulting roles, or provided personnel to be employed in the administration. Yet this concerted governmental effort did not mean that the political and economic values tied to productivity became any more coherent. In fact, the Productivity Program brought together contradictory values. On the one hand, it called for free enterprise—that is, the freedom to manage without interference from government regulation or union influence; on the other hand, it demanded collaborative labor relations—that is, worker representation by trade unions and plant-level collective bargaining as practiced in the United States.[2]

Although the officers of the Productivity Program were united in their urge to achieve social change in Europe, Bellows stood out among his colleagues for his "eagerness and enthusiasm" for his work.[3] One of his first tasks was to create an operations manual for hiring people in the new administration; he later bragged about cutting red tape and forging new ways, usually with the intent of doing good for the individual employee. For example, he stood up for his secretary when she was threatened with release because of suspected communist activities; he sent the family of a US Marshall officer in Italy back to the United States early for medical treatment, although dependents were typically allowed to travel only with or after the officeholder; and he contracted with a physician and hired a psychologist for the support staff of the Marshall administration in Paris because many of them, unlike foreign service personnel, were not used to the conditions of living abroad and were experiencing what he called "culture shock."[4] As the director of the productivity division in Paris, Bellows needed to navigate American needs and goals as well as European sensitivities about American dominance; he and his fellow officers did so with sometimes almost missionary spirit, driven by a belief in the uniqueness of their economic model (figure 3.1).

How far the Marshall Plan actually helped European economic recovery has long been a subject of debate among historians.[5] More recently, historians of the Marshall Plan have pointed out that it was based on voluntary agreement of the European side, and they have viewed the Marshall Plan primarily as a program to encourage European integration for

Figure 3.1
Everett Bellows (second from left) and US productivity officers at a two-day conference on the Productivity Program in Paris in March 1952. While the officers sought to overcome long-standing European social relations, they met in the splendor of Chateau de la Muette, a place reminiscent of absolutist France that by then served as the headquarters of the Organization for European Economic Cooperation (OEEC).
Source: Everett H. Bellows Personal Papers (public domain). Courtesy of Harry S. Truman Library and Museum.

economic as well as national security reasons. These scholars generally agree that the United States dominated by consensual hegemony and acted as an "Empire by invitation."[6] At the same time, the major intent of the Productivity Program was to bring American-style productivity to participating countries, and in this sense it can be seen as an Americanization program, making European economies more similar to each other and the US model by implementing the American values of productivity.[7] Despite this intent toward homogenization, American administrators took care to adapt the Productivity Program to the political and economic conditions in each European country. Taking European conditions and sensitivities into account, Bellows and his colleagues sought to convert—rather than coerce—Europeans to the creed of American productivity.

This chapter introduces the Marshall Plan's Productivity Program along with its projects and political objectives. After a brief discussion of New Deal labor policies, it begins by providing the historical background of the program, explaining its major goals, and looking at the definition and measurement of productivity within the program that BLS officers helped develop. It then analyzes the political and economic values attached to American-style productivity within the Marshall Plan as they are exhibited in the Productivity Program and the demonstration projects. It concludes by exploring the contradictions of the program: the officers respected the Europeans and worked to convince them (rather than force them) to adopt American business values, while at the same time, the program aimed at making European companies and markets more like the US model—a fact that would help introduce US productivity technologies, such as production lines and electronic computers. This chapter lays the foundation for examining, in chapter 4, the public-private partnership of the Marshall administration with business and labor groups that supported the Productivity Program, and in chapter 5, the German reception of the Productivity Program.

Consumerist Labor Policies

In order to understand US productivity after World War II, it is important to first briefly review New Deal labor policies, and their shift from welfare capitalism to a form of industrial democracy that gave workers a say in corporate matters through union representation, collective bargaining, and grievance procedures.[8] In the wake of the Great Depression, workers shifted from expecting support from their local ethnic and religious communities as well as their welfare capitalist employers to seeking support from national industrial unions and an expanding state. Both white and black workers began to vote in increasing numbers for the Democratic Party—rather than Socialist or Communist alternatives—placing their faith in Franklin D. Roosevelt and New Deal legislation. While the Federal Emergency Relief Administration and public works projects addressed the most immediate needs, and put many unemployed people to work, other programs—such as Social Security benefits after retirement, the insurance of bank accounts through the Federal Deposit Insurance Corporation, and a national minimum wage—expanded state benefits to the general population. Far from seeking a revolutionary overthrow of the capitalist system, workers supported private ownership of property and sought to work within the two-party system toward a form of capitalism that promised everyone, owner or worker, a fair share.[9]

At the same time, workers started organizing in the national unions of the emerging Congress of Industrial Organizations (CIO), eschewing the older and more socially exclusive American Federation of Labor (AFL). While the AFL had focused on organizing mostly white male, skilled craftsmen, the CIO embraced women, workers of other ethnic backgrounds, and unskilled workers. Workers began to organize in company after company, backed by union-friendly legislation, such as the National Labor Relations Act (commonly known at the Wagner Act), which guaranteed the basic rights to organize in trade unions and engage in collective bargaining. But rather than demanding a fundamental redistribution of power—such as a role in hiring and firing, work and wage assignments, and production decisions—workers sought to improve their conditions through collective bargaining. Considering themselves partners within the capitalist systems, they sought increased wages, reduced works hours, and better job protection through seniority rights and grievance procedures.[10]

Ford, Filene, and other welfare capitalists who argued for low prices and high wages had foreshadowed consumerist reasoning that came into full swing during the New Deal years. Low prices and high wages were intended to create "purchasing power," which "was a shorthand for redistributive economic policies designed to enable the working and middle classes to buy basic necessities and still have enough left over to shop at Filene's and even drive a Model T."[11] The combination of housewives protesting rising prices and workers demanding higher wages on the bargaining table and in picket lines led to a "new mass-consumption political economy," with a powerful alliance between middle-class consumers and organized labor. Addressing both the problems of prices and wages, the new economic reasoning built on the assumptions that underconsumption was at the root of the economic problems and strengthening consumerist demand—through employment programs and legislation to increase wages—would address the economic problems. This consumerist reasoning set the stage for the Marshall Plan's Productivity Program after World War II in which the plant-level collective bargaining that emerged during the New Deal years would play a central role.

The Marshall Plan's Productivity Program

In June 1947 in a commencement speech at Harvard University, US secretary of state George Marshall announced a systematic program to help European recovery. He promised a program "directed not against any country or doctrine but against hunger, poverty, desperation and chaos."[12] Circumstances had convinced Marshall that the United States—instead of providing emergency help on an ad hoc basis—ought to develop a systematic aid program that would allow Europeans to help themselves, stabilize the region, and eventually build up a strong trading partner for the US economy. At the time, the economic situation in Europe was again deteriorating; after production had almost reached prewar levels in the initial postwar years, it again fell precipitously. Following a hard winter and failing crops, urban populations faced starvation, and a public health crisis threatened. Under this burden, Great Britain pulled its troops from Greece and Turkey, leaving both countries vulnerable to Soviet influence, and leading Harry S. Truman to formulate his doctrine of Soviet containment. Marshall came away from a two-month conference in Moscow with the conviction that the Soviets had sabotaged rather than collaborated with European recovery. A bipartisan group of policy makers under the leadership of Senator Arthur Vandenberg eventually passed the Marshall Plan—formally known as the European Recovery Program—which would funnel the equivalent of some $13 billion to participating countries between 1948 and 1952.

Paul Hoffman, a former president of the Studebaker Company, became the head of the new agency created to administer the Marshall Plan, formally called the Economic Cooperation Agency.[13] Hoffman was a core member of a politically influential group of internationalist business progressives to which IBM's Thomas Watson Sr. also belonged. In 1942, Hoffman founded the Committee for Economic Development in anticipation of postwar economic problems and challenges. In this function, he also directed a report on the Marshall Plan written by a committee of prominent citizens including many of the same business progressives who were involved with a report by the president's Committee on Foreign Aid, the so-called Harriman Report. Both reports identified three main goals for the Marshall Plan: humanitarian—to meet the moral obligation to end the crisis in Europe; economic—to build a strong European economy; and political—to strengthen Western Europe against the Soviet Union. Under Hoffman, the administration of the Marshall Plan continued the public-private partnership forged during the New Deal years.[14]

While Hoffman was concerned about the Soviet threat, like Marshall, he saw the solution to European misery not in ramping up European defenses but rather in strengthening Europe's economy, particularly its civilian and consumer industries. Funding during the first year of the Marshall Plan was mostly spent on emergency shipments, especially agricultural goods to relieve the European food and fuel shortages. In addition, the Marshall Plan funded the purchase of machinery considered essential for industrial reconstruction such as turbines for energy production and machine tools. From the beginning, however, a small part of the Marshall funding was devoted to technical assistance to help increase the productivity of European economies. This program was dear to Hoffman's heart from the outset; he and other Marshall Plan administrators hoped it would help increase prosperity and lure European workers away from the promises of Communist parties.[15] The emerging Productivity Program, sometimes also called the Productivity and Technical Assistance Program, soon became the programmatic core of the Marshall Plan.

In its first two years, the Productivity Program was a still-limited program for technical assistance between the United States and Great Britain, the so-called Anglo-American Council on Productivity—aid that was also extended to other European countries such as France. This assistance focused on the reform of civilian industries through exchange visits, paid from what were dubbed counterpart funds. These funds originated from the need to address the alleged dollar gap: European governments raised funds in return for aid shipments through the Marshall Plan. Since US vendors had no use for European currencies, the US government paid the vendors in US dollars. Of the counterpart funds, the US government received 5 percent for administrative costs, and the rest was used for infrastructure projects and economic improvements, including productivity projects. In 1950, with the outbreak of the Korean War, pressure rose for immediate European rearmament for North Atlantic Treaty Organization defense, and the formerly independent Marshall administration was subsumed by the State Department.[16] From 1950 to 1953, productivity measures were intensified in a productivity drive that sought to more systematically increase productivity in all participating European countries; also, Cold War imperatives for immediate European rearmament now competed with the goals for economic reform. Congressional amendments demanded that European countries create national productivity centers to raise awareness for productivity in their countries, and that productivity projects be tied to cartel restrictions and collaborative labor relations in hopes of raising European standards of living and combating Communist promises. In addition to such projects in small and medium firms in civilian industries, military procurement projects sought immediate production increases in larger companies while waiving some of the political goals attached to productivity.

Defining and Measuring "Productivity": From Production Increase to Factory Performance

The meaning of "productivity" was settled only over the course of the Productivity Program, with the help of officers from the BLS. Initially, Marshall officers and other persons involved with the Productivity Program used the notions of production increase and productivity interchangeably. Production increases could be achieved by simply employing more workers and raw materials for a given task; a production increase meant a higher amount of goods produced. As we saw in chapter 2, Köttgen had prescribed such a production increase for the recovery of Weimar Germany from World War I when he promoted longer hours and a higher work pace to achieve higher production. Marshall administrators, however, eventually saw it as their goal to promote more efficient work in European countries by raising the output per hour or the output per man hour; in other words, Marshall officers sought to increase the *rate* of productivity, not merely the *volume* of production. BLS officers were involved in this decision, settling on productivity increases over production increases.

BLS officers brought the seeming objectivity of their quantitative measures to the Marshall Plan's Productivity Program. During the war years, the BLS had been embroiled in a dispute with labor unions over the objectivity of its cost-of-living index, with labor unions charging that the bureau disregarded the decline in the quality of goods and the disappearance of cheaper items from the shelves. Labor unions eventually began to publish their own, competing index.[17] Emerging from this experience, statisticians emphasized the importance of quantitative information and that it be objective. Former BLS commissioner Isador Lubin argued in his 1947 presidential address to the American Statistical Society that the ability to meet the challenges of European recovery would "to a large degree be determined by the availability and intelligent use of pertinent data," and "never before has adequacy of data and statistical integrity been so essential.[18] Clague, who had conducted the calculations of the first productivity indexes for the BLS in the late 1920s and had ascended to the position of BLS commissioner in 1946, would be successful in again establishing the objectivity of the BLS's work and its statistical measures. Under his leadership, the BLS lent significant support to the Marshall administration's productivity efforts.[19]

The BLS had continued its productivity calculations during the New Deal and World War II years, although cost-of-living calculations had taken priority at this time. During the New Deal years, concern over unemployment had lessened attention to productivity research, and with the outbreak of World War II, BLS officers turned to concerns over scarce manpower during wartime. In 1940, Congress had passed legislation that established the BLS's Productivity and Technological Development Division, charged with providing government and private agencies "with current information on productivity, technological developments, and factors influencing productivity; and to maintain files and issue reports on technology and other topics relating to utilization of materials and human resources in peace or war."[20] The division continued both the quantitative and qualitative traditions of productivity studies within the BLS, compiling indexes of output per person hour of labor and unit labor cost, and reports on labor requirements per unit of output in specific—often militarily relevant—industries that required factory visits. James M. Silberman, the chair of the new division, was aware of the sensitivity of productivity calculations and strove for balance in his office's support for the Productivity Program.

Silberman became involved with transatlantic productivity exchanges in the early days of the Marshall Plan. Like BLS officer Clague, Silberman had studied economics and public administration at the University of Wisconsin. And like Clague, Silberman left an unfinished PhD dissertation to join the federal administration. Still in his early career, he appears to have become the BLS point person for support of the Productivity Program, conducting productivity surveys in Great Britain and France, and later also in Austria and other countries. In the context of European postwar recovery, Silberman realized that the appearance of objectivity would be important because Americans and Europeans disagreed about the level of productivity on either side of the Atlantic.

Silberman traveled to Great Britain in spring 1948 to assess the level of productivity in British industries; his report became one of the foundations for the Anglo-American Council on Productivity, a predecessor to the Productivity Program.[21] Being aware that the level of productivity was a sensitive topic in Anglo-American relations, Silberman wanted his (handwritten) notes on British productivity to be kept confidential because "an observation on a difference may so easily be twisted into a criticism by one who wished to do so, and there are some who would wish to do so," and because he feared that it would destroy the "chances of effective work if analysis becomes twisted into destructive criticism." At the same time, he noted clearly that "differences in productivity are real. They are not statistical fictions, accounted for by differences in quality of production or made wholly necessary by market considerations."[22]

Silberman produced a descriptive report that resembled the BLS's industry studies of the early 1920s more than the quantitative indexes of the later 1920s because, as he observed, "productivity studies carried out in the past have proved that it is impossible to measure the multitudinous factors affecting productivity, especially for large groups of plants and industries."[23] With regard to technological aspects, Silberman remarked that British companies often used the same production technologies as their US counterparts, but used them in different ways. While British companies may have used modern presses and other machines, for example, they did not transport materials from one workstation to another in efficient ways but instead frequently used wicker baskets. And with longer work cycles, where single workers did many things, they lost the benefits of division of labor. In addition, there were fewer hand-powered tools such as pneumatic screwdrivers, nut runners, and abrasive wheels. Even between US and British plants of comparable size, big British plants resembled collections of smaller operations, with less specialization and standardization, and less in-house production of many parts, such as hinges for refrigerators in a refrigerator factory.[24]

At the same time, Silberman related productivity to tenets of the US economic system: high wages, low prices, and cooperative labor relations—the same values that Ford had already promoted in the 1920s, as we saw in chapter 2. Silberman reasoned that the availability of cheap labor was at the root of the poor organization of production; he found it "hard to bring home the point that human labor is never cheap, and today is the most valuable commodity in Britain, since it is all she has to export or sell." In addition, he observed that noncollaborative labor relations—which he called a "caste system"—had pervading effects, with some European workers even refusing promotion to foremen. He briefly noted that there was "no disposition to create [a] large market through planned low price." Moreover, the "British"—it is unclear whether he meant industrialists, workers, or both—believed that the US market was huge and theirs was small, that Americans "get by on low quality—'regimentation,' not standardization," and that the United States had "huge plants" and a "dizzy work pace."[25]

The BLS later produced so-called *Factory Performance Reports* for the Marshall administration—that is, reports based on cost accounting data on the performance of individual US factories. Productivity, the reports implied, was an outcome of how the manufacturing process was organized—how the manufacturing floor was organized, which machines were used, how materials were transported between workstations, and how work was planned and controlled. Other productivity factors, such as marketing and labor relations, found only passing mention in the reports, if they were mentioned at all.[26] Altogether, between 1951 and 1954, the BLS produced forty-eight of these reports, which were intended for use in Europe to compare European production with American production, evaluate operations, identify areas of good and poor performance, and make the necessary improvements to increase productivity.[27] The reports focused on consumer industries such as textiles, canning and food, furniture making, and radio and television manufacturing. Based on more comprehensive data, the reports also mostly resembled the BLS productivity reports of the early 1920s. For each industry, the BLS developed questionnaires in collaboration with management as well as labor and trade associations. The reports were typically based on voluntary reporting by between one and two dozen companies—a significant increase from the handful of reporting companies in the early 1920s. The reports followed the case method—that is, for each reporting company, they provided the labor input for essential production steps rather than averaging across companies. Also, the reports were technical, giving detailed descriptions of the production process with photographs and technical drawings, and supplying sample forms for production planning and control.

The BLS *Factory Performance Reports*, however, were not without criticism. In particular, labor members of an advisory board to the Productivity Program, the Advisory Group on European Productivity (AGEP), critiqued the reports as too technical and not properly communicating the labor goals of the Marshall administration. Almost like an add-on, all reports included the same introductory paragraph that stated that productivity depended on many factors, "most of which are difficult, if not impossible, to isolate and measure," such as the skill of individual workers, workers' health conditions, and their attitudes. The reports, however, noted that productivity was due

> primarily to the efficiency and condition of the equipment with which the workers are supplied, the layout of the factory which determines the efficiency of the flow of raw materials and goods-in-process, the skill of management in planning and directing production processes, the aggressiveness of management in creating consumer demand and developing effective marketing procedures, the design of the product, and the quality of the raw materials.[28]

After a review of the reports, the AGEP head, John W. Nickerson, concurred that they did not contain statements about wages per hour or wages per unit of production, nor did they mention union efforts toward production increases—aspects that would be instructive for European unions in their outlook on productivity technology and approach to collective bargaining. An AGEP labor member argued that "the studies in their present form would be more harmful than helpful in the psychological effect they would have on European labor," presumably because their reduction of productivity to technological and organization improvements could suggest that economic factors such as wages and prices were irrelevant.[29] Therefore, the AGEP sought to minimize the use of BLS *Factory Performance Reports* in European countries. The seemingly objective productivity measures did not meet the larger political and economic goals as well as values of the Productivity Program.

The Values of American Productivity

Marshall administrators intended to instill a new sense of productivity in European workers, industrialists, union representatives, and politicians. They aimed to shape European firms and economies after the model of American productivity—in effect, they wanted to Americanize them. This section will look more closely at the economic and political values that the Productivity Program brought to Europe. The overall goal of the Productivity Program was to create dynamically expanding economies in European countries, where increasing productivity would raise the standard of living of European workers (figure 3.2). Productivity within the program was to "cover not only technological improvements but also business practices and labor practices related to distribution of the benefits of high productivity," and "be aimed not just at productivity improvement, but at share-out productivity improvement, with equitable shares of the results accruing to ownership, labor and consumers."[30] In other words, the objective was to "stimulate and assist a general conversion of the European economy to a more dynamic system geared to high and efficient production, high wages, lower prices, mass distribution and mass consumption with a high level of employment."[31] The Productivity Program thus was intended to implement a complex of social, economic, and political ideas that encapsulated the American capitalist system: mass production, mass distribution, free enterprise, and collaborative labor relations.

American companies developed and refined mass production methods in the late nineteenth and early twentieth centuries. British firms had initially forged the changes of the first industrialization, replacing manual and horsepower with steam, and dividing work processes into individual steps and between workers. Adapting British developments to American conditions—for example, by using abundant waterpower rather than steam power, which led to the rural versus urban location of plants—American inventors and entrepreneurs took over. Beginning in the federal armories in Springfield, Massachusetts, and Harpers Ferry, West Virginia, in the early

Figure 3.2
This 1956 IBM advertisement appeared in different languages in US and European newspapers and magazines. The caption presented the company's data processing machines as tools for companies to take advantage of "expanding markets everywhere" as well as "sustained and ever-increasing productivity," incorporating core productivity values.
Source: IBM World Trade News (August 1956): 14. Courtesy of International Business Machines Corporation, © International Business Machines Corporation.

nineteenth century, American manufacturers produced interchangeable parts in sequential operations with special-purpose machinery, replacing the work of skilled artisans with precision measurements that could be used by less skilled workers—the American system of manufacturing. They allowed for the production of large numbers of consumer durables, such as sewing machines, bicycles, and reapers by the second half of the nineteenth century, although handwork and skilled machine work prevailed in parts of the production process.[32] Improvements in the flow of materials on the shop floor, replacing hand-drawn carts in multistory buildings with moving belts and chains in one-level canning factories and Chicago slaughterhouses, eventually culminated in Ford's manufacturing line in the early twentieth century—the realization of mass production.[33]

As US manufacturers produced more goods, they also developed distribution methods to create new markets that helped avoid overproduction.
Manufacturers labeled their productions, creating a direct relationship with
consumers nationwide who demanded Ivory soap or Diamond sugar rather
than bulk products. Packaging also promised guarantees of quality, clean-
liness, and the correct weight and measure of goods. Manufacturers used
new distribution channels as well. While most manufacturers had previously worked with commission agents, many began dealing with regional
wholesalers who purchased goods outright, and took responsibility for
shipping and other costs; other manufacturers who needed to handle perishable goods, such as biscuits, or provide instruction and repair services for
technical products, such as sewing machines or typewriters, created their
own distribution networks at significant costs. Store credit and, by the early
twentieth century, installment purchase plans and consumer credits made
goods available to growing numbers of Americans. While manufacturers had
often pursued their own advertising in the late nineteenth century, proudly
showing their factories' smokestacks in colorful posters, professional firms
took over advertising by the early twentieth century, displaying products
in the (imagined) contexts of consumers' lives to create demand instead of
technical awe. Finally, new retail outlets such as department stores, mail-
order houses, and chain stores bought large quantities of merchandise
directly from manufacturers, bypassing wholesalers and efficiently organizing the flow of large quantities of goods within their companies.[34] Mass
markets for branded goods allowed manufacturers to reach growing numbers of consumers nationwide, creating stable demand for labeled products
with prices more constant throughout business cycles, thereby protecting
manufacturers against seasonal and economic fluctuations.

Free enterprise distinguished the US capitalist system from the state-run
economies of Soviet fashion as well as from cartel-dominated economies
in European countries. In the late 1940s and throughout the 1950s, the
United States was undergoing a major political sea change in favor of the
free enterprise system. During the 1930s, in the wake of the Depression,
the labor movement had gained strength, and the federal government had
intervened in the economy on behalf of labor through legislation like the
Social Security Act, the National Labor Relations Act (guaranteeing the
right to organize in trade unions, engage in collective bargaining, and pursue collective action, including strikes), and the Fair Labor Standards Act

(introducing a minimum wage and the forty-hour workweek). After World War II, however, government interventions in the economy again became contested, as did the proper size of the welfare state and the scope of union power in factories. A conservative business coalition, which included the National Association of Manufacturers (NAM) and the Chamber of Commerce, pushed through the Taft-Hartley Bill, which repealed the prolabor legislation of the National Labor Relations Act, and new legislation, such as the Full Employment Act, passed only in watered-down versions, defining employment as a responsibility of private business. The free enterprise system came to embody the conservative business ideology as the conservative business community embarked on a public campaign to counter the prewar rise of the labor movement and the economic interventions of the federal government. In gatherings on factory floors, in communities, and in institutions such as churches and schools, free enterprise was used to appeal to the values of individualism, freedom, and productivity, and win over American workers.[35]

Likewise, collaborative labor relations were thought of as typical for the US system, differing from the antagonist approach of socialist and communist labor movements. The core of the collaborative industrial relations model that Marshall Plan administrators hoped to bring to Western Europe was plant-level collective bargaining. In the United States, unions began to negotiate wages, hours, and other work conditions with employers in the late nineteenth century, replacing competition in the labor market with collective bargaining between workers and employers. Based on a decentralized system of labor representation, local unions negotiated labor conditions with employers at the plant or company level, and a single union had exclusive representation of a group of workers.[36] Some business conservatives may have claimed that their labor relations were harmonious in principle, blaming (antagonistic) unions for their struggles with workers; in this sense, free enterprise may be understood as in accord with "collaborative" labor relations. Yet within the Productivity Program, the idea of collaborative labor relations was generally interpreted in ways that went beyond such appeasement; it aimed at strengthening the free labor movement in European countries and increasing the general standard of living.

These two values—strengthening the free labor movement and free enterprise—are of particular interest because while each of them was thought to distinguish the US capitalist system from other economic systems, they were also at odds with each other. In addition, both were hotly contested in the United States during the postwar years.[37] Integrating both in the Productivity Program allowed the administration to enroll a bipartisan coalition in support of the program, and contemporary politicians brushed over the potential contradictions between these values, making them seem as if they were in harmony. For example, a 1951 amendment to the US foreign aid program by a bipartisan group of young senators under the leadership of William Benton, a Democrat from Connecticut, combined the two objectives. The so-called Benton Amendment supplied additional funding for the Productivity Program to "provide the incentives for a steadily increased participation of free private enterprise," and "discourage cartel and monopolistic business practices." But the amendment did not stop there; it called for funds to "encourage where suitable the development and strengthening of the free labor union movements as the collective bargaining agencies" too.[38] Integrating the two values also enabled Marshall administrators to assemble a broad group of business and trade union representatives to implement the Productivity Program, as we will see in chapter 4.

Debates surrounded the Productivity Program about whether uniquely American conditions enabled high productivity. Addressing a group of European industrialists, Hoffman, the first head of the Marshall administration, observed, "You always point out to me that we have much greater natural resources. And you point out that in the United States we enjoy a great single market. I've had these things pointed out to me innumerable times."[39] Marshall administrators generally admitted that there were constraints on European productivity, such as inadequate natural resources, a lack of capital goods including power plants, modern factories, and machinery, and small national markets. Yet they were quick to emphasize the European "lack of will and ability to undertake rapid technological improvements," which went "deep into the Western European social and economic situation and the customs, habits and attitudes of both business and labor."[40] As Marshall administrators complained, "The basic system on which the European economy runs is largely a moribund form of capitalism geared to low productivity, low wages, low consumption, high prices and high unit profits protected by restrictive business practices. Under this system actual production and productivity are far below what existing natural resources, capital equipment, market opportunities and technology would permit."[41] Warning that it was "very dangerous to kid oneself" and "say something because it sounds pleasant to oneself," Hoffman concluded that Europeans could certainly increase their productivity if Americans produced "one-third of the total goods in the world and one-half of the total manufactured goods with one-fifteenth of the land area of the world, one-fifteenth of the people in the world, and one-fifteenth of the natural resources of the world."[42] Productivity administrators therefore sought to remedy three crucial problems: the lack of technological skill, the lack of business, managerial, and industrial know-how, and the lack of will to produce and distribute efficiently.

The Productivity Program adopted a variety of methods, which included showing productivity films in Europe, awarding grants-in-aid for research supporting productivity ideas, organizing studies of European products to assess their productivity features, conducting industry surveys in the United States and European countries for comparative purposes, running a technical question-and-answer service, and organizing exchange programs that brought European managers, engineers, workers, union representatives, politicians, administrators, and students to the United States to study and experience American productivity, and brought US consultants to Europe to offer consultancy services and seminars (figure 3.3). The signature program comprised the so-called demonstration projects that aimed to exhibit the features of US productivity in Europe. I will look at these demonstration projects more closely here because they highlight important features of US productivity within the Marshall Plan.

Devising and Selecting Demonstration Projects

European companies could receive moderate financial aid through the Productivity Program for demonstration projects that improved productivity and exhibited the values of collaborative labor relations and free enterprise. By definition, demonstration projects were "an action program at the individual plant level" that provided management and labor "with technical advice, methods and techniques, and any other assistance necessary, to obtain increased productivity." Demonstration projects also required "a mutually accepted working agreement to share equitably the benefits of such increase and the resulting savings between management and labor in terms of increased earnings, and consumers in the terms of lower prices."[43] Demonstration plants also had to refrain from restrictive business

Figure 3.3
The Marshall Plan's Productivity Program included a mobile productivity film program.
Source: Productivity Film Program, box 5, entry 1203, RG 469, NARA.

practices—that is, the formation of cartels. Finally, they had to welcome outside visitors to help disseminate productivity methods. Increasing productivity and disseminating US values, demonstration projects thus combined the technical and political goals of the Productivity Program.

Persons associated with the Productivity Program used anti-Communist rhetoric to describe the projects. Mrs. Wallace Clark, a member of AGEP advising the Productivity Program, suggested that the United States "should have 'cells' in other countries as the Communists have" in Western countries. Only US cells "could not be underground, secretive and destructive" like, for instance, US Communist Party members, who had worked through labor unions and other organizations without revealing their party affiliation during the Popular Front period of the 1930s. US cells, though, "would be open and above-board examples of the democratic way," and "should be model factories set up and operated for teaching and training in the American know-how and the way of thinking that has gone into it." She believed that "such democratic 'cells' would be the most effective answer" to anti-democratic propaganda.[44] The demonstration projects were introduced in 1950 when the Marshall administration faced pressure to aid the North Atlantic Treaty Organization rearmament as well as show results; the goal for the demonstration projects was immediate impact in order to provide "quick and rapid methods for actually applying practical productivity" without having to await longer deliberations.[45]

While Marshall administrators debated the implementation of demonstration projects, they imagined that the projects would lead to ripple effects in their industries through higher wages and lower prices.[46] They imaged that as demonstration plants achieved higher efficiency and lower prices, they would exert competitive pressure on other firms to adopt the same practices in order to stay competitive.[47] For example, Carl R. Mahder, the US head of the Productivity Program in Germany, explained:

> When we have succeeded in reaching this goal, we will then have realized the establishment of the economic machinery that will make it possible for a shirt manufacturer—or a producer of any other goods—to take his shirts that he can now produce at cheaper costs, and sell them on the market at a cheaper price and take the market from his competitors, and then he will have demonstrated to his competitors that if they do not want this market to be taken away from them, they had better try to find out what is going on. The competitor will then be placed in a position—he will be forced if he wants to keep the market—to go about doing much the same type of thing. He, too, will then have to figure out the steps that he must go through in order to bring his costs down, with the result, eventually, we hope, that we will have an analogy in Germany with what we have had in the United States for the past five or ten years.[48]

As proof of these dynamics, Mahder cited the decreasing prices for television sets in the United States from $700 to $900 to by then $160 to $200.

While the Marshall administration internally used the term "demonstration project," it allowed local missions to use other terms "for political and psychological reasons."[49] In Germany, for example, the mission used the terms "pilot plant" and "pioneer plant" in public statements. The more frequently used term, "pilot plant," highlighted the experimental character of demonstration projects, capturing the administration's intent that demonstration projects serve as a "workshop laboratory, where measures to improve productivity are introduced and considered to develop and to test new methods and procedures with the results analyzed and evaluated on a current basis." By contrast, the term "pioneer plant" emphasized the projects' exploratory character, and their serving as a "living demonstration that the possibilities really are unlimited for greater and closer cooperation and work between management and labor; and that together they can develop joint undertakings in their mutual interest providing to each returns far larger than was believed practicable and realistic."[50]

The shifting terminology may have been an indication of the initial local resistance to the idea of demonstration projects. For example, officers in Italy noticed that they got "into trouble" every time they used the term and instead began to refer to these projects as "productivity programs." Similarly, US officers in the Netherlands reported that the "word, 'Demonstration' Plant, is a red flag," and they considered alternative notions such as "cooperative plants"—despite its ambiguous meanings since it could be understood as either a plant with cooperative labor relations in the sense of the Marshall Plan or a plant with cooperative ownership in the sense of the nineteenth-century cooperative movement.[51] The reasons for this resistance remain unclear. It may have been because "demonstration project" made Europeans feel like students on the school bench, being taught by the US master. But the resistance may have also been the result of European political sensitivities to American attempts at working around their national sovereignty. The initial announcement of the demonstration projects in June 1951 forcefully stated that the Marshall administration would "modify its general policy of working only with and through governments and will be in direct touch with trade unions, individual firms, individual managers, trade associations, labor leaders, and, especially, will be working with individual plants." The administration hoped to "be working with the concurrence and, hopefully, with the help of the governments of the participating countries, but not exclusively through those governments."[52] A year later, by November 1952, the Marshall administration sought to "avoid the appearance of heavy-handed American intervention" in relation to demonstration projects that were now to be initiated locally, with the local US mission providing discrete advice and other assistance in the background but not having a public role.[53]

Yet the selection of firms for demonstration projects reveals that Marshall administrators did not seek model plants that would exhibit high labor efficiency. They instead prioritized political over technological goals, choosing plants where management and labor were willing to implement share-out agreements—even if the firms were barely competitive. Indeed, a number of the firms under closer consideration were not profitable, not to mention efficient, and had serious deficiencies in corporate organization and accounting. Some of the firms even appear to have been close to going out of business and seemed to have grabbed the demonstration project funding as a last effort to stay afloat.

One demonstration project, the family-run textile mill Spinnerei und Weberei Wiesental, was a typical example of a firm with significant organizational challenges. With the owner and director making all hiring and promotion decisions, there was no human relations department, an indication of a general lack of organization and delegation of responsibilities. Also, the company lacked a proper cost accounting system, and a productivity survey of the company revealed that its "pricing policies are dictated by outside influences rather than by . . . [its] own management"—a possible reference to cartel-like collusion.[54] Productivity administrators strongly recommended introducing a cost accounting system and work standards; they saw these measures as more important than the introduction of new machinery.[55] For the plant to be considered for the demonstration project program, Mr. Nestel, the director, who had participated in a productivity visit to the United States, and the works council must have agreed to a share-out of productivity benefits. Nestel indicated that he wanted to pass on his gains to consumers in the form of better material for the same price rather than price reductions, and he hoped that higher wages would solve his problem of workers leaving for better-paid construction work during the summer. These considerations appear to have satisfied the requirements to become a demonstration plant for a company that certainly was far from a model in efficiency.[56]

At the same time, administrators turned away plants that were more profitable. One such example was the shoe manufacturer Dorndorf Schuhfabrik, which submitted an application for DM 1.4 million for technical improvements, such as new cutting tables and new means of transportation as well as an electronic computer to be used for payroll calculations (one of the rare applications for a computer). While the scope of the application may have been far outside the typical project size, Dorndorf was a typical German version of a patriarchal welfare capitalist firm. The color-printed annual report accompanying the application emphasized the company's concern for the "Dorndorf-family," expressed in the hiring of a female social worker to complement "male reason and objectivity with motherly sympathy" for a holistic care of the employee's physiology and psyche. Instead of higher salaries, the company proposed to pass on productivity benefits to employees in the form of additional welfare capitalist measures, such as an extension of the suggestion system and training course, the creation of a new employee magazine, youth camps along with a youth group and house, clubs and choirs, a library, and a training school for apprentices. It is unclear what disqualified the company from the demonstration program, but it might have been because the company's requests surpassed by far the amount of funding typically granted under the demonstration program, or because of the company's unwillingness to change labor relations and sign a share-out agreement.

Another productive plant that was not considered for the demonstration program was the Homann Stove Works in Wuppertal. When the current owner took over the company from his father in 1933, he reduced the number of types of household stoves produced from over 450 to 40, and integrated his production under a single roof. Producing 350 stoves a day, the company held a 10 percent market share in Germany, and Homann aimed at increasing production to 1,000 stoves a day. The company had good relations between labor and management, with 68 percent of the employees organized in the industrial union IG Metall, and Homann kept an open mind regarding share-out agreements under the demonstration program and said he was eager to produce a "modern, well built, efficient household stove," which would sell for a low enough price to create demand. Also, Homann had a "modern outlook," drawing on US stoves to create his own stove designs, and used consultants for design and human relations questions. Homann, however, expressed the fear that sharing his company's knowledge and experience with the entire industry would lead to overproduction, thereby greatly intensifying competition. His marketing director, Mr. Zähringer, alleviated this fear by remarking that with the help of their envisioned marketing plan, they would far outpace their competitors.[57]

The Homann factory appears to have been an immediate candidate for a productivity program: it was modern, efficient, and successful, and had good labor relations. An open question is why the initial inspection never led to a further survey and consideration of the plant for the program. Did Homann decide not to pursue the opportunity because of his concerns over sharing his knowledge and experience? Or did Marshall administrators not consider the company suitable? In a number of other cases, Marshall administrators felt that a well-run company would receive funding through the banking system. Mahder, the head of the Productivity Program in Germany, even explained this reasoning at a Paris conference on demonstration projects. Describing the example of a successful German lumber company that had developed a process to manufacture boards from sawdust and wood chips that would significantly reduce the cost for lumber, he said, "This man has such a good thing he should be able to find his financing elsewhere," and the administration would only reconsider its initial denial if the manufacturer had not been able to secure independent funding within six months.[58]

Overall, the Marshall administration, through the Productivity Program and its demonstration projects, aimed to change social and class relations. As John W. Nickerson, the head of the AGEP advisory group to the Productivity Program, emphasized, if the Marshall administration was to achieve its main objective of "keeping millions of people in the free world from drifting toward Communism," it had to "give them hope that through participation in an industrial economy similar to ours they may obtain freedom from want and a chance for a better standard of living."[59] Likewise Harold B. Maynard, one of Nickerson's colleague on the advisory board, hoped that the availability of "more of the good things of life" through the Productivity Program would "help to awaken the desire of the individual to improve himself and may tend to develop a more fluid society."[60] Yet Maynard thought that the question of social change was one for the social scientist, not for the advisory group.

The demonstration projects were an essential part of the Productivity Program—for programmatic reasons, because they exhibited the core of the American economic system of high salaries and low prices, and for political reasons, because congressional support for Marshall Plan appropriations hinged on the success of the program. On the ground in Europe, the Marshall administration encountered opposition mainly from the Communist left and the cartelized right.[61] A major problem was that unions in Germany insisted on regional or national collective bargaining, which made it difficult to generate higher salaries from productivity benefits—an issue that will be further discussed in chapters 5 and 6.[62] Despite such resistance, Marshall administrators asserted with almost missionary zeal that the Productivity Program had "sown the seeds of Productivity in many areas," and emphasized that "while some may have fallen on stony ground, much fell on fertile soil, had taken root, and is flourishing."[63] Or as another Marshall administrator said, demonstration projects were to convince through "work rather than words"; they were to shine like the biblical light on top of the hill.[64] Such religious rhetoric in the assessment of demonstration projects alluded to the early puritan vision and ideas of American exceptionalism; it also indicated the identity of some Marshall administrators as missionary cold warriors. Inspired by their spiritual experiences as evangelical Christians, they were driven by an almost missionary desire for Europeans, through their own convictions, to adopt the American productivity model.

US Productivity Officers as Marshall Missionaries

As we saw in chapter 2, Weimar Germans had developed their own rationalization movement, creating institutions, such as the Reichskuratorium für Wirtschaftlichkeit, and introducing machinery and production planning in their companies. Germans thus were familiar with the ideas of technological rationalization. But the Productivity Program's social and political objectives, in particular free enterprise and collaborative labor relations, went beyond such technological rationalization, posing a challenge to German and European ways of thinking. Marshall administrators believed that to convince Europeans of the American way, they would have to change "basic attitudes—attitudes which are deeply implanted in European cultures and which in part stem from ancient institutions such as the medieval guilds."[65] In order to accomplish the objectives of the Productivity Program, they felt the need to make an "emotional impact," "catch the imagination and hopes of people," and replace "tradition, fear, ignorance, pride, hopelessness, and selfishness . . . with their opposites: progressive ideas, courage, intelligence, humility, hope and selflessness."[66] Such a change could not be achieved by force; instead, Marshall administrators thought they had to "win people over" so that the people would "see the righteousness of the position, and . . . what they are going to get out of it." Administrators believed that it was "impossible to take an idea out of anybody's head" and they could "only give him an idea [that] he thinks is better; he accepts it and it then replaces the other idea."[67] Marshall administrators thus became missionaries of the American gospel of productivity.[68] This religious dimension of their work shaped their efforts at convincing Europeans of the American productivity model.

To be sure, Marshall administrators also used other metaphors to describe their work. They used medical metaphors, for example, and talked about supplying "a kind of 'blood transfusion' to inject into the Marshall Plan countries the production methods and ideas that have built America's economic power."[69] A member of the AGEP advisory group used the image of a doctor prescribing medicine without asking "the patient what he wants, though [the doctor] may sweeten the medicine somewhat."[70] Occasionally they used the sales metaphor, thinking of themselves as being in "a 'retail' business," having "to change the attitude of a person somewhere, or a pattern of action on a farm or in a factory."[71] The religious metaphor, however, seems to most adequately capture their idealism and in some cases even their spiritual identity.

A few administrators of and advisers to the Productivity Program, including Nickerson and Myron Clark, the director of the Productivity and Technological Assistance Division, were members of an evangelical Christian breakfast group. Part of the Christian Leadership Movement, these breakfast groups were organized to help believers integrate their religious values and professional lives. Nelson Cruikshank, the labor adviser to the Productivity Program, stated at the International Christian Leadership Conference in Washington, DC, in February 1953 that "offering aid to a personal friend is one of the most dangerous things which can happen to a friendship. Unless exceptional care is taken, [the] friendship is bound to break up. The same is true of nations. We want to help our allies but we want to keep them for friends. What we need in the Mutual Security Agency [Marshall administration] is people with the spirit of St. Francis and the political skill of Machiavelli."[72] Evidently, Marshall administrators were sensitive to the challenges posed by their mission in Europe; their religious beliefs provided guidance on how to address these challenges.

Marshall officers realized that it was essential to find the right people for the job. "While technical competence is extremely important," said Arthur McLean, chief of the Productivity Program's technical staff branch, "there is a danger of placing too much emphasis on that and neglecting other qualifications such as a sincere belief in and enthusiasm for the program, proper attitude toward Europe and Europeans, etc."[73] Some Marshall administrators found inspiration in the testimonies of their fellow believers. At the International Christian Leadership Conference, for example, C. F. Hamann of Western Electric stated that changing others' attitudes depended on changing one's own attitude. He described how he went through different stages, from initially giving up non-Christians as lost, to embracing all fellow men with a peaceful attitude unless they were "ornery," to finally embracing all fellow men. At this last stage, he stated, "he was able to change attitudes more easily, pleasantly and successfully," because "attitudes can be changed by sincerity of purpose and that sincerity can only come from carrying out God's purpose and from seeing God's truth."[74] Discussions on the floor suggested that officers should "live simply and unostentatiously" in the countries that they were sent to, and must have a "strong spiritual sense which causes people to sacrifice their ego and be humble."[75] Marshall administrators suggested that the proceedings of the International Christian Leadership Conference be made part of the orientation program for government employees and consultants sent to Europe.

In addition, Marshall administrators worked closely with cultural anthropologists to better understand the European mind-set and improve their method of persuading Europeans. As anthropologist David Rodnick explained at a conference on the Work-Study Training Program within the Productivity Program, "Europeans are much like Americans," but they know that the United States is "a big, rich country, a powerful country," and they are dependent on the United States, and therefore "they are very much on the defensive." In order to prevent European workers—who already "have a sense of inferiority"—from feeling "helpless," and thus "resent and react rather strongly against" Americans, Americans had to "lean backward and unroll the carpet for them—make them feel they are wanted." Given that Americans had to cut down on their own "sense of importance," this seemed like an insurmountable task, particularly because this attitude had to be instilled not only in Marshall administrators in Washington, DC, and Europe but also in other organizations related to the Productivity Program, such as in companies that welcomed European experts, and colleges and plants that hosted European work-study students.[76]

Finally, Marshall administrators also infused with religious meaning George Marshall's initial direction that the initiative for the aid program was to come from Europe.[77] They insisted that the Productivity Program would not work unless the Europeans made productivity ideas their own and believed in it. For example, William H. Draper Jr., then the US special representative in Europe, asserted that he did not "believe that we as Americans operating in Europe can foist on the Europeans our concept of industrial techniques, management and labor relations, nor do I think that we as outsiders can remake their social system or industrial practices." Draper said that Americans should "keep on preaching the gospel of productivity . . . but should turn over the responsibility for really increasing productivity to the Europeans themselves." In his view, "much good work has been done, but unless Europeans are ready now to take the active leadership in achieving better technical methods for increased and cheaper production," he doubted that "direct efforts on our part [the Marshall administration's] will have great effect."[78] Other Marshall administrators agreed. Quoting the assistant administrator for production, William H. Joyce, they asserted that "it is one thing for us to talk about ourselves—it is many times more effective when a European worker says to his own people: 'I was there, I saw it for myself; here are the facts!'"[79] Similarly, the chief of the Productivity Division, Mr. Foerster, claimed, "We, as Americans, cannot have a productivity programme in Europe. It must be a European one."[80] Motivated by their spiritual background, Marshall administrators thus tried to get Europeans to adopt their creed of American productivity as their own.

Conclusion

Inspired by their experience of religious conversion, Marshall Plan officers sought to *convince* Europeans of American business values rather than *force* the American way of business onto Europeans. While the missionary spirit may have instilled in US officers compassion and seeming respect for Europeans and European circumstances, a closer look reveals that it did not grant Europeans a *choice* in adapting or debating the American model. Like missionaries, Marshall Plan officers were convinced that their own gospel was the correct one; they displayed little willingness to learn from Europeans, and expected Europeans to adopt their creed as a wholesome system. There was no room for transatlantic debate or criticism. The Marshall Plan officers' unfailing belief in the American model astonishes, given that even the meaning of productivity was uncertain at the Productivity Program's start, and the Productivity Program was tied to contradictory political values, such as the free enterprise system and collaborative labor relations, which were at that time under heated debate in the United States. Yet despite their strong convictions, Marshall Plan administrators did not simply force productivity onto Europeans. On the one hand, they insisted on a few universal objectives—to be implemented through demonstration projects—that were to reshape European economies after the US model; in this sense, the Productivity Program was a homogenizing program. On the other hand, Marshall Plan administrators acknowledged that European economies differed from each other as well as from the United States, and they allowed national Productivity Programs to be adapted to local political, economic, and social conditions; in this respect, the Productivity Program could be expected to have heterogeneous effects.

Besides the Marshall Plan administrators' considerate rhetoric and religious overtones, one should not forget their ultimate, hegemonic objective: to instigate a social revolution—however benign—in Europe. Not only eager rank-and-file officers, like Bellows, promoted this goal; it was upheld by the highest ranks inside and outside the Marshall administration. Joyce, the head of the Productivity Program in Europe, angered the French public when he called the French economic system "more feudalistic than capitalistic," indicating the need for social change.[81] Further removed from the operations in Europe, Senator Benton used even stronger verbiage in his address to Congress: the Productivity Program assisted the "Marshall Plan countries to undergo a constructive revolution," and he did not mean "the kind of revolution that starts out with mass executions and winds up as a cold-blooded dictatorship—but the healthy kind of revolution which can bring life and vigor and vitality to the economic system of the countries we are trying to aid."[82]

While no electronic computer was funded through the Marshall Plan, electronic computers were to become the iconic productivity technology of the postwar decades. By reshaping European companies and their economies after the US model, the Marshall Plan's Productivity Program created a fertile ground for the later adoption of electronic computers. For example, the demonstration projects incentivized the introduction of formal cost accounting procedures that would lend themselves to be computerized. In addition, the executives of the leading US computer manufacturers in the postwar years, IBM and Remington Rand, were part of the progressive business community that supported the Marshall Plan and its Productivity Program. Both IBM and Remington Rand opened their doors to Europeans studying productivity culture and technology in the United States through the Productivity Program, and introduced European visitors to new electronic productivity technologies. Free enterprise and collaborative labor relations, the two contradictory values tied to the Productivity Program, allowed Marshall officers to form a diverse coalition in support of the program; it included business and management associations, executives from companies like IBM and Remington Rand, and trade unions and labor advisers. This diverse coalition will be looked at more closely in the next chapter.

4 US Management and Labor Debate Political and Economic Values of Productivity

The Marshall Plan's Productivity Program published a brochure with captivating visuals detailing the differences in productivity between the United States and European countries. One graphic showed that an American worker needed to work 2 hours to purchase a pair of overalls; a British worker needed to work more than five times as long, or 10.5 hours; a French worker needed to labor more than seven times as long, or 14.5 hours; and an Italian worker needed to work even more, or 48.5 hours (see figure 4.1). The graphic also showed similar comparisons for shoes, bicycles, and butter. The brochure's graphics hid its questionable conflation of wages and prices with the statistical definition of productivity—that is, the output per hour or per worker. It implied that if only the Italian worker was more productive, he could buy more. But if the Italian worker was paid more, or if goods cost less, he would be able to buy goods using fewer work hours, regardless of how high his *output* was during those work hours.[1] And wages and prices were influenced by other factors aside from a worker's output, such as collective bargaining and the strength of labor unions, the unemployment rate and its pressure on the job market, and competition and cartel agreements to set prices. Brushing over these complex relations, the brochure used the politically loaded definition of "productivity": productivity no longer merely denoted output per worker or per work hour, but now it also meant free enterprise and collaborative labor relations—two ideas infused with specific values as well as the goal to shape the setting of wages and prices after the US model.

The brochure could easily be mistaken for one of the countless brochures and films that the Marshall Plan's education program showered on participating European countries. Yet this brochure was addressed to a different

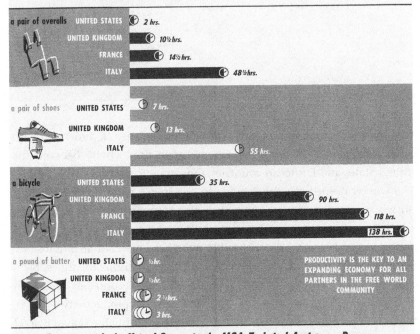

Figure 4.1
This illustration from a Marshall Plan brochure demonstrates that Americans needed to work significantly less time than Europeans to afford daily commodities. Figures are for 1950.
Source: Mutual Security Agency, "Waging the Peace . . . on Your Factory Floor," box 20, entry 172, RG 469, NARA.

audience: managers and workers in US companies. Titled "Waging the Peace . . . on Your Factory Floor," the brochure claimed that global peace began at home by "imparting to overseas visitors the know-how behind US productivity."[2] It aimed at convincing US employers and employees to open their factory gates to European visitors, and explained to them why it was important to impart what was presented as a matter of technical "know-how"—rather than a matter of economic and political values—to them: it was a matter of mutual security in order to strengthen the free world in the Cold War competition. To do so, the Productivity Program's administration forged an unlikely network of US corporate and management associations along with labor unions that supported the program ideologically and practically.

The Productivity Program's administration co-opted private citizens and private organizations, relying on what could be called a public diplomacy network to support the goal of instilling productivity mindedness in Europeans. It enrolled corporate and management associations, such as the National Association of Manufacturers (NAM) and the National Management Council, for logistical support of the many European study groups traveling to the United States through the Productivity Program. It also looked to these organizations for people to serve as consultants traveling to Europe to advise European companies on productivity measures, and for the leaders of the NAM and National Management Council to convince member companies to open their factories to European visitors. The Productivity Program encouraged labor union officers, such as those in the American Federation of Labor (AFL) and the Congress of Industrial Organizations (CIO), to join the Productivity Program administration as labor officers in Washington, DC, and European locations to give the Productivity Program legitimacy so as to convince European labor unions and workers of US-style productivity against Communist promises to redistribute wealth between owners and workers, and generate support for the program among US workers who could see Europeans as competitors. This practice of co-opting private citizens and organizations went beyond public diplomacy efforts in other areas, such as the space program, which communicated US technological prowess and superiority to peoples around the globe.[3] In a clear departure from what historians have called the cooperative state of the 1920s and 1930s, the Productivity Program formed a network of private organizations that were to work toward goals formulated by the administration—albeit goals that took the interests of these private organizations into account.

The Productivity Program administration embraced the kind of conflicts of interest that tore apart the New Deal coalition. Through the New

Deal, producers such as rural farmers and organized workers sought higher remuneration for their work. Their higher wages meant higher prices, which made urban shoppers and middle-class consumers recoil. Defying similar conflicts, the Productivity Program administration formed a public-private partnership that included organizations that were usually at odds—companies and labor unions—and did so by incorporating goals that each partner could identify with. On the one hand, the Productivity Program administration encouraged free enterprise—a goal that private corporations identified with and promoted in a nationwide campaign. On the other hand, the administration encouraged cooperative labor relations, thereby drawing in the labor movement, which fought for plant-level collective bargaining. This tolerance for internal incoherence and contradictions enabled the Productivity Program administration to assemble a broad coalition in support of its program. Discounting the potential conflict, they argued that productivity increases would eliminate the class inequalities under conditions of economic scarcity. Their view differed from an antagonistic understanding that productivity increases would only benefit the rich and a redistribution of wealth was needed to benefit the poor. As Charles Maier has compellingly stated, Productivity Program officers turned the issue of inequality from a political one—the redistribution of wealth—into an economic one—the increase of average standards of living; they talked the "economics" of productivity increases instead of the "politics" of redistribution.[4]

The goal of this chapter is to explore the interests and interest groups brought together by the Marshall Plan's Productivity Program, and reveal some of the program's internal tensions. The chapter first discusses the signature project for European entrepreneurs, Project Impact, which brought almost three hundred leading European industrialists to the United States to observe US-style productivity through company tours and participate in the First International Conference of Manufacturers. A closer analysis of the conference presentations reveals that free enterprise took on different meanings in domestic and international contexts. The chapter then examines collective bargaining and looks at Clinton S. Golden, a CIO union leader who, in the late 1930s, developed the idea of collaborative labor relations and was a leading labor adviser, shaping the Productivity Program from the labor perspective. In conclusion, it discusses how the Productivity

Program was to some extent an Americanization effort, with the goal of making European firms and labor relations more like the US model.

Project Impact and the First International Conference of Manufacturers

On the morning of December 3, 1951, the former Marshall Plan administrator and then president of the Ford Foundation, Paul G. Hoffman, gave the opening address at the First International Conference of Manufacturers. Some three hundred leading European industrialists from eighteen countries and one hundred top executives from the United States had assembled in the ballroom of the Pierre Hotel. Located across from Central Park, the hotel offered old-world distinction with its grand decor, French haute cuisine, and a mansard modeled after the Versailles royal chapel. While the Europeans might have felt at home in these surroundings, Hoffman had a provocative message for them. He emphasized to his audience that "those of us who believe in a free economy have got to prove that a free economy can not only offer the freedom which is the basis for human dignity, but we must also offer a better standard of living." Warning that "the Kremlin's proposal [that] can achieve a better standard of living by taking away all the goods from the 'haves' and giving them to the 'have nots' is very appealing to hungry and desperate peoples," Hoffman charged that such a redistribution of existing wealth was "only a method of distributing poverty." In Hoffman's eyes, it was "not enough to promise freedom"; a free economy had to provide a better life for all people through increasing prosperity.[5]

Organized by the National Association of Manufacturers, the First International Conference of Manufacturers was the culmination of a productivity project targeting leading European industrialists. From the US side, the top executives of large multinational corporations attended, including General Electric's Philip D. Reed, General Motor's Charles E. Wilson, IBM's Thomas J. Watson, and Remington Rand's Marcell N. Rand, as well as the presidents of smaller, nationally operating companies, such as Roy Fruehauf of the Fruehauf Trailer Company, H. W. Prentis of the Armstrong Cork Company, and William C. Stolk of the American Can Company. They represented mass production companies mostly, while representatives from labor-intensive industries, such as textiles—which opposed Marshall funding because they feared international competition—were missing. The

European delegations comprised both large and small family-run companies, from mass production companies as well as labor-intensive industries. The German delegation, for example, included members from large corporations, such as Dr. Heinrich Tramm, the chairman of the board of Ruhr-Chemie AG, and Hans Boden, vice chairman of the board of the Allgemeine Elektrizitätswerke (AEG), along with owners of smaller firms, such as the textile manufacturer Hellmut Krengel of the Eschweiler Bekleidungswerk in Bielefeld and the glass manufacturer Otto Seeling from Fürth. The conference was the final event of Project Impact.[6]

Before, the European participants had toured US corporations for one to two weeks in groups of forty to fifty persons from different countries and similar industries. Organized by the National Management Council, the tours roughly followed the itineraries of the Weimar German visitors: through the American Midwest with the obligatory stop at Ford's factories in Detroit. In addition, they toured companies that few visitors before World War II would have seen, including one of IBM's plants in Poughkeepsie, New York, and Remington Rand's factory in Elmira, New York (figure 4.2). At every stop, the European visitors received warm welcomes and ample local attention. When one of the groups arrived by train in Elmira to visit Remington Rand's plant for calculating machines and typewriters, for example, it was greeted by popping flashbulbs from the local press, and Remington Rand's vice presidents chauffeured the visitors in private cars to the company's plant. In addition to a tour of the manufacturing facilities,

Figure 4.2
Twenty European industrialists from ten European countries visited the IBM plant in Poughkeepsie in November 1951 through the Productivity Program's Project Impact. *Source: IBM World Trade News* (January 1952): 2. Courtesy of International Business Machines Corporation, © International Business Machines Corporation.

a Thanksgiving turkey was served for lunch at the company canteen to celebrate the holiday, and the mayor, the women's association, and union representatives greeted the group.

At the end of the tours, all participants met in Washington, DC, for discussions with top government officials, including Richard Bissell, then the head of the Marshall administration.[7] A chartered train finally brought everyone to New York City for a three-day conference with their US counterparts. The conference program focused on six major topics surrounding productivity: production, marketing, employee relations, finances, competition, and public relations. The organizers of the conference may have aspired to an ideal of transatlantic exchange, but the highly regimented discussions prevented any real possibility of persuading anybody.[8] Although conference presenters from both sides of the Atlantic expressed diverging and converging viewpoints, the conference laid the groundwork for continued transatlantic exchanges between top executives.[9] In addition, the US manufacturers' willingness to open their plants to European competitors as well as their conference participation demonstrated the US business community's dedication to the Marshall Plan's Productivity Program, and marked the administration's reliance on this support, both ideologically and practically.

Mostly progressively minded and internationally oriented executives, organized in business associations such as the Business Advisory Council and the Committee for Economic Development, had supported the Marshall Plan since its inception. After George Marshall's announcement of a European aid program, they backed two reports, one headed by Hoffman for the Committee for Economic Development and one by Averell Harriman for a presidential committee, that both strongly supported aid to Europe, emphasized the need for higher European productivity, affirmed the commitment to capitalism and free enterprise, and even accepted inflationary consequences for US consumers from shipping abroad materials that were in short supply in the United States. Progressive businessmen, such as Hoffman, Reed, and James D. Zellerbach of the Crown Zellerbach Corporation, also backed the Marshall Plan by giving congressional testimony. Once the legislation passed, members of the business and academic community served in leadership positions in the Marshall administration because of Congress's reluctance to have the State Department control the aid program. Hoffman, initially an unwilling recruit for the administrator's position, studded his administration with businesspeople, including William C. Foster, president of Pressed and Welded Steel Products, as deputy special representative in Europe; Bissell, a professor of economics at MIT, as assistant deputy administrator; and Harriman as special representative in Europe.[10]

In addition, the Productivity Program administration relied heavily on business associations such as the National Management Council to act as subcontractors to organize the visits and company tours of European productivity groups. People in these associations knew which plants were suitable to visit in different industries and for different groups; they had the contacts to arrange the visits; and they had the trust of member companies, which made the companies willing to open their doors to foreign visitors. Moreover, a large number of companies, particularly in the automobile industry, the agricultural implements industry, and the electric mass manufacturing industry, welcomed European study groups, showed them their production sites, engaged in discussions, and often hosted them in generous ways.[11] While companies may have simply been altruistically motivated, they may also have been interested in making connections with European partners in hopes of future orders or licensing agreements, or mutual exchanges that would allow them to learn from European knowledge and practices.

Among the companies receiving study groups through the Productivity Program were IBM and Remington Rand, both of which would soon market their products as productivity machines. IBM advertised its "machine tools for office work" as a way for office productivity—which increased only 40 percent between 1850 and 1950—to catch up with shop floor productivity—which increased 1,400 percent during the same time.[12] Remington Rand similarly offered its machines as productivity technology, depicting them as the "oil" that "keep[s] the wheels of production, distribution and administration turning . . . efficiently, economically; . . . no matter what the job, there's a Remington Rand system or machine that does it *better!*" (figure 4.3).[13]

But the US business community was not unanimous in its support of the Productivity Program, nor was it united in its willingness to welcome European visitors; as the need for the brochure in figure 4.1 indicates, some companies had to be convinced to participate in the program.[14] Large, international mass production firms tended to support the Marshall Plan,

Figure 4.3
This Remington Rand ad was titled "Efficiency Is the Lubricant." The company marketed its office machines—from typewriters and card catalogs to calculators and punch card machines—as the lubricant that allowed companies to operate efficiently. *Source: New York Herald Tribune* (Paris), April 6, 1953, 9.

possibly in hopes that it would help close the dollar gap—that is, the inability of foreign companies to import from the United States because of their lack of access to US dollars. If foreign companies were more competitive through higher productivity, they would be able to export their own products and accumulate foreign currency reserves that would eventually create foreign markets for US manufacturers. But nationally operating firms with high labor costs, such as agricultural firms and the textile industry, tended to fear competition from abroad and instead sought protection through high tariffs. They opposed the Marshall administration's goal of increasing European exports to the United States with the aim of closing the dollar gap. Companies in these industries as well as in the chemical, steel, rubber, and aluminum industries refused to host European study groups, or they hosted groups only reluctantly.[15] Their voices were channeled by conservative business associations such as the NAM.

While the NAM generally proclaimed its support for the goals of a European aid program, it also sought to minimize the government's role in administering such a program. The association called for regular reviews of the program and demanded conditions for the reception of US aid. In October 1947, several months after Marshall's initial proposal for a systematic aid program, the NAM still called for replacing government loans and credits to foreign countries with international financing through private sources and international institutions—similar to the recovery after World War I.[16] During the congressional hearings for the Marshall Plan, Herbert Schell, a textile manufacturer and member of the NAM's international relations committee, testified on behalf of the association. He now declared the NAM to be "all for the Marshall Plan," provided that proper safeguards were included. He expressed the NAM's general support for the goals of the aid program—to prevent the spread of Communism—but at the same time, he used the aid program to call for domestic conservative political projects, such as cuts in government spending, reductions of individual income taxes, a plan to pay off public debt, an end to inflationary monetary expansion, and government control of interest levels to bring the US "economic house in order."[17] For other countries to receive US aid, he called for them to commit to currency reform, a balanced budget, and rejection of nationalization programs. He also stated that any extension of aid should be based on performance, and that an independent agency, not a government department, should administer the program. Several months after the Marshall legislation had passed, the NAM still argued in internal memos against aid financed by the US government and declared that international aid could "best be served through humanitarian agencies."[18] For the 1949 and 1950 appropriations of the Marshall Plan, the NAM generally supported the appropriation but urged "continuous critical examination" of the operation of the program, and that immediate relief effort be reduced in favor of "long-range objectives of improving and expanding the production facilities of the participating countries."[19]

Progressive and conservative members of the business community could all rally behind the value of a free economy, as Hoffman laid it out in his opening statement at the manufacturers' conference. They often advanced different meanings of freedom, though. For example, Thomas R. Jones, chairman of Daystrom Inc., a management holding company, stated that it was "the philosophy of freedom—political freedom, which means not just the freedom to vote for a government or a government official of your choice, but freedom from excess of interference from government itself"—that together with religious, educational, and other forms of freedom formed an "encompassing *freedom of choice*." In Jones's view, freedom included the free choice of education and occupation, and through it the opportunity for individual uplift—somewhat different from Hoffman's promise of uplift for all from increased prosperity. But both acknowledged the existence of class and class differences. And both defined a free economy as one that was devoid of cartels and subsidies, given that, as Jones continued, "the thoughtful and far-seeing American business man is deathly afraid of these devices because they inevitably lead to controls." In his view, "government controls and a high standard of living are mortal enemies and business which welcomes government subsidies and asks for government controls will eventually become the servant of government."[20] This is part of the definition of free enterprise—the freedom from government control. The other part of free enterprise, the freedom from interference by labor unions, was also a unifying factor for Project Impact and the US business community.

Free Enterprise Domestically and Internationally

Free enterprise became the rallying cry of the US business community in the years immediately following World War II, a time of labor uproar with a wave of sit-down strikes and work stoppages protesting wage and price freezes. Free enterprise meant private, competitive business, free from government regulation and union intervention, and a NAM-orchestrated campaign aimed at swaying public opinion in its favor. On company time, employees watched films and discussed values, including patriotism, freedom, individualism, competition, and abundance through higher productivity. They and their families were exposed to these values through mass media articles, posters, billboards, advertising, radio and TV programs, and business-sponsored education programs at schools and churches. In the postwar years, rolling back union representation was at the top of the agenda for business representatives. Business associations formed coalitions with conservative legislators to repeal union-friendly New Deal legislation. Thus, for example, the 1935 National Labor Relations Act that had eased union organization was replaced by the 1948 Taft-Hartley Act that made labor organization more difficult. In forceful strides, management claimed its right to manage.[21]

Free enterprise also became a central value in *international* interactions surrounding the Marshall Plan. Here, US manufacturers criticized not union representation per se but instead the particular form of industry-wide collective bargaining practiced in many European countries, which they likened to cartel and monopolistic practices. US manufacturers argued that rather than taking an individual company's competitive situation into account, agreements in industry-wide collective bargaining coalesced on the lowest wage and production standards in the industry to protect the least competitive firms.[22] Labor relations also were the most controversial topic at the First International Conference of Manufacturers, with disagreements between US and European participants as well as among European participants.

In his position statement on labor relations, Clarence B. Randall, the president of the Inland Steel Company in Chicago, promoted a number of things that were attractive to labor: high wages; the three-way division of productivity gains between investors, consumers, and workers; a transparent promotion policy; active communication policies between management and workers through employee magazines and suggestion programs; and the increasing encouragement of workers to invest in company shares. Yet he strongly rejected industry-wide collective bargaining, stating it was one point on which US manufacturers held "deep convictions." He explained that "for the preservation of the full benefit of the free enterprise system, it is important that the problems of each company be dealt with on their own merits, uninfluenced by the different circumstances which will surely prevail in other geographical areas and in competitive plants." In his view, industry-wide bargaining had "many vices: it breeds cowardice in management; it protects the inefficient and thus reduces competition; and it affords a ready tool for socialization."[23] Even more strongly, Ira Mosher, the director of the Russell Harrington Cutlery Company near Boston, likened industry-wide collective bargaining to monopoly and cartel practices. He decried the "stultifying effects of wage agreements covering an entire industry in the nation" and claimed that industry-wide bargaining had "a seriously depressing effect on productivity," calling it "another dangerous aspect of that whole complex problem of monopolies and cartels."[24]

On the European side, Carl Jacobsson, the president of the Swedish AB Svenska Metallverken, observed that free enterprise was "an economic creed" that had "permeated all layers of the population" in the United States. He believed that it stemmed from the "dynamic character of the American society, where free competition is the governing principle." He assured US manufacturers that European employers shared "this faith in free enterprise . . . wholeheartedly," and that only the demands for security by workers and labor parties led to "a mixed economy which often lacks the vigorous spirit of the American society."[25] Likening free enterprise to a religious belief, Jacobsson took it to mean free competition—rather than freedom from government regulation and union representation. Also, he blamed labor for the lack of competition in Europe versus the manufacturers' desire for cartel agreements or monopolistic advantages. In its position paper on labor relations on behalf of the European participants, however, the Swedish delegation also defended industry-wide collective bargaining because it usually stipulated minimum wages, and companies were free to pay above-tariff rates. Industry-wide collective bargaining allowed pursuing goals beyond changes in a single corporation too, such as federal legislation and control of corporate mergers. While the Swedish delegation admitted that higher wages from productivity improvements would increase workers' interests in collaborating in rationalization measures, they stated that such a practice would disrupt the European system of collective agreements, which is "based on the principle that wage setting shall be approximately similar for labour groups within one and the same branch," because "there is really hardly any reason why workers within a department which has been rationalized should enjoy higher wages than workers in the same business employed in a department in which rationalization had not yet been introduced and where the conditions of labor are in reality inferior to those in the rationalized department."[26] In the ensuing discussion, however, European statements on labor relations were all over the place, generally highlighting differences between European countries.[27]

US participants legitimized their right to manage as professionals. In his position statement on labor relations, Randall, of Inland Steel, said, "Management must manage," because "only confusion and loss of efficiency" will follow if there is no "single voice of authority and responsibility." For example, Randall explained, if worker representatives with veto power were included on the board of directors—a demand that German labor unions fought for as the manufacturers gathered in New York—then this could lead to inefficiency and undermine management leadership. Randall claimed that management was "rapidly becoming a profession in the United States," with entry "determined solely on merit," without "barriers of birth and social stratification to overcome," and "open to every worker." Talent, according to Randall, was "the only requisite for participation in our management group."[28] Likewise, Jones, of the management holding company Daystrom Inc., asserted in his position statement on production policies that access to the managerial "class" was "open to anybody with native ability," even without a college education. Notably, Jones used the Marxist term "class," which usually denotes class conflict between the working class and the bourgeoisie, although he distanced himself from the term because he recognized that it "can be so readily misunderstood." Making gendered assumptions about management, Jones stated that a person may get into management "through influence but he stays there only through ability, a strong and healthy body, and a willingness to work hard," implicitly admitting that social background may at least impact access to management.[29]

William H. Joyce, the Marshall administration's assistant administrator for production, picked up the topic of the managerial profession in his speech at the concluding banquet, giving it a special spin. Joyce claimed that management was "increasingly regarded as a profession in the United States," with the typical attributes of a profession, including a focus on efficiency, sense of social responsibility, and basis of expert knowledge. Joyce distinguished professional managers from those who derived their managerial prerogative from their social position. The latter "may look upon himself as the owner of a business, or as the agent of the owner," Joyce stated, implying that the interests of owners may conflict with those of other stakeholders such as workers and consumers, but a professional manager—"whether an entrepreneur or a salaried executive"—"looks upon himself as the directive personality who integrated and balanced the interests of the shareholder, the employees, and the general consumer."[30] In other words, Joyce claimed here that professional managers in the United States, deriving their position from merit, were able to distribute the gains from productivity, while the typical owners of small, family-run companies in Europe, deriving their position from their social standing, had interests opposed to those of workers and consumers.

Free enterprise, then, was a malleable concept that took on different meanings. Domestically, it primarily meant freedom from government regulation and union intervention. Internationally, it meant freedom from cartel agreements and monopolistic practices, and primarily served to oppose industry-wide collective bargaining, almost accepting the form of company-wide collective bargaining practiced in the United States. In its malleability, it managed to appeal to members of the business community, like IBM's Watson, who was internationally minded but opposed union organization. With its domestic ring of opposition to union organization, free enterprise suggested to businessmen that the Productivity Program carried values they identified with. Free enterprise thus served as an identifying concept for the Marshall Plan's Productivity Program, allowing the administration to enlist US businesses to support the program in various ways, from conceptualizing and administering it, to welcoming European visitors in their corporations as well as convincing them of the US model of capitalism, competition, and labor relations.

US Labor Unions and the Marshall Plan

In addition to working with US executives and business associations for the Productivity Program, administrators also included US labor unions in these efforts. Having labor unions on board signaled that the Productivity Program was not only for the wealthy but rather for everyone; it gave the Productivity Program legitimacy in the eyes of European workers. Within the Productivity Program, unions coalesced around the notion of collaborative collective bargaining. The idea of collaborative collective bargaining was developed by two CIO leaders, Clinton S. Golden and Harold J. Ruttenberg, during the late 1930s, and publicized in their 1942 book, *The Dynamics of Industrial Democracy*.[31] While industrial democracy had initially, in its nineteenth-century origins, intended to carry democratic principles into the growing corporate sector through participation in managerial decisions and socialist nationalization schemes, by the 1930s, it had come to mean contractual collective bargaining within the confines laid out by New Deal legislation. Golden and Ruttenberg argued that this system of collective bargaining could lead to collaboration between labor and management in their mutual interest to improve their company's competitive position through higher productivity. When Golden was appointed by Hoffman as

the codirector of the Marshall administration's Labor Office, together with
the AFL representative Bert Jewell, he brought the concept of collaborative
bargaining with him. Before looking closely at this concept in the next section, I will turn to Golden's personal background as well as the role of US
labor unions in the international labor movement.

When Golden joined the Marshall administration in 1948, he was
both an elder statesman and an intellectual forerunner of the labor movement, looking back at a thirty-plus-year career of union organizing and
leadership. Independently minded, he rebelled against school and authorities as a youth, and as an adult, pursued ideas beyond taken-for-granted
assumptions even in the US labor movement. For example, leading a strike
against the Cramp shipyard in Philadelphia in 1921, Golden denounced
the "extremes to which the master class will go in their mad efforts to crush
the workers" by maintaining a "system that permits a parasitic minority to
control the lives and destinies of so many by the ownership of their jobs,
as well as the press and State"; this radical accusation posed a stark contrast to the moderate position that he would develop in the 1930s. Golden
built a wide network in the labor movement, traveling extensively as a field
representative for Brookwood College, a residential labor college in Philadelphia in the 1920s, and in the 1930s, as a government mediator and steel
organizer under CIO president Philip Murray. During World War II, Golden
was appointed vice chairman of the War Production Board and the War
Manpower Commissions, two agencies charged with enormous planning
prerogatives as they diverted civilian production to military production—a
task in which they had to balance management and labor interests. By the
end of the war, Golden intended to retire from government work and take
a position as a lecturer on labor problems at Harvard University, but was
reluctantly drawn into government service.[32] In 1947, he was appointed
chief labor adviser to the American Mission for Aid to Greece, and in 1948,
became chief labor adviser to the Marshall administration.

For Golden personally, his appointment to the Marshall Plan was a logical consequence of his experience in Greece. American aid to Greece, he felt,
was severely hampered by the failure to consider the concerns of working
people. The United States advanced aid to Greece after the British had with-
drawn from the country because, recovering from World War II, Britain's
own precarious economic situation did not allow it to further maintain its
engagement in the region. Based on the Truman Doctrine, which aimed

to stop the expansion of Soviet influence, the United States began extending aid to Greece. As chief labor adviser, Golden became convinced that the United States needed to develop an alternative vision of prosperity to lure working people in Greece away from Communist promises. When the Greek assembly passed the death penalty for participation in labor strikes, however, Golden sought to resign because he felt that US aid only left Greek workers the choice between an autocratic government and Communist influence. On his return from Greece in 1948, he worked to help prevent similar mistakes through the Marshall Plan, the new and even more comprehensive US aid program to Europe that was then in the making.

US labor unions had by the late 1940s rebuilt international relations after the wartime interruptions. The AFL, the older of the two major unions in the United States, enjoyed more established international relations and pursued staunch anti-Communist policies. The CIO, founded only in 1936, had not been able to secure solid international relations before the war, and in the immediate postwar years, the union included a small but vocal Communist left wing, which the CIO expelled under public pressure in the late 1940s. In the last days of military action in Europe, in February 1945, union leaders from the United States, Britain, and the Soviet Union agreed to form a new unitary international organization, which resulted in the foundation of the World Federation of Trade Unions (WFTU) in Paris in September. Only one union from each country was to be admitted to the organization, though—that is, only the AFL or the CIO from the United States. With the Soviet Union already pulling the strings behind the scene, the CIO was admitted, and the AFL found itself excluded from the major international union organization.[33]

US labor unions supported the Marshall Plan domestically and internationally. In September 1947, the annual AFL convention wholeheartedly endorsed the Marshall proposal and called for an international union conference on the European aid program. The AFL primarily saw in the aid program a means to combat Soviet influence and reinsert itself into international affairs. A month later, the CIO passed a general resolution in favor of an aid program to Europe, although CIO president Murray had pushed for a stronger endorsement. Only in January 1948 did the CIO executive board formally declare its position in favor of the Marshall Program.[34] Golden called the labor cooperation in the Marshall Plan essential, likening it to the allegorical nail that caused the loss of a battle. In the proverb "For

Want of a Nail," an important message that would decide the outcome of a
battle was not delivered by the messenger because his horse lost a shoe due
to a missing nail.[35] While the allegory could be read as an indication of a
usually minor role of labor—a horseshoe nail—Golden meant to assert its
crucial function. The backing of labor, like the cooperation of the business
community in the Marshall Plan, was essential in three important ways.

First, through their committee work and congressional testimony, labor
leaders assisted the legislative effort to pass the Marshall Plan. Union executives, such as the AFL and CIO secretary treasurers George Meany and
James Carey, were members of the committee on the Harriman report,
which outlined the goals and principles of the Marshall Plan. Union testimony to Congress was mixed, though. Representatives of the left-leaning
International Longshoremen's and Warehousemen's Union criticized the
aid program, saying that it would keep agrarian areas from moving forward
technologically, help the recovery of German heavy industries, and lead to
extensive US control over European economic relations. Murray, however,
emphasized that the program would aid Europeans in a destitute situation,
help them rebuild their own economies, and allow them to solve their internal problems without foreign interference because no political or financial
conditions would be attached to the aid. Walter Reuther, president of the
CIO United Automobile Workers (UAW) union, unabashedly spoke in favor
of the aid program, which in his view would lead to prosperous and socially
responsible economic relations in the United States and Europe.[36]

Second, labor union leaders convinced their own rank-and-file members
of the benefits of the aid program. Many US workers feared that Marshall
aid would be to their own detriment for two reasons: they worried that
Marshall monies would not be used for the sale of US goods in Europe and
therefore would not boost the US economy, and that the Marshall Plan created European competition in US markets. Concern over unemployment
was rising in the United States, with a national average unemployment
rate of 5 percent, and labor advisers warned that pockets of higher unemployment, which reminded some of the Great Depression, were "likely to
affect the thinking of the rank-and-file union member toward the European
Recovery Program."[37] Leaders countered this fear with a statistical analysis,
obtained through the BLS, that proved the positive employment effects of
the Marshall Plan. Worry over foreign competition came particularly from
the textiles, lumber, and shoe industries in New England and the middle

South as well as on the West Coast; labor officers feared that workers would demand higher tariffs and lower imports. Union leaders recommended an expansion of world trade to create demand for US products. They also recommended ensuring a domestic market through employment policies, and suggested educational campaigns through the labor unions and agricultural, business, veteran, African American, and women's associations.[38] Work by union organizers combatted workers' fears and ensured that these fears did not rise to the level of political decision making. The union organizers prevented the Marshall Plan from becoming a scapegoat for US economic problems.

And, third, union members took positions in the Marshall administration in Washington DC, in Paris, and in the missions of participating countries, which meant that union members were involved in decision making at all levels. As Golden emphasized, never, in all his years of government service, "has American labor had the opportunity for such full participation in formulating policies and decisions" as in the Marshall administration.[39] By 1950, the Marshall administration employed 65 union officers in its labor and information divisions. Labor officers vouched for the intentions of the aid program, combatting Communist propaganda in Europe that said American labor was opposed to the aid program because it was "allegedly a device of 'American Wall street war-mongering imperialists to destroy the independence and sovereignty of European countries.'"[40] Critics have charged that the Marshall administration's labor service was hampered by animosities between AFL and CIO officers who sometimes devolved into personal infighting; that appointments were often haphazard—partly because the unions were unwilling to guarantee officers a return to their previous jobs, leading to appointments of available rather than best qualified persons; and, finally, that Golden as chief labor officer, and Bert Jewell, his AFL counterpart, were both of advanced age and in ill health and not involved in the work with their full efforts.[41] Widely networked and trusted in the labor movement, however, Golden was well positioned to select officers for the labor service. Together with Jewell, Golden usually appointed a CIO and an AFL officer for each mission, ideally at the same level. Golden insisted that labor advisors be appointed based on their experience in the labor movement and that their primary function be the interaction with European labor movements. The responsibilities of labor advisors were to be different from those of labor information personnel, for example, with the former helping strengthen European labor unions, and the latter working for communication programs.[42]

Within half a year of the Marshall Plan's start, however, labor officials criticized the aid program for its technical orientation, with little attention being paid to social reform, because the Marshall administration prioritized economic recovery over social justice. One of the first to launch this criticism was Jay Krane, a CIO official appointed to the WFTU office in Paris who was in daily contact with Marshall officers. By early 1950, labor information officers Harry Turtledove and Lemuel Graves wrote a series of reports arguing that European workers would not believe in the Marshall Plan unless their material conditions improved. In response, the CIO opened its own office in Paris, headed by Victor Reuther, Walter's brother, to independently reach European workers, outside of Marshall channels, and the AFL and CIO launched "Operation Bootstrap," a labor education campaign financed and conducted by the unions that sent US labor officials to Europe.[43] When the Marshall administration strengthened its Productivity Program in 1950 and onward, labor officials hoped that now social reform efforts would be more adequately addressed. The concerns of labor officials were then mostly about procedural issues, such as appointing labor officers to European exchange groups, particularly labor groups, visiting the United States.[44] An open question is whether their concerns changed because labor officials were in programmatic agreement with the Productivity Program, or whether they figured that their unions were reaching European labor independently, through their own programs, to achieve the goals they were after.

Internationally, union cooperation with the Marshall Plan had precipitated the breakup of the WFTU, and it split the international labor movement. With Soviet officers delaying a WFTU meeting to discuss the Marshall proposal, the British Trade Union Congress eventually organized an international labor conference in London in late 1948. Fifty-eight non-Communist unions from fifty-three nations, including both the AFL and the CIO along with social democratic unions from participating Marshall countries, attended the conference. Soviet and East European unions as well as the Italian and French Communist ones remained absent from the conference, which turned into the founding gathering of a new international labor organization, the International Congress of Free Trade Unions (ICFTU), rivaling the WFTU. Overall, the ICFTU program presented a compromise between American and European union traditions, leaving out controversial issues such as capitalist free enterprise versus social democratic goals of transforming society, and free market and private property versus government planning and nationalization. Yet the unions were united in their antitotalitarian stance, cementing the split between East and West within the labor movement.[45]

In this larger context, the Productivity Program's value of collaborative labor relations was a declaration of the kind of labor relations that the Marshall administration sought to avoid: the antagonist labor relations that Communist labor unions claimed were the result of a conflict of classes derived from the ownership of the means of production by the capitalist class and that could only be resolved through communal ownership of the means of production. In addition, collaborative labor relations presented a value that US labor unions could rally around; it signaled that the Productivity Program was not only a business program but a program by and for working people in the United States and Europe too. In the United States, it also signaled an embrace of New Deal policies and their promotion of (company-based) collective bargaining, in opposition to free enterprise. Most specifically, as will be discussed in the next section, collaborative labor relations indicated a form of collective bargaining that Golden and Ruttenberg had been developing since the 1930s. Based on the idea of industrial democracy, they argued that labor and management could and should work together to increase the productivity of their companies, and raise the standard of living for everyone. Despite the decline in Golden's health and his at times half-hearted involvement in the Marshall administration, his appointment as chief labor adviser suggested that the administration favored this particular form of labor relations. It also indicated a possible divide between the administration's and the labor movement's goals.

Industrial Peace through Collaborative Labor Relations

Golden's extensive network within the labor movement and his previous experience in government service qualified him as chief labor adviser. But by the late 1940s, not everyone in the labor movement shared Golden's positive attitudes toward management. Even Murray, Golden's CIO boss, "never trusted an employer," as Meyer Bernstein, a labor adviser in Germany, observed. Golden, by contrast, "had more respect for them as individuals," and "thought they were decent individuals who represented a point of view, an economic interest," and in whose own interest it was to be rational.[46] With Golden, Hoffman thus appointed a chief labor adviser who promised to work well with corporate representatives. Golden's goal was to achieve "industrial peace"—that is, a situation that went beyond the mere absence of labor conflict, in which management and organized labor "coexist, with each retaining its institutional sovereignty, working together in reasonable harmony in a climate of mutual respect and confidence."[47]

Golden and Ruttenberg laid out their views in their book *The Dynamics of Industrial Democracy*. Based on their experiences with the Steel Workers Organizing Committee, they developed thirty-seven principles of union-management relations, among them that workers organize in unions for economic as well as psychological and social reasons; that "management's assumption of sole responsibility for productive efficiency actually prevents the attainment of maximum output"; and that union-management cooperation increases productive efficiency, improves the competitive position of a company, and raises earnings for both workers and owners. While their list of principles is longer than an astute rhetorician would recommend for effective communication, rich case stories supported each principle—often plausible, if not real, stories from Golden and Ruttenberg's experience in the labor movement. For example, there is the story of "Big Mike," a worker at a Pittsburgh railroad company who stormed into his boss's office, still in his work clothes, to demand a free rail ticket. In the middle of a meeting, his boss ordered Big Mike out, and Big Mike then told his boss "to go to hell." The authors explained that this desire to tell the boss to "go to hell" was intertwined with the psychological and social motives for union membership. Or there was the story of an educational union play, *The Innocent Upsetter*, in which a new worker lost his job after proposing an idea to set up the company's machines to allow for three times higher output. While the play may have been intended to encourage union members to explore which kinds of conditions would allow them to freely share their ideas for technological improvements, a discussion of the play at a management seminar resulted in the president of a steel company ordering a union handbook on how to achieve a free exchange of ideas in his plants—something that he assumed already existed in his company.[48] Many managers appreciated Golden and Ruttenberg's book for explaining workers' motivations and shop floor dynamics, even if they did not necessarily agree with the authors' goals and arguments.

Labor leaders, however—particularly the militant leaders essential for controversial union organization drives—often found it hard to meet management with respect and trust. Like Murray, to whom Golden and Ruttenberg dedicated the book, they maintained a more adversarial attitude toward management. Golden and Ruttenberg had little use for adversarial labor leaders if collective bargaining was to develop through three phases: first, a primitive phase when management and the union initially met to negotiate wages; second, a phase when the bargaining partners discussed other issues such as vacations, working conditions, a grievance system, safety, severance pay, pensions, and arbitration; and third, a phase that did not emerge in all bargaining relations—the phase of "enlightened" or "supercharged" bargaining. This final phase was instilled with trust and mutual respect, and allowed both parties to collaborate for their common interest in the success of the company, such as through technological improvements.[49]

Golden and Ruttenberg laid out the conditions for collaborative collective bargaining. Most importantly, companies needed to agree to hire only union members in order to give workers security and free unions from the need to continuously organize, and instead devote energy to causes such as productivity improvements.[50] Also, they repeatedly emphasized that decisions needed to be based on democratic discussions—between equal partners—rather than made unilaterally, even if they were made with benevolent intentions. Railing against paternalistic decisions in welfare capitalist companies, Golden and Ruttenberg charged that Ford's opinion that "all men want to be told what to do and get paid for doing it" was "dictatorship."[51] Even well-intended decisions, such as wage raises, they warned, would be met with suspicion and resistance if unilaterally decided by management because workers would complain about perceived injustices in the wage scale. Only through discussions between labor and management along with mutually approved agreements could such perceptions be avoided. And through discussion, even more controversial decisions could be made, such as those about the introduction of new technologies.

Productivity played an important although not central role in Golden and Ruttenberg's book, and they recognized productivity as a bargaining chip for unions. For example, if a company claimed that its competitive situation did not allow wage increases, the union could help increase productivity to lower costs and enable wage increases without price raises. The authors argued that workers, from their daily experience on the shop floor, often knew about sources of waste and inefficiencies, and had ideas for remedies. Implementing these ideas would give workers a sense of accomplishment and liberate their creative capacities—crucial aspects since the authors claimed that psychological recognition, in addition to economic security, was one of the motivations of workers to organize in unions. To take advantage of the workers' creative potential, companies were to set up committees of management and labor representatives that solicited as well as assessed proposals for improvements. Technological improvements would be discussed with workers in a transparent mode, without suspicions, if the union collaborated, and no worker would be threatened with job loss. Job replacements, if necessary, were to be achieved through the usual attrition, and adequate replacement positions were to be found for all displaced workers. The costs for these measures were to be charged against the investment for the improvement, and the proceeds from technological improvement were to be shared between owners and all employees, thus avoiding wage inequalities that could lead to further labor unrest.[52] Golden and Ruttenberg emphasized that only collaborative labor relations allowed companies to take advantage of their workers' intimate knowledge of the production process and their creative capacity for solutions, because the union protection freed workers from concerns over negative effects for themselves and their colleagues, such as job losses and speedups. By contrast, they railed against welfare capitalist programs for employee suggestions for improvements because these programs only threatened workers—who didn't have proper protection of their jobs and working conditions—with social isolation for collaborating with management.

While some of Golden and Ruttenberg's arguments, such as the emphasis on productivity, were amenable to the Productivity Program, others were not. In particular, the Productivity Program promoted free enterprise and plant-level collective bargaining, while Golden saw district and industry-wide collective bargaining and national planning as a "natural outgrowth" of plant-level collective bargaining—a position strongly opposed by conservative business associations such as the NAM.[53] Golden welcomed that some industries that were almost completely unionized, such as the US railroad and bituminous coal industries, were able to advance to industry-wide collective bargaining at least at the regional level. Industry-wide collective bargaining would allow addressing problems beyond the corporate level, such as cooperation on federal legislation, transportation costs affecting the whole industry, the problem of wage differences between workers, and the issue of marginal producers by having organized labor and other interest groups control ongoing concentration processes in an industry. Eventually, industry-wide collective bargaining would allow sharing the proceeds from productivity improvements between owners, workers, and consumers, with consumers being represented by the government.[54]

In 1948, when Golden was appointed as labor adviser in the Marshall administration, he also directed a case study program on the causes of industrial peace—that is, harmonious labor relations with mutual respect and trust—under the auspices of the National Planning Association, a New Deal era think tank with labor, business, and agriculture representatives. The case study program reportedly originated from Golden's observation that the media dramatized strikes in their reporting, even when during the post–World War II strike wave of 1946, nine out of ten collective bargaining contracts were signed without industrial conflict. The individually published case studies investigated the conditions for collaborative labor relations and were widely circulated in the United States as well as in Europe, distributed through the Marshall administration.[55] The first case study of the Crown Zellerbach Corporation, a paper manufacturer on the West Coast, found especially wide attention, and the company president, James D. Zellerbach, served as the head of the Marshall mission to Italy. Unlike some of the later studies, which included incidences of strikes and labor strife, Crown Zellerbach represented a case of "astonishing" labor peace, without a single day of work interruption in fourteen years of collective bargaining. Management quickly recognized the right of union organization, and the union accepted private ownership and acted out of industry consciousness, not craft or class consciousness.[56] Crown Zellerbach also was an example of regional collective bargaining between an association of pulp and paper manufacturers on the Pacific Coast and two AFL unions, which was not a type of collective bargaining favored within the Productivity Program.

The concept of collaborative labor relations thus provided the basis on which union officers worked in the administration of the Productivity Program. Their participation was testimony to the labor movement's

endorsement of the program, although not everyone in the labor movement necessarily agreed with Golden's conciliatory approach to collective
bargaining relations, and although the labor movement's goals for industry-
wide collective bargaining and national planning did not exactly align with
those of the Productivity Program. The labor unions' participation in the
Productivity Program may have convinced European workers and their
unions about the intentions of the Marshall Plan and the Productivity Program. At the same time, labor advisers like Golden knew of the emerging
concerns in the labor movement about unemployment or international
competition, and they warned the Productivity Program administration
and sought to alleviate those fears.

Conclusion

The Marshall administration brought members of the business community
and the labor movement—whose interests and goals were often at odds
with each other—into the Productivity Program. The business community
identified with the value of free enterprise based on private ownership and
market competition, while labor representatives identified with the value
of collaborative labor relations based on plant-level collective bargaining.
Support by business and labor leaders along with their organizations was
essential for the program; their endorsements and congressional testimonies helped pass the Marshall Plan, and they convinced their communities
of the benefits of the program and addressed concerns when they emerged.
Corporations and business associations assisted with the logistics of European exchange groups as hosts and contractors, and labor advisers joined
government service and played an essential role in convincing European
workers and unions of the benefits of the program. This strong public-
private coalition assembled behind the Productivity Program based on the
notion of productivity.

US management and labor continued to negotiate the values of productivity technologies as the Marshall administration brought them to the
European continent. This chapter showed that unresolved questions and
disagreements remained among Americans, such as over whether collective bargaining should be conducted at the company level or the regional
or industry one, and whether national planning should have a role in the
larger economy. The Productivity Program, brushing over these unresolved

questions, did not offer a coherent model of corporate and labor relations but instead supplied an assemblage that different social and economic groups could identify with. Productivity thus was also an inherently inconsistent concept, tied to the contradictory values of free enterprise and collaborative labor relations. This observation is in line with recent critiques of the argument that American aid reshaped European corporations and economies after the US model.[57]

While it was not a stated goal of the Marshall administration to reshape European firms and labor relations after the American model, European corporations during the postwar years came to resemble the American model, preparing them for the adoption of electronic information technologies. In the early 1950s, only one in four firms in the German and French manufacturing sectors had over five hundred employees, compared to 46 percent of large firms in the United States at the same time. But by the late 1960s, over 40 percent of German and French firms had over five hundred employees, and corporate or mixed ownership increasingly replaced family ownership.[58] Also, most large companies adopted the multidivisional structure, or M-form, which was characterized by decentralization and organization along product lines—a form of organization that emerged in the United States in the early twentieth century and had been virtually unknown in Europe before World War II. In these larger, multidivisional corporations, information was important for managerial decision making; the corporations commanded higher capital resources and had larger data processing needs. Therefore, they were more conducive to acquiring computers than small family firms.

The postwar years thus marked the beginning transition from Ford's production line to electronic computers. Ford's factories were still a stop on the Project Impact tours; so were the factories of Remington Rand and IBM now. While Ford executives remained conspicuously absent from the First International Conference of Manufacturers, executives from both IBM and Remington Rand supported the Marshall Plan and its Productivity Program as members of the progressive business community. Both companies were leading office machine producers in the early postwar years, selling mostly electromechanical technologies such as calculators, typewriters, and punch card machines; they also ventured into electronic computer technology, and both companies soon marketed their machines as productivity machines.[59] The Productivity Program thus created the conditions for the acquisition of electronic computing machines in Europe: the program encouraged corporate growth, incentivized the introduction of cost accounting systems, and spread productivity mindedness.

Productivity, finally, allowed Marshall officers to redefine the political issue of inequality between social classes in economic terms by promising that higher productivity would raise the standard of living for everyone. As John W. Nickerson, the head of the Productivity Program's advisory board, emphasized, "sharing" was central to the idea of productivity, and he "would want nothing to do with it" if sharing wasn't part of the program. "Productivity without sharing" was "worthless" to him, and "sharing without increase of productivity" was "futile." Nickerson urged, "We must share the results of this increased productivity, we must share it with Labor and we must share it in reduced prices with the consumer, and we must not have unemployment." In a departure from the BLS officers' view in the 1920s, Nickerson believed that there needed to be a "definite relation" between higher productivity, higher wages, and lower prices.[60] This promise of a larger piece of a larger pie appealed to US labor unions, and was why they supported the program. The question of productivity technology and unemployment, which Nickerson alluded to, will be the focus of the last chapter. In the next three chapters, I turn to productivity in West Germany. In 1952, Democratic senator Paul H. Douglas still warned the Marshall administration that he was "getting a bit fed up at the relative failure of the program to help the great masses of the people," and in his next annual appropriations vote, he would take into account whether the funding reached "the little people of Europe."[61] The next chapter looks at how West Germans of all social classes viewed American productivity.

5   German Perceptions of US Productivity

Productivity administrators were convinced that "seeing is believing," so they worked at exposing European workers and managers to American productivity technologies. They wanted Europeans from different social and economic backgrounds to see for themselves the mass production and distribution of consumer goods. More important, they aimed at exposing Europeans firsthand to larger political and economic values, particularly free enterprise and collaborative labor relations. US Productivity Program officers hoped that Europeans would turn into missionaries of US productivity and unleash in their own countries the social productivity revolution that American administrators aspired to, leading to dynamically growing economies with higher standards of living based on high wages and low prices. As we saw in chapter 3, demonstration projects were to implement productivity technology and collaborative labor relations in European firms so that they would serve as shining examples; Productivity Program officers, however, encountered problems in identifying and approving suitable firms. Therefore, exchange programs became the most important vehicle of exposing and convincing Europeans and Germans of American productivity.

The Marshall Plan's Productivity Program entailed three different exchange programs. First, groups of Europeans visited the United States for four to eight weeks to study specific production technologies at US companies, attend seminars on collaborative labor relations, and meet with labor and industry leaders.[1] Second, US industrial engineers and other experts traveled to Europe as consultants of European firms. And third, young Europeans workers came to the United States for a yearlong student exchange program, the Work-Study Training for Productivity program (WSTP).

The goal of the WSTP was "to help strengthen the free trade union movement in Europe so as to offer an affirmative program in meeting the trade
union and production problems."[2] The participants were to learn technical and organizational skills for individual improvement and professional
advancement as well as experience plant-level collective bargaining and
local union organization in hopes that they would later take on leadership
positions in European labor movements.[3] Created in late 1951, the program
already brought an astounding number of over 550 European participants
to the United States in 1952, its first year of operations. For the first two
months of their stay, the students lived on college campuses, where they
took full-time classes in American history, economics, and labor relations as
well as English-language classes. Afterward, they were placed with companies as paid employees, where they experienced the "famously high union
wages" and observed collaborative labor relations. They joined their local
union, paid dues, and attended union meetings, while continuing their
classes at local colleges in the evening.

In June 1953, the employee magazine of Studebaker, an automobile
manufacturer and the former company of Paul G. Hoffman, the Marshall
Plan's first administrator, reported on a group of WSTP students working for
the company in the South Bend, Indiana, area. The article was titled "They
See America," and showed young men—most of the exchange students
were men—participating in the daily lives of their local community, such
as at work, dinner with their host families, the local YMCA where some of
the exchange students boarded, the Studebaker company's choir practice,
a church service, a game night playing checkers, and a "coke date" with
local women.[4] An image of the group at an evening class at the University
of Notre Dame shows well-dressed young men in suits and ties—and some
even with bow ties and pocket squares—attentively listening to an instruc-
tor. With their fine leather shoes and fashionably styled hair, the work-
study students certainly no longer resemble the typical European working
class from which they originated; they appear to be partaking, at least
visually, in the American style and standard of living. The article seems
to suggest that participating in American daily life allowed the students
to experience the benefits of American productivity and surpass European
class boundaries—regardless of union activities, which remained conspicuously absent from the narrative. How European visitors, and in particular
German exchange visitors, viewed American productivity along with social
and economic relations is the topic of this chapter.

Figure 5.1
Working-class Europeans participated in the yearlong Work-Study Training for Productivity program in the United States.
Source: Studebaker Spotlight (June 1953): 3. Courtesy of Studebaker National Museum.

Not surprisingly, the program participants returned from their American visits with a broad range of impressions and wide-ranging opinions about American productivity. While technological improvements constituted the part of the program that most visitors eagerly embraced, collaborative labor relations proved a more difficult sell. Most WSTP students appreciated what they perceived of as good labor relations. For example, the Austrian student Josef F. Waclena observed that relations between labor and management were "far more cordial and friendly" than in Austria; his compatriot Fritz Reihinger felt that he did not "have any supervisors but only good friends, helpful and kind"; Peter Kojaager from Denmark complained that "we in Europe still have something left of the old spirit: 'I am the boss and you are my workers.' The Americans are not so 'snobbish' in that way"; and Domenico Cotto from Italy reported that there was "marvelous cooperation

between workers and foremen, as well as between workers and management," and "both foremen and managers behave like friends of the workers
and not like supermen, as unfortunately too often happens in Italy."[5] By
contrast, Bert Wessel, the owner of a German ceramics company and participant in Project Impact, dismissed collaborative labor relations as mere
advertisement, with previous slogans, such as "Milk from contented cows,"
now being replaced with "Products, made in America, by the freest people
of the freest nation of the world."[6]

Socioeconomic background appears to have been one of the factors
influencing the perceptions of European visitors. Industrialists tended to
be more skeptical of US productivity, pointing to a lack of natural resources
in Germany, and a less beneficial climate, smaller markets, and the scarcity
of capital as explanations for lower productivity in Germany and other
European countries. There is little indication that Project Impact convinced any of the German industrialists of the values of US productivity;
to the contrary, while they generally appreciated the warm welcome into
US companies and homes only several years after fighting the war against
each other, some also felt reproached by their hosts. Another important
factor was the visitors' attitude and receptiveness to new ideas. In an
open-minded study group, for example, in which industrialists and laborers decided to share hotel rooms for the duration of the trip in a social
experiment that would have been unheard of in the stratified postwar
German society, participants returned with approving opinions of US productivity.[7] By contrast, another group with a strict focus on technological
improvements seems to have barely noticed the social and political factors
enabling the high US productivity.[8]

The WSTP program was the most successful of the exchange programs
in convincing young Germans and Europeans of US-style productivity.
While some students had better experiences than others, depending on
how well their job placement matched their skills and interests, most students returned with positive impressions of US productivity, lauding US
labor-management relations. As one US officer observed, the length of the
students' stays may have been decisive in achieving this positive result:
"the students did not receive very favorable impressions of our [US] social
and industrial system in the first month or two," but "their impressions
improved throughout the year."[9] In addition, the students' age may have
mattered. While the participants in Project Impact and the study groups

were frequently established industrialists or experts at the height of their careers, and thus at a more advanced age, the WSTP students were much younger and were at a formative age. Spending a year abroad, in any situation, is often a highly impressive and sometimes life-changing experience. Returning students, however, enthralled with the American model, sometimes became frustrated with their inability to advance professionally or change their social and working environment. More important, exposure to US unionism frequently estranged them from their own unions because the students began to wish for more participation in union decisions and more budgetary transparency.

German travel reports provide insights into the reception of the Productivity Program and productivity ideas in Germany. The chapter first looks at the Productivity Program in Germany, implemented by US officers of the Marshall Plan mission to Germany together with officers of the Rationalisierungskuratorium der Deutschen Wirtschaft (RKW), because it reveals what Germans identified as the major hurdle to higher productivity: the lack of capital for investment in new technologies. By contrast, US officers viewed antagonistic German labor and social relations as an obstacle to higher productivity. The chapter then discusses German travel reports. German and European WSTP students recognized differences in social and labor relations in the United States and their own countries, but their experiences in local US unions estranged them from the top-down hierarchical organizations of their own unions, rather than motivating them to take on leadership positions. By contrast, many members of shorter study trips to the United States focused on technological production methods, and disregarded the social and economic values of US productivity; they did not agree that dynamic economic growth would allow raising the standards of living for everyone, and changing social and labor relations. This chapter shows that local reception and perceptions determined how far American productivity shaped European and German societies and economies.[10]

The Productivity Program in Germany

When the US Marshall administration announced its Productivity Program in June 1951, US administrators invited participating European governments to submit their own individual proposals for implementing a Productivity Program in their country, based on their national economic situation and needs. Consequently, the Productivity Program took on different forms in different countries, indicating which problems local administrators identified as the major hurdles to higher productivity. The Danish program proposal, for instance, centered on the following industries: metal casting, clothing, textiles, and shoes; the Italian program pursued a community approach to increase the productivity of selected cities; the Belgium government, receiving only negligible Marshall funds because of its strong export position, devised a minimal productivity program aimed at disseminating the idea of productivity without any concrete supporting projects; the British proposal focused on productivity research and education; and in both France and Germany, the governments proposed loan programs for small and medium-size companies.[11]

In July 1953, after two years of negotiations, the German Productivity Program was finally announced—ironically, at the time as appropriations for Marshall funding ended.[12] The difficulty in securing the cooperation of German trade unions, particularly the German trade union umbrella organization Deutscher Gewerkschaftsbund (German Trade Union Confederation, or DGB), caused part of this delay. As the US administration emphasized, one of the major reasons for the Productivity Program was "the necessity of achieving a social impact on the target countries of Western Europe by passing on to labor and consumers the benefits anticipated from a productivity drive before our efforts are defeated by political instability"—that is, by Communist agitation.[13] Reaching and convincing European workers and their trade unions of the advantages of the American economic model thus was a major goal of the Productivity Program. German trade unions and their leadership were, at least in principle, positively inclined toward productivity improvements. As we saw in chapter 2, the Weimar trade union delegation to the United States had set out with skeptical questions and returned with positive views of rationalization in the United States; in particular, it called for more work organization in German companies. The German trade union leadership maintained this positive attitude toward rationalization after the war, and at the 1949 DGB founding congress in Munich, passed a resolution in support of rationalization and technological improvements.[14]

Still, union approval for the Productivity Program was forthcoming only slowly. One reason for the reluctant public trade union support for the Productivity Program was the attitude of rank-and-file workers who feared that productivity improvements would lead to a higher work pace and unemployment. For German labor leaders, publicly supporting the Productivity Program meant risking disapproval from their base at an essential time when they fought for codetermination in German corporations, as I will discuss in more detail in chapter 6. In addition, German unions and employer associations practiced regional, industry-wide collective bargaining, which was diametrically opposed to the Productivity Program's model of collaborative labor relations with plant-wide collective bargaining. Such industry-wide collective bargaining, German labor leaders argued, allowed addressing problems such as controlling concentration processes in an industry or cooperation on federal legislation that went beyond individual companies. In contrast, US Productivity Program officers and labor leaders held that industry-wide collective bargaining would end up setting wages at the lowest level to allow the least productive companies to survive, rather than encouraging competition and productivity increases.

Partway through the two-year negotiation process, in April 1952, the DGB finally agreed to participate in the Productivity Program under certain stringent conditions. The most important condition was that "in all other fields of the economy the same equal cooperation must be established"—a veiled demand that codetermination be introduced in all German industries, although the term "codetermination" was not used. Another key condition was that the "results achieved through rationalization" were, "in agreement between the social partners"—that is, labor and management representatives—to be "used for wage and salary increases and price reductions as well as for economically necessary investments."[15] In other words, the DGB demanded that productivity benefits be divided between workers, consumers, and industrialists, in accordance with promises by US Marshall administrators. Other conditions were the formation of an inquiry committee to broaden the Productivity Program, a reduction in the number of types of products, the sharing of experiences between plants to promote rationalization, and—another potential stumbling block—the opening of company books to the collective bargaining parties.[16]

The German Productivity Program included a DM 100 million loan program—of which DM 70 million were devoted to loans to small and medium-size enterprises, and DM 30 million for demonstration projects—and an additional DM 17 million to be spent for grant-in-aid for research and other supporting projects. In the German view, the loans to small and medium-size enterprises were the core of the Productivity Program. They focused on the sector of the economy that German administrators perceived as constitutive—small and medium-size, often family-run companies—and addressed the lack of capital that German administrators and entrepreneurs saw as a chief detriment to German economic development.[17] As Germans occasionally reminded Americans, credit scarcity was an economic condition partly created by the 1948 currency reform under the US administration that eliminated cash reserves for investments in addition to protecting the wealth of the owners of property and capital goods—which remained untouched by the currency reform—while small bourgeois and working-class households lost their main wealth, which had been in the form of savings.[18] Building up capital was one of the primary goals of West Germany's first economic minister, Ludwig Erhard, and the Productivity Program addressed this obstacle to plant improvements through a credit program for small and medium-size enterprises.

In the view of US Marshall administrators, by contrast, the demonstration projects were the core programmatic objective, as we saw in chapter 3, because they aimed at eliminating restrictive business practices in Europe and introducing US-style plant-level collective bargaining practices. As such, the demonstration program was open for applications from all German firms, including large corporations.[19] The German Productivity Program thus was a compromise between American and German administrators (figure 5.2). Americans saw the demonstration projects as the core of the program, while Germans saw the loan program as the core and accepted the demonstration projects only as a necessary nuisance.[20] These disagreements would soon result in renegotiations of the program.

The Marshall administration authorized the RKW to implement the Productivity Program. The RKW was the successor organization to the Weimar RKW, the umbrella organization of the German rationalization movement, on behalf of which Köttgen had studied US productivity technologies in the 1920s.[21] Both organizations coordinated the work of more specialized associations, such as the investigation of apprenticeship and management training by the German Institute for Technical Vocational Training (DINTA), and the development of incentive pay systems by the Imperial Committee for the Determination of Working Hours (REFA). Yet while the Weimar RKW had a centralized organization, focused on its Berlin headquarters, the West German successor RKW had a federal structure, formed

Figure 5.2
Presentation on management training by US productivity officer Harold C. Zulauf in August 1953 in Mehlem near Bonn. US officers William Fradenburg, Frederick Scheven, and Carl Maher observed the presentation from the second row, while the German attendees were seated in the front row.
Source: Box 16, entry 1202, RG 469, NARA.

by state rationalization boards. Also, labor unions were a full sponsor of the West German RKW, with equitable representation on all committees. Why the Marshall administration designated the RKW as the agency responsible for the implementation of the Productivity Program remains unclear. With its technical and organizational experience in the German rationalization movement, the RKW may have offered itself as the natural conduit for the Productivity Program. For the RKW, the designation brought financial security, and for the next two decades, productivity programs were the organization's major focus.

Authorizing the RKW to administer the Productivity Program had its drawbacks, however. First of all, US administrators often complained that the RKW was too theoretical—which could have been a stereotypical criticism by pragmatically minded Americans against a systematic German organization, or might have been a veiled criticism of other aspects. The RKW brought the baggage of its collaboration with the Nazi government, having voluntarily offered itself to the Nazi government as a corporatist instrument to control the German economy. From 1934 onward, the government installed a Nazi leader to run the RKW, ousted critical thinkers— like Köttgen—and Jewish employees, and restructured the organization. The RKW became instrumental in Nazi Germany's economic mobilization and collaborated in the expropriation of Jewish merchants and real estate owners in Austria and Poland.[22] The acronym and personnel carryover indicate continuities with the organization's past.

Second, the RKW brought to the Productivity Program the traditional German understanding of rationalization, with its emphasis on technological improvement, as we saw in chapter 2. By contrast, US Productivity Program officers strove for social and institutional rather than technical change. Thus Carl Mahder, the US officer in charge of the Productivity Program in Germany, stated in October 1952 that German technical efficiency needed no improvement from the American side; instead, the goal of the Productivity Program in Germany was to "create the framework within which an industrial democracy can function and in which we can give play to those things which we hold valuable in a free economy"—in other words, the goal was to create collaborative labor relations in a free enterprise economy.[23] And third, the Weimar RKW had been founded as a business initiative with government encouragement; only a small number of labor advisers were included on a pro forma basis when the RKW received public funding in 1925. Although after World War II, the RKW embedded parity governance in its organizational structures, and a National Productivity Council with labor and management representatives oversaw the RKW's Productivity Program work, the labor movement suspected that the RKW still privileged management over labor concerns.

Within a year after launching the German Productivity Program, conflicts emerged again. In March 1954, after German administrators had dispensed the first half of the loan program for small and medium-size enterprises, they requested that American officials release the second half. Productivity measures were paid through counterpart funds—that is, European funds raised in return for Marshall shipments; US officers therefore needed to approve the disbursement of the funds. While the loan program had quickly caught on in Germany, however, the demonstration projects, which US administrators had insisted on for programmatic reasons, got to a slow start. Almost six hundred small and medium-size enterprises had received loans by February 1955, and proposals for twice the amount of the second half of the loan program had already been processed, while only twelve demonstration projects had been approved during the same time, for a total of DM 5.3 million.[24] The head of the US mission to Germany, John W. Tuthill, explained that the two parts of the Productivity Program, the loan program and the demonstration plants, were "closely related in intent and in purpose," and even though the loan program had been satisfactory, the demonstration plant program "has not been marked with comparable success."[25] Thus, American officials refused to release counterpart funds for the second half of the loan program for small and medium-size enterprises as long as the demonstration program wasn't moving ahead. Holding back the funds was their leverage to make Germans implement the demonstration projects that, in the eyes of US administrators, had been neglected.

After more than a year of negotiations, Marshall administrators in Washington, DC, finally agreed to refocus the German Productivity Program, and shifted part of the funds from the demonstration projects to the loan program for small and medium-size enterprises in West Germany and a similar new program in West Berlin. As Tuthill released the second part of the counterpart funds in May 1955, he asked that the nontechnical aspects of the Productivity Program—read, free enterprise and collaborative industrial relations—be emphasized more strongly and the consultant activity be pointed in this direction.[26] Obviously, Americans were still wary that while the German program would increase productivity through technical measures, it would not reduce market control through restrictive business practices or change existing patriarchal labor relations.[27] With the programmatic results of the loan and demonstration projects limited, exchange visits from Germans to the United States became an essential part of convincing Germans of US-style productivity.

Views of Labor and Class Relations in Germany

US productivity officers saw German labor and class relations as antagonistic, and this perception informed how they organized the exchange programs. Although US administrators acknowledged that "both management and labor have lived in an economy of restriction and scarcity for so long that they now both accept it as normal," and that "both are apathetic toward change and regard the inevitable transition period with foreboding, primarily because of the threat of temporary technological unemployment," they still thought that "it is the European workers who have felt the pinch of an economy of scarcity and high prices."[28] Similarly, Harold C. Zulauf, a US Productivity Program officer in Germany, stated that German "labor does *not* receive its *merited* increase in the present GNP"; he cited the examples of a well-run plant with manufacturing costs of 5 percent for direct labor, and textile workers striking because they received lower income for working than they would get from the government if they were on relief or unemployed, which he called "unhealthy conditions."[29] For Zulauf, this was a problem primarily because low wages meant that workers' purchasing power could not be realized to dynamically grow the economy. This US view of German labor relations as antagonistic may be because German labor unions were organized much differently than their American counterparts and were often affiliated with political parties, unlike in the United States.

The older US labor union, the American Federation of Labor (AFL), came from a craft union tradition, while the younger Congress of Industrial Organizations (CIO) organized workers in mass production industries, such as the automobile and electric industries, and originally included small but vocal socialist and communist wings. Both unions operated independently of political parties. While the CIO accepted that unions needed the government to set a legislative framework to protect the organization of labor unions, both unions rejected the idea that the government should set labor conditions through legislation.

In Germany in the late nineteenth century, labor unions were formed with different ideological orientations and affiliated with different political parties across the whole spectrum. The largest were the socialist unions, the General German Union Association (ADGB) and the Afa-Bund. Called "free unions" in the German and Austrian context to distinguish them from Hirsch-Dunckersche Vereine and Christian unions, they comprised a broad spectrum of socialist perspectives based on the idea of class conflict and an atheist worldview; also, both were close to the Social Democratic Party (SPD). The Hirsch-Dunckersche Vereine was close to the left-liberal German Democratic Party (Deutsche Demokratische Partei, DDP) during the Weimar years. Following the model of British trade unions, the Hirsch-Dunckersche Vereine fought for social reform through a reconciliation of interests along with the cooperation of workers and management. While it accepted strikes as a means of last resort, it rejected class conflict as non-democratic and nonunion. Yet the Hirsch-Dunckersche Vereine played a minor role in the German labor movement, with only 122,000 members in 1910, compared to the 2.5 million members of the free unions. A Christian union movement emerged in the 1890s, based on the Catholic social doctrine of the 1891 papal encyclical *Rerum Novarum*. Close to the Catholic Center Party, its three core ideas were personality (the invulnerable dignity of every human), solidarity (the social principle of human togetherness), and subsidiarity (the social principle of responsibility and self-help in small social units, starting with the family). While there were smaller Protestant unions, Christian unions were dominant in Catholic areas, such as the Rhineland, Westfalia, Ermland, southern Germany, Palatine, Saarland, and Upper Silesia. In addition, there was the German National Association of Commercial Employees (Deutschnationaler Handlungsgehilfen-Verband), which organized mostly white-collar employees and civil servants, and was close to the national conservative German National People's Party (Deutschnationale Volkspartei, DNVP) and eventually the Nazi Party. The Free Workers Union of Germany (Freie Arbeiterunion Deutschlands) was the only union organization not aligned with a political party.

The oldest German unions, the free socialist unions and the Hirsch-Dunckersche Vereine, were formally organized in the 1860s, at the same time as the SPD. While the unions devoted themselves to improving working conditions, the political parties sought to influence legislation. The SPD was the only labor party in Germany until the end of World War I, when the Communist Party broke off. The SPD remained committed to social reform and continued to work for improvement through the legal process, becoming one of the major democratic forces in the Weimar years, when it formed coalition governments and furnished the first president, Ebert. The Communist Party, by contrast, devoted itself to revolutionary causes, and in the late 1920s, formed its own union, the Revolutionary Union Opposition (Revolutionäre Gewerkschaftsopposition). During the Nazi regime, all unions were brought into line in the German Labor Front (DAF), and Communists faced special repression. After World War II, Hans Böckler formed the trade union association DGB, uniting the free unions in different industries, such as the metalworking and chemical industries. Christian unions played a more limited role, and the Hirsch-Dunckersche Vereine filled a negligible role in the postwar years.[30]

US productivity officers may have seen German labor relations as antagonistic because German unions did not resemble US labor organizations, where class consciousness played a negligible part, and unions operated independently of political parties and the government. By contrast, the most important German union organization, the DGB, espoused the doctrine of class conflict, although the union worked for reform through the legislative process rather than for a revolutionary overturning of ownership principles. In addition, the DGB closely affiliated itself with the SPD. Ideologically, the Hirsch-Dunckersche Vereine most closely resembled the US model of collaborative labor relations as it was advanced by Golden in its effort to reconcile labor and management interests, but the influence of the Hirsch-Dunckersche Vereine had declined precipitously by the end of World War II. In terms of independence from political parties, the Freie Arbeiterunion Deutschlands resembled US unions, but it only played a minor role in the history of the labor movement in Germany too. US Productivity Program officers eventually worked with the largest union association, the DGB, to strengthen the "free" union movement, in the sense of being free from Communist influences. But the DGB's class-conscious ideology differed from the US politics of productivity, which sought to raise the standard of living for everyone in a dynamically growing economy based on rising productivity rather than redistributing existing wealth in a stagnant economy.

When talking about antagonistic labor relations, US officers may have also referred to the strict boundaries between social classes that shaped German work relations and daily life. At a seminar for returning German WSTP students, the psychologist Ernst Korff captured this concern in his story of a painter in Germany who had contracted with a small company. When the owner of the company noticed a luxury car parked in his company's lot and found out that it belonged to the painter, he ceased doing business with the painter.[31] The car was perceived as a transgression of social class boundaries. Other items that marked social classes included individual possessions, such as a radio set in the living room or a man's Sunday suit (particularly in the nineteenth century); habits and clothing distinguishing workers from their superiors; and the often-ornate and intimidating design of corporate offices in Germany marked differences between the shop floor and corporate administration.[32] A photograph of workers leaving a shoe factory at the end of their shift shows them in front of the company gate, walking individually, without engaging or talking to each other. They have empty, tired faces, and wear nondescript hairstyles and coats that mark them as workers (figure 5.3). This image presents a stark contrast to the spruced-up, young WSTP students attending an evening class after work (figure 5.1).

Figure 5.3
Employees at Dorndorf, a shoe manufacturer in Zweibrücken, Germany, leaving after work.
Source: "Dorndorf Geschäfts- und Sozialbericht," 1952, box 18, entry 1202, RG 469, NARA.

Cars, consumer items, clothing, style, and habit marked social classes in daily interactions. They were embedded in a culture that believed people were born into the social position that they belonged, and that a transgression meant a violation of social order. Sociologists in postwar Germany debated the economic basis of these social distinctions. On the one side, in an ethnographic study of displaced families in West Germany, Helmut Schelsky coined the term "leveled middle-class society" (*nivellierte Mittelstandsgesellschaft*), a social structure that was formed by counteracting processes of social mobility. The working class had been ascending to its social position since the 1920s by participating in the welfare of the industrialized society, while higher social groups faced social descent through wartime displacement and wealth destruction. Together, both processes resulted in the formation of a leveled middle class in Schelsky's analysis.[33] On the other side, Ralf Dahrendorf argued, in a modified Marxist class analysis, that interest groups continued to exist in postwar Germany, and now solved their conflicts in moderated and arbitrated processes, such as industrial democracy, which enabled organized business and labor to settle their conflicting interests nonviolently.[34] Different from the United States, where over 80 percent of consumers identified as middle class and aspired to a middle-class lifestyle through bargain shopping, distinct consumption patterns persisted in West Germany. Working-class households spent a higher ratio of their income on food and luxuries, such as tobacco, coffee, tea, and alcohol, and invested in entertainment goods, such as radios and television sets, before other social groups did. Different from American consumers, German middle-class consumers maintained a frugal restraint on buying and emphasized quality over price. Compared to German working-class households, they spent a lower ratio of their income on food, and more on housing, high-quality durable goods—such as refrigerators, washing machines, and vacuum cleaners—and educational services and goods, such as books. Class differences and values thus determined consumption patterns more than income levels.[35]

Sociologists agreed that markers of social distinction continued in German society—be it as indicators of real or aspired social difference.[36] Members of the empirically based Cologne school of sociology around René König developed statistical markers of social status based on occupation, income, and "prestige." Erwin K. Scheuch, a young and rising member of the school, knew conditions on both sides of the Atlantic through a

Fulbright scholarship at the University of Connecticut in 1951, and later as a Rockefeller postdoctoral student. He argued that American Marshall officers and sociologists like Talcott Parsons mistook expressions of respect in Germany as markers of social inequality.[37] Markers of social distinction were not indicators of class consciousness, he purported; rather, they were remnants of preindustrial—feudal—rules in a highly industrialized society based on mutual rights and obligations. The father's occupation—not goods or ownership—determined a family's social status, Scheuch contended. Germans sought proficiency in their occupation. The respect for skills rendered social ranks less divisive; thus, Germans aspired to perform better in their given job instead of seeking to qualify for the next higher position. While social mobility rates in Germany were comparable to other Western societies, Germans believed in it less and aspired to it less.[38] In accordance with these sociological analyses, WSTP students returning to Germany observed that a person's occupation was a formative element in German society, providing craft-consciousness and moral ties that formed an aim in life. In the United States, by contrast, a job was merely a means to achieve other goals in life—through high salaries—and a person's social status did not derive from his (or her) occupation.[39]

Showing US Labor and Class Relations to German Visitors

When US Productivity Program officers brought German and European exchange groups to the United States, they wanted to expose them to collaborative rather than antagonistic labor relations. But this was more easily said than done, partly because collaborative labor relations were an ideal versus a reality, and partly because of rising labor strife. US unions sought to participate in economic gains after wartime wage and price freezes through a postwar strike wave, and conservative business associations tried to roll back New Deal labor gains through a campaign for free enterprise as well as the 1948 Taft-Hartley Act that restrained union organization, purged the radical union wing, and curbed interunion solidarity. These larger social tensions also emerged within the WSTP program.

Some colleges participating in the WSTP program, such as Syracuse University, found that their advisory committees transcended special interests between labor and management, but other colleges, such as Bridgeport University, experienced contentious meetings of the labor-management committee and blamed union members for unreasonable demands. Some labor officers stressed the essential role of the labor movement in the development of a consumerist policy and dynamically expanding economy. H. Carl Shugaar, a CIO officer, emphasized that unions demanded consumerist policies because "people are weak as individuals, strong as groups," and it was an "unlikely day when a group of benevolent employers call in workers at the end of the year when they've made too much money and offer to share the overflow."[40] Labor officers within the Productivity Program forcefully spoke for the right to unionize to advance consumerist policies—a controversial position in the United States during the postwar years.

US labor officers, like Franklin G. Bishop from the CIO and Everett Bellows, put these domestic controversies in a larger perspective: while union organization may have been controversial within the United States, this was not the case in US foreign policy. They alluded to the supportive role that US unions played in US consumerist policies in Europe, swaying European workers away from the promises of communism, and urged that this role be recognized. More pointedly, Bishop suggested that if companies "fight to prevent [union] organization, they are fighting the United States throughout the world."[41] These tensions provided the larger backdrop for the WSTP and other exchange programs.

The program soon encountered difficulties. One of them was the problem of placing students in suitable jobs because, initially, students were supposed to only work in plants with established labor unions to ensure that students would be exposed to plant-level collective bargaining and be able to participate in local union activities. Local colleges conducted job surveys to match students to the needs of local industry.[42] But possibly due to transatlantic miscommunication, the skills of the first students were different from the ones expected, and the 1952 economic downturn in the United States meant that local companies no longer needed additional workers. WSTP officers found that not placing students in jobs, or placing them in positions that poorly matched their skills, led to high levels of frustration. Students were already dealing with language problems, insecurity, isolation, and home sickness; therefore, satisfaction with their jobs and the prospect of personal advancement were major factors in making students happy. Unhappy students, in contrast, tended to see their own situation as well as conditions in the United States more negatively, and thus defied the goals of the WSTP program of convincing students of the model of US productivity and labor relations. As such, WSTP officers removed the requirement of placing students only in unionized companies, thereby allowing students to be placed in smaller, typically nonunionized companies, for example, with the understanding that students would not be placed in companies with contentious labor relations.

This was an anxious and much-debated decision because it meant less control over what kind of labor relations the WSTP students would be exposed to—at a time when labor relations were a hot-button issue in the United States. And indeed, soon one student was found to have participated in a strike—an expression of hostile labor relations—supposedly without even understanding the reasons for the strike. Some of the students, though, had been able to earn high US union wages, emphasized Nelson Cruikshank, one of the Productivity Program officers, allowing them to participate in the positive effects of collaborative labor relations. With Europeans experiencing both sides of the productivity coin, Cruikshank argued, Americans could "afford to let them see the whole picture, if we really believe that the good out-weigh the bad"—which he clearly did.[43]

Positive German Views of US Labor and Class Relations

US Productivity Program officers made sure that European exchange visitors returned to their home countries with the intended views of American productivity as well as labor and class relations. At the end of each shorter study trip, groups met with an officer for a well-documented evaluation meeting to assess their views, and should these evaluation meetings raise any red flags, officers followed up with additional meetings after a group's return in a process that resembled indoctrination more than persuasion. The groups composed elaborate reports on their visits, which the RKW published for the benefit of others in the industry.[44] WSTP students reflected on their exchange year in reports to the International Council for Youth Self-Help (Internationaler Rat für Jugendselbsthilfe), the agency organizing the exchange program on the German side. Within a year, the council organized and meticulously documented multiday seminars for returning students, partly to aid with their reintegration, and partly to analyze and interpret their observations and experiences in light of American productivity ideas (figure 5.4).[45] With few exceptions, study groups returned from the United States with positive opinions about labor and class relations in

Figure 5.4
Participants of the Work-Study Training for Productivity program at a debriefing
conference in April 1955.
Source: Box 5, entry 1202, RG 469, NARA.

the United States. Likewise, WSTP students often held glowing views of
their experiences abroad, reporting on their first sights of New York, their
ability to purchase cars and take trips across the United States, and in one
case, even obtain a pilot's license. Yet the US experience also estranged students from their own labor unions, unintentionally derailing the goal of
forming future leaders of free labor unions. I will first discuss the views of
the WSTP students, and then turn to the ones of the participants of the
shorter study trips and Project Impact.

German and European visitors to the United States witnessed the higher
standard of living firsthand. One of the German WSTP students created a
list of "what you could afford if Dad had a job here [in the United States]
similar to the one he has in Germany: A house (six to ten rooms and central heating), a car, a complete television set, a trip to Europe every two
or three years, two children and complete vocational training for them,
including university training."[46] While differences persisted in the United

States between households of different income levels, homeownership became affordable with the postwar move into suburban developments, and families began acquiring second cars as well as shifting their spending from laundry services and domestic help to household durables, such as refrigerators, washing machines, air conditioners, and television sets.[47] The students' families in Europe, by contrast, still faced postwar deprivations that delayed the acquisition of consumer durables. For example, in West Germany, for the first purchase of household appliances, people bought four out of five of their washing machines, dish washers, television sets, and freezers after 1958.[48] Working-class families in European urban centers still took advantage of public transportation or often purchased motorcycles rather than cars, although car ownership grew with 22 percent of working families owning a car by 1962. And while in 1950 only 6 percent of dwellings were owned by workers, the number had climbed to 32 percent by 1968.[49]

WSTP students often told compelling stories of their insights into labor relations in the United States.[50] For example, Klaus Kesselhut, a young architect who stayed in Saint Paul, Minnesota, described a situation that required the employees of the small company where he worked to do overtime without extra pay shortly before Christmas. The four employees did the work out of a collegial spirit toward the owner, who worked closely with them and did not earn considerably more than them. Kesselhut stated his own attitude was that "the employer was my friend, and in the case of my friend I don't bother with union regulations."[51] Lacking visible signs of class distinctions, such as coats or white collars, employers and employees in the United States often addressed each other by first names and pursued collegial relations without reducing managerial authority. While the students noticed that differences existed between skilled and unskilled workers, they did not comment on racial or ethnic divides, unlike the union travel group in the 1920s. The students stated that there generally were less distinctions in the standard of living between different occupational groups; in their eyes, this led to less social tensions, a higher sense of collaboration, and higher overall satisfaction among workers.[52]

The students perceptively noted that this more collaborative atmosphere relied on a different attitude of workers toward their work and superiors. One student, Olaf Beckmann, a mechanic, said that workers, including union members, created a climate in which everyone strove to do their share of work to the end of a shift rather than extracting a few idle minutes from the employers. He had been in the United States only two weeks when he stopped working a few minutes before the end of the shift and began to wash his hands, since the habit in Germany was to pack up ten minutes before the end of the shift. A colleague—a union member, Beckmann emphasized—immediately took him to task, saying that it wasn't fair to stop early since workers were paid to the end of their shift. Beckmann saw this episode as a sign that workers themselves contributed to creating a fair work climate and good relations between labor and management.[53] He experienced local union members as a disciplining force in the company—a role that had only been held by his supervisors in Germany, not by his colleagues. The union members nearly violated what Beckmann perceived as solidarity between colleagues, yet Beckmann himself also exhibited a similar interest in the success of the company—a hallmark of collaborative labor relations in Golden's view.

Exchange Students Estranged from German Trade Unions

The major political goal of the WSTP program was to form future leaders of free unions in Europe.[54] Indeed, the exchange students returned from their year in the United States with deep admiration for the American system and American productivity. But their experiences, especially their participation in local union affairs, also estranged them from their own unions in Germany. The students particularly appreciated the democratic decision making and budgetary transparency in their local unions in the United States. One German student placed in Syracuse, New York, for example, reported on his union's decision to send delegates to a conference in Rochester, New York, about a hundred miles away. Since the expenses were to be covered from union fees, members voted on how many delegates to send—they chose three rather than two—and voted on the delegates' per diem rate—high, to represent the local union. In the student's view, the decisions reflected the members' pride in their local union. Another student remembered receiving his union fees back at the end of the year because they hadn't been spent; in his eyes, this was an indication of the budgetary consciousness and transparency in his union.[55] These stories also expressed the accountability and trust between rank-and-file union members and their local functionaries.

The German students generally liked plant-level collective bargaining because they thought it allowed union members to be more involved in economic affairs and better understand their rights. The students also thought that the close relations between local union officials and rank-and-file members had a moderating impact on collective bargaining negotiations. Students reported that managers and union delegates—imbued with their members' trust as well as close community relations—met each other in wage negotiations not as opponents but instead as partners, and that American union representatives acted more flexibly and collaboratively in contract negotiations than their German counterparts. For example, if a union in the United States demanded unreasonably high wage increases, management might give union leaders access to the company's financial reports. By looking at these records, union leaders often would convince *themselves* that their demands were too high and could then defend this conclusion to union members. Herr Karp, the DGB delegate with whom the returning WSTP students discussed labor relations, admitted that there was a larger chasm between rank-and-file members and union functionaries in Germany. Union functionaries feared being considered unfit if they did not achieve salary raises and worried about being called "fat cats" because of their salaries.[56]

But Karp also argued that regional collective bargaining, as it was practiced in Germany with negotiations between regional delegates of employer associations and unions, imbued labor unions with more momentum (*Stosskraft*) because it could build on the solidarity of workers across companies. While he claimed that German employers preferred leaving tariff negotiations to their associations, he argued that the DGB needed to be involved in political debates to represent the interests of its members. While half the exchange students agreed, the other half held that such political involvement negatively impacted the climate in corporations. In addition, students felt more removed from their unions in Germany; they argued that a success for unions in Germany always remained "a success of the institution, its policies and its functionaries"—that is, of the union and its staff—rather than an achievement with which they could personally identify.[57]

Productivity Program officers brought WSTP students to the United States in hopes that as future labor leaders, the students would implement plant-level collective bargaining after the US model. The returning students certainly found many advantages in the ways that US local unions operated and were enamored with the model of plant-level collective bargaining. But these experiences also estranged them from their own union leadership in Germany. Indeed, unions were reluctant "to [send] their leaders for training to USA to make _____ out of them."[58] Rather than strengthening the labor movement by building its leadership, the WSTP program thus had the unintended effect of splintering it.

Critical Views of US Productivity

Other visitors such as the participants of Project Impact or shorter study groups, however, were not to be convinced of American productivity as easily as the WSTP students. They focused on technological methods, and either ignored the larger social and economic factors, or dismissed them as a public relations ploy. Many visitors arrived with preconceptions, some of which could be overcome, and others not. Also, some visitors felt misunderstood and therefore resisted the US model, although US Productivity Program officers sought to overcome such mutual misunderstandings. The industrialists participating in Project Impact, in particular, were critical, regardless of whether they came to the United States with more progressive or more conservative attitudes. They resisted the idea that shared-out productivity would solve the problem of inequality by promising everyone a part of higher profits versus having to distribute limited profits between workers, industrialists, and consumers.

Some travelers openly rejected the US model, taking a skeptical, if not cynical, stance toward interactions between workers and managers. For example, Otto Seeling, the conservative director of a glass manufacturing company in Fürth, on returning from his two-week trip to the United States under the auspices of Project Impact, stated that "the jovial tone between American workers and their bosses should not be considered unerring proof" of successful collaborative labor relations; rather, that sort of informal friendliness was part of a mode of interacting commonly found in the United States.[59] Similarly, Hubertus Wessel, a junior partner at Wesselwerk AG in Bonn, which then produced sanitary ceramics such as toilets and sinks, marveled at the reception of his Project Impact travel group at Remington Rand's typewriter and computing machines plant in Elmira, New York. Corporate directors in their own cars picked up the four dozen

European visitors from the local airport and drove them to the plant, where they were greeted by the mayor, one of the company's vice presidents (probably Marcel N. Rand), an employee representative, and a speaker from the local women's association. In addition, the local press reported extensively on the European visit. Impressed, Wessel called this kind of reception "applied and demonstrated 'public relations.'"[60]

Wessel was probably involved in his company's technical development because five years after his trip, he filed a US patent for applying a ceramic surface to building elements. His travel report largely concentrated on technological production methods.[61] At Remington Rand, for instance, he noticed artificial lighting and air-conditioning in the one-story production facility and was astonished by the small number of production lines in use. He also commented on the well-dressed and good-humored employees, who were seated while working; some were even smoking. Likewise, at Ford's River Rouge plant, Wessel remarked on the effective organization of the shop floor and the arrangement of tools. He also noticed the flexible way that materials were transported to allow for changes in production design: instead of loading them onto conveyor belts, individual vehicles replenished small material storages at the assembly line. Technological and organizational efficiency was Wessel's main focus and, while he did not discuss labor relations, his designation of employees as members of an allegiance (*Gefolgschaftsmitglied*)—versus members of a team, for example— implicitly revealed his views of labor relations.[62]

Likewise, the German study group on steel heat treatments had a singular focus on technological improvements. For every hour at a company, the group spent an hour in the evening taking copious notes on what they learned. Productivity Program officers accused the team of a nationalist-imperialist attitude, being "driven by a singleness of purpose: to extract from American manufacturers and researchers the utmost in manufacturing secrets, cost of production data, and up-to-the minute laboratory research discoveries for the greater good of the Vaterland and its conquest of world markets as a means of economic and political prestige."[63] Although the itinerary included visits to American homes for dinner and everyday activities, such as lawn mowing and dish washing, productivity officers noted that "the management representatives and technical experts manifested little interest in the American way of life, in shared-out productivity, in labor-management or human relations in American industry."[64] While the group's singular technological focus may have been an extreme outlier, it is an example of the German visitors' continued emphasis on rationalization technologies that began with the Weimar travelers of chapter 2.

Not surprisingly, the initial study group reports centered on technological improvements and workplace organization. A shift occurred in 1953, with a new, stronger look at "questions of marketing and distribution, corporate organization, human relations and general factors affecting productivity."[65] Now, technically oriented study groups began to consider the social and political aspects of productivity. A 1955 study group on productivity and manufacturing, for example, reported that Americans had consciously and impressively exposed Germans to the "fundamentals of productivity," pushing them to look beyond technological improvements at the "economic humanism that, in the American view, explained the technological-economic success of the United States more than any technological revolution of process engineering." The eleven group members— engineers, toolmakers, and company owners—admitted that as "matter-of factly thinking technicians," they would have ignored the social and political aspects of productivity in their continual focus on daily operations. Instead they devoted the entire first half of their report to human relations, dynamic economic growth, relations between management and workers, collective bargaining, work safety, public relations, management training, and advertising and marketing.[66] The second half was devoted to the architecture and layout of manufacturing buildings and machine tools. It remains an open question as to how much attention German readers paid to the first half of the report.

By 1954, a leading delegate from the German ministry for economy, Walter Hinsch—who had previously pushed for the German small and medium-size loan program to proceed independent of the demonstration project program—promised to advocate that the focus of study groups be shifted to the "intangible factors of productivity," such as "human relations, labor-management relations, advanced personnel management, executive and supervisor training, communication between top management and the manual workers, [and] industrial accident prevention." He now believed that "the introduction of the American intangibles in German industry would do more to increase efficiency than technological improvements" since Germany was already ahead of the United States in some technological aspects such as standardization.[67] Notably, Hinsch excluded share-out agreements—the formal agreements between management and labor to share the benefits from productivity increases—from his list of intangibles. The US evaluation officer was convinced that Hinsch's was yet another German group seeking higher productivity as a means to lower production costs to conquer European and global markets. Still, the RKW began to organize study groups with a stress on the social and economic values of productivity, such as groups on human relations as well as teamwork and productivity.[68]

Many other study groups arrived with preconceptions about the United States. Often, these could be countered. For instance, a German study group on mine safety expected to find "an industrial waste of unsightly factories, slums, slag heaps, and deserts, brought on by reckless farming and tree felling," and instead were surprised by the "cordial welcome everywhere," with Americans being "kinder and more helpful to each other than Europeans."[69] In other teams, misunderstandings may have caused resistance to the American model. For example, the German machine and precision tool team members had been lauded for their enthusiasm, and for management and labor representatives deciding to room together during their stay. But in a follow-up meeting, they complained that the industrial psychologist, Herbert E. Krugman, who led the human relations meeting, "did not understand the Germans, nor the German culture," and team members suspected Krugman had never traveled outside the United States.[70] They felt insulted by suggestions that Germans were less enthusiastic and took fewer initiatives. Team members stated that German workers feared that higher productivity would lead to overproduction and unemployment; in their view, the capacity of the German market could not be grown, and the opening of exports markets, particularly in the United States, was the only solution to avoid overproduction. Despite their open attitude, the group obviously was not to be convinced of the idea of a dynamically expanding economy.

Likewise, the all-labor study group members from the German leather industry came with strong preconceptions. Not only did they arrive in New York in the sweltering July heat with long underwear because they had been told about cold spells in the United States, but they also believed that human lives were not valued in the United States. They had been told there were high murder rates due to gang violence, and high death rates due to road accidents and industrial accidents. They believed that in fast-paced American factories, human lives counted less than raw materials and machines. Also, possibly based on Communist propaganda, they expected that their tour would be controlled day and night by guides. In addition, the members believed that theirs was a purely technical mission "to study machinery, equipment and production processes from the worker's point of view," and they had been asked by their bosses to "keep their eyes and ears open for new ideas," steering them away from the nontechnical aspects of US productivity.

One of the group members, Paul Lucius, a union official from Offenbach, wondered why, "with wages so high, retail prices can be so low." He primarily explained the higher productivity by lower craftsmanship that resulted in "shoddy, crude-looking goods," and took higher consumption figures as an indication of lower quality. Willi Bier, the works council head of Dorndorf—the patriarchal shoe manufacturing company whose workers are depicted in figure 5.2—theorized that the higher number of workers under age eighteen lowered German productivity. Overall, the team members appeared to expect that "their bosses would feel proud to learn that all things being equal their plants were as efficient as the vaunted American factories," suggesting more subservient than antagonistic labor relations in their companies.[71] Strong preconceptions and a technical focus kept them from recognizing the role of high wages and high purchasing power.

Heinrich Krumm, a participant of Project Impact and leather manufacturer from Offenbach—like Lucius—complained that the European delegation of successful industrialists "often had the feeling that they were questioned and reprimanded with the somewhat strict attitude of a superior teacher who had done better in life." Defensively, he debated the validity of productivity as a statistical measure. Americans were specialists "in thinking in figures," and tended "to use simple comparisons of figures and to evaluate them accordingly." Doing so, however, Americans were "often unable to see the much more complicated structure of the European economy." Krumm charged that productivity measures did not account for wartime destruction, nationalization, the dismantling and prohibition of production, the loss of middle-aged men, and the refugee crisis in Germany.[72]

Historians nowadays agree that the German economy had already begun its recovery by the time that Marshall funding started, partly due to the nature of wartime destruction. While urban centers and transportation infrastructure had been subject to bombing attacks that destroyed office buildings as well as dwellings, and impeded travel and communication, postwar Allied analyses showed that air raids frequently missed manufacturing plants—including those in the armaments industry—due to a lack of precision, leaving mostly intact the German manufacturing infrastructure located outside urban centers.[73] But in the contemporary perception, wartime interruptions were a convenient argument to which Krumm turned, disgruntled by his inability to engage with American counterparts eye to eye. Productivity Program officers realized how detrimental such misunderstandings were. They worked with historians, such as David Landes, to understand European conditions, and hired psychologists and anthropologists as consultants to help them "lean over backwards" to accommodate Europeans and avoid such misunderstandings.

Krumm, industrialist and self-described progressive, should have been an easy convert to US productivity. Suggesting that possible political antagonism in German companies had turned into cooperative labor relations, he professed that in his thirty years of experience, he had never met a trade union secretary who "would have refused to take up any ideas and suggestions concerning the plant, and to support any improvements for its further progress" as long as employers openly discussed with union representatives questions related to labor and technology. Yet Krumm also attested to a disjunction between union officials and rank-and-file members when he suggested that workers opposed technological improvements because they "may often be unsufficiently [sic] informed."[74] Krumm's company practiced the methods to assess work and determine payment recommended by the Imperial Committee for the Determination of Working Hours (REFA), one of the German rationalization committees associated with the RKW. He saw these methods as comparable, if not superior, to US methods, although higher profits allowed more generous profit-sharing arrangements in the United States.

Krumm was extremely upset, though, by European industrialists being called "backward, conservative, conscious of their class and thereby unsocial" by their American hosts. He was one of the rare industrialists who admitted to the different nature of social relations in the United States rather than simply dismissing them. He acknowledged US labor relations to be "quite normal and friendly, though perhaps lacking a little of the warmth which we are used [to] in European plants," and thought US trade unionists to be "less politically inclined but very well informed about economical [sic] and technical matter." Observing that "old-fashioned and class-conscious attitudes, as they can be noticed in old Europe towards plants or rather employers, are unknown," he saw the welcome by mayors, sheriffs, and other notables who proclaimed their "appreciation for the decision to visit just *this* plant of *this* community" as an expression of local interest in a company's success. He thought the composition of the management was one of the reasons for this local embrace, with executives, technical heads, and foremen forming the management team regardless of their capital share in the enterprise, preventing "envy towards the success of the enterpriser." The head of his family firm, Krumm led a privileged life. Yet he was extremely concerned about what he saw as an American "prejudice" against European industrialists because he thought it a "dangerous argument in the discussion of the broad public."[75] Krumm apparently feared that unions and others social forces in Germany might use the US rhetoric to strengthen their position. Thus he was not to be convinced of US productivity and may have shaped local opinions in the Offenbach leather industry, contributing to the critical attitude of Lucius and other members of the leather labor group.

Many other manufacturers had deeply settled conservative views and would have been difficult to convince of US-style productivity. For example, the glass manufacturer Seeling held a patriarchal welfare capitalist worldview despite a biography that resembled an American rags-to-riches story. Orphaned at an early age, Seeling began working young, and because he was unable to complete his formal schooling, continued to study in evening classes. He held progressive opinions as a young man and joined the SPD before World War I, but in his later years moved to more conservative parties. Visiting the United States through Project Impact, he saw American productivity through patriarchal welfare capitalist eyes. Thus he criticized the lack of washrooms at Ford's River Rouge plant because in his view, it forced workers to wear gloves on their way home so that they wouldn't soil their cars.[76] Not only did Seeling neglect the fact that American workers owned cars—in US eyes, an indication of their higher standard of living compared to their European counterparts; he also ignored that gloved and protected hands, free of callouses, would mean that the workers would not be marked as blue collar in their daily interactions.

The only thing that Seeling appears to have envied Americans for was the lack of an economic debate, which in his view, saved a lot of "friction." Seeling did not consider American proposals of higher productivity and share-out agreements as solutions to the distribution conflicts he so deplored; rather, he thought workers and consumers needed to be convinced of the existing German system, and that the goal of the economy was to meet their needs "as good and as cheaply as possible." Seeling wished that a "large unified conception" could be reached.[77]

Likewise, the staunch conservative Wessel rejected outright the idea of share-out agreements. During Project Impact's international conference of manufacturers, Wessel socialized with Eldridge Haynes, a NAM official whom he had met earlier that year when Haynes had traveled to Germany to support the opposition against codetermination legislation, a form of corporate government pursued by the DGB. This acquaintance is indicative of Wessel's position on labor relations. In lieu of collaborative labor relations or codetermination, Wessel argued that those who wanted to participate in corporate decision making—socialists calling for the control of "capitalists," in Wessel's words—also had to share the risk and responsibility for these decisions. Collaborative decisions were thus to replace calls for control. Since both "capital and labor"—Wessel used the terminology of class conflict—had the same goal of a higher standard of living, he suggested that corporatist bodies (*Körperschaften*) be created with representatives from both sides. These bodies were to make decisions on the basis of votes, with the amount of capital invested per employee determining the voting rights. This arrangement would solve two problems. First, majority decisions would help overcome conflicts of interest among employers as well as among employees, such as between small and large corporations, or unionized and nonunionized employees. Second, in Wessel's view, employees would gain an interest in technological investments—or automation—since it would increase their capital and thus increase their voting rights; as he put it, they would begin to think in entrepreneurial terms.[78] Such corporatist bodies, however, favored more capital-intensive companies—some of which may have invested more in rationalizing equipment and machinery. While such decision making may have favored technological improvement, it did not address the question of who was to benefit from higher productivity through new technology—one of the central tenets of the Productivity Program's promises of shared benefits between industrialists, workers, and consumers. These industrialists were upset that Americans viewed German labor relations as antagonistic, and failed to recognize a difference between labor relations in Germany and the United States.

Therefore, they saw no need for dynamic economic growth to share profits from higher productivity as well as change labor and social relations in Germany.

Conclusion

These views and perceptions of German visitors were not fully aligned with the US views; as such, it is unclear how far Germans applied US productivity technology and values on their return home. While those with critical perspectives could not be expected to implement productivity ideas, it remains an open question as to how far those who returned with positive views tried and succeeded in employing US productivity. Students as well as workers may not have been in a position to effect significant change. For example, the Dutch WSTP student A. J. Seip urged the Productivity Program administration to *"not* bring over plain workers from Europe" because management would not accept the young workers' suggestions and would dismiss the worker by saying, "Oh, he has been in America for a year, and now he thinks he knows it!" Instead, Seip encouraged the administration to select young managers who would soon rise in the corporate hierarchy and could then begin to implement changes in labor-management relations.[79]

Seeking to change productivity ideas in European countries, US officers evaluated study groups for their effectiveness. They tried to increase the impact by selecting the right participants and providing effective follow-up assistance. For example, Productivity Program officers found that study groups were most effective if two or more participants, including labor and management representatives, came from the same company. Through follow-up meetings, they supported study group members in their efforts to effect change, and realized that as Seip had indicated, workers were vulnerable in the European social and corporate context. If, for instance, a worker experienced a poor reception to his presentation about his visit to the United States—be it because of a lack of respect for his experience due to his social rank, or because of a lack of training and practice as a public speaker—Productivity Program officers would assist him in preparing for his next one or even collaborate with him. This way, the officers hoped to prevent "psychologically very negative" experiences and "considerable economic loss" for the Productivity Program administration should the visitor cease disseminating US productivity ideas.[80]

The Productivity Program also created opportunities for European and American workers to get to know each other, bridge differences, and create approval for the Marshall Plan. Labor adviser Nelson Cruikshank reported that US union members were initially indifferent to their organization's "transpheric"—as they said—"flights about the importance of international relations" that didn't impact the local situation in their midwestern towns and elsewhere across the United States. But then union members found that "these international relations are beginning to mean something to us and to our home-town"; once "these birds from Europe, from France and Norway and other countries, come to our local union meetings, we work with them at the bench and at the shop and they talk about their problems." Cruikshank himself admitted to having at first been extremely skeptical of the WSTP exchange program, and concluded that although "every possible mistake" had been made, the program had turned around and started to create a series of "good" and "heart-warming" experiences. For example, a group of Italian students in Minnesota planted a tree with a hand-carved plaque that expressed their appreciation of the hospitality of the local community as a sign of "friendship," "fraternity," and "solidarity among free people."[81]

European WSTP exchange students, like the ones portrayed in the Studebaker journal, may have been the most promising missionaries of American productivity.[82] Their long exposure to American production technologies and a culture of collaborative labor relations frequently convinced them of the American model. Due to their youth and lower social status, however, they often faced difficulties when trying to introduce American ideas in their home countries. The shorter visits of European study groups and visits through Project Impact yielded mixed results. German visitors, with their tradition of the rationalization movement, were intent on studying manufacturing technologies and likely did implement technological ideas in their companies. But they remained more skeptical with regard to labor relations; conservative as well as progressive industrialists proved largely resistant to the American model, and some labor representatives also returned with reservations. Many resented feeling lectured to, and argued that American productivity was not suitable to Germany's lower labor costs, fewer capital resources, and smaller markets. Likewise, the concept of a dynamically growing economy proved a hard sell. Critical industrialists resented sharing the profits of higher productivity, and while WSTP students and labor delegates liked US labor and social relations, they did not connect them to a dynamically growing economy.

Thus, neither young workers nor technical experts or industrialists embraced the politics of productivity that promised to distribute increasing profits instead of redistributing existing wealth. Still, the Productivity Program created a long-lasting ideological and institutional legacy in transatlantic and even global international relations. For example, the WSTP created a student exchange program in Germany that still exists, nowadays under the auspices of the Carl Duisberg-Gesellschaft, which provides international training and education for young student workers from around the globe. And the European organization coordinating productivity projects, the Organization for European Economic Cooperation (OEEC), transformed into the Organization for Economic Cooperation and Development (OECD), which among other things, provides technological assistance to members worldwide. In the next chapter, we will see that in Germany, corporate governance took a different direction from the one advocated by the Productivity Program: instead of plant-level collective bargaining, German legislation implemented codetermination.

6 Codetermining German Labor Relations

On April 4, 1951, the *Wall Street Journal* informed its readers that "a revolution in labor-management relations is in the making here in the West German capital this week."[1] But the paper was not reporting on the kind of social revolution that US Productivity Program officer Everett Bellows and his colleagues were seeking to instigate in West Germany and Western Europe. Rather, it was reporting about legislation before the West German parliament, the Bundestag, that sought to implement codetermination, a form of corporate governance that gives workers a say in corporate decisions and limits industrial power. Codetermination presented an alternative to the form of collaborative labor relations that Marshall Plan officers wanted to employ in Germany through their Productivity Program. To some American observers, the West German rejection of collaborative labor relations in favor of codetermination was a step on the "road to socialism." This led two American organizations—the National Association of Manufacturers (NAM) and the National Foreign Trade Council (NFTC)—to intervene in the German debate on codetermination; they wanted to protect managerial freedom in corporate decisions. Their delegates, Eldridge Haynes and Gordon Michler, held a press conference in the West German capital of Bonn a day before an important parliamentary debate on codetermination and threatened that US companies would withdraw their business from Germany should the legislation pass.

On both sides of the Atlantic, many were unhappy with Haynes and Michler's intervention. The German association of labor unions, Deutscher Gewerkschaftsbund (DGB), protested that their statements unduly interfered in foreign affairs. Even those in Germany who were opposed to a strong form of codetermination—such as industrialists and employer associations, whose position Haynes and Michler sought to strengthen—refrained from taking advantage of the Americans' statements. In the United States, some voices, such as the writer and cultural adviser Lewis Galantiere, had publicly warned the NAM for months not to interfere in the German debate. And labor officers in the Marshall administration, finally, protested that Haynes's dual involvement in the codetermination debate and a planned productivity seminar for industrialists from the Ruhr region threatened to raise suspicion among German union leaders about American intentions, and thus endanger the overall success of the Productivity Program.

The US intervention in the German codetermination debate is a fascinating story of the intermingling of public and private diplomacy, and the negotiations of boundaries between the two. It also is the story of the competition between different economic visions and orders, both between and within the United States and West Germany. The NAM and the NFTC were two conservative business associations that, in the United States, supported free enterprise. In particular, the NAM had opposed the New Deal and, before World War II, fought the acceptance of labor unions. After the war, it had largely acquiesced to the new legislation and urged its members to make the best of collective bargaining where it existed, but still supported the repeal of labor-friendly New Deal legislation, and opposed any extension of collective bargaining beyond individual plants or companies. On the German side, the closest to this position was Chancellor Konrad Adenauer's coalition government's embrace of Ludwig Erhard, the economic minister, and his vision of the social market economy. The social market economy was an economic system based on private ownership and free competition in which the state was to guard the competition by preventing counteracting forces, such as monopolistic tendencies. Therefore, both social market economy and free enterprise built on free competition, but they differed in the role of the state. In Erhard's social market economy, the state was to guard the economic framework without interfering in economic processes—like the state setting traffic rules without directing the traffic.[2] In a free enterprise economy, corporations sought freedom from state interference. In this sense, Erhard's vision of the social market economy was an alternative to free enterprise.

Indeed, the social market economy is often hailed as a third way, between American free capitalism, on the one side, and Soviet-style state planning, on the other. A closer look, however, reveals that the realities were more complex. Occasionally, Germans criticized US officers for favoring planning over free competition, despite the strong antitrust legislation of the 1890 Sherman Act. Planning was an aspect of New Deal economic policies, and progressive US business associations, such as the Council for Economic Development and the National Planning Association, envisioned a more active role for the state than did free enterprise proponents.[3] Similarly in Germany, the DGB fought for codetermination. While codetermination harkened back to the Weimar precedent and its socialist vision of overcoming class conflict through economic democracy, it did not on the surface limit free competition. But through backdoor channels, it introduced a corporatist element into Erhard's free market economy. Corporate interest groups, such as unions and business associations—which were to send delegates to corporate boards of directors—gained a say in corporate decisions rather than individual managers or the state making decisions. In other words, the realities of economic systems were more convoluted than easy models or visions suggest. The codetermination debate provides an opportunity to open the black boxes of competing visions on both sides of the Atlantic and take a closer look at their complex interactions across the Atlantic.

Among all the countries participating in the Marshall Plan, West Germany was the one that Marshall Plan officers might have expected the most from as far as implementing their economic and labor relations ideas. After all, with political and economic elites delegitimized after the demise of the Nazi regime, West Germany had to shape and rebuild its political and economic system anew. Many labor and union leaders, however, were not socially and politically delegitimized, because they had been part of the resistance against the Nazi regime and had suffered from the regime's suppression, or had passed the time in exile—a fact that may have helped them pursue their own vision of corporate governance instead of bowing to the American model. Yet large industrialists, particularly in the coal and steel industries, like Krupp and Thyssen, were delegitimized through their association with Nazi rearmament. Threatened by Allied deconcentration and de-cartelization efforts, they still fought for their concerns, and in a weakened position, were willing to compromise in the codetermination debate. Erhard, along with the ordoliberal circle that he was a part of, saw the post–World War II break as a chance to implement the economic vision they had developed during the 1920s and 1930s, partly in response to what they conceived of as the problems of planned economies in the Soviet Union and the Nazi regime.

Political economists have categorized the United States as a liberal market economy and Germany as a coordinated market economy.[4] Labor relations—aside from modes of finance, vocational training, and technology transfer—are one of the defining characteristics of each variety of capitalism; therefore, it is fitting to investigate the emergence of corporate governance in West Germany and the codetermination debate. While political economists categorized different varieties of capitalism, economic historian Werner Abelshauser has probed the historical origins of the West German variety.[5] Building on his work, this chapter takes a closer look at the emergence of the West German economic order in the post–World War II years, revealing the complex interactions of liberal visions and social pressure from corporatist groups. The German social market economy is not the social market economy that Erhard imagined in 1949; it is an intricate amalgam of his vision and traditional elements in the German economy, such as codetermination—which has been called an "essential component of the German market economy"—as well as a continued tolerance for cartel-like cooperation in which firms agree on prices and technological standards rather than competing on these.[6] While the codetermination debate can easily be seen as a failed transatlantic exchange—which is one reason why German narratives exclude reference to the American side—a closer look reveals the multifaceted economic visions and conditions on both sides of the Atlantic.

The goal of this chapter is to show the complex historical roots and amalgamations behind the different varieties of capitalism in the supposedly homogeneous capitalist Western bloc. It begins by laying out the situation of corporate governance and labor relations in the early days of the West German republic as well as Erhard's vision of the social market economy. The chapter then takes a close look at the codetermination debate in Germany, in which labor unions sought worker participation in corporate decisions. While Haynes and Michler, on behalf of the NAM and the NFTC, intervened in the German legislative process and sought to shape the legislation as well as prevent it from passing, the US government took a neutral position in the debate, and US labor unions gave, at the most, cautious support to the quest for codetermination. The chapter concludes by assessing the outcome, the intricate amalgam of Erhard's economic vision as well as the corporatist influence of labor unions and business associations.

Corporate Governance and the Social Market Economy

Labor relations in the early Federal Republic of Germany were a patchwork because the occupation authorities and the German states had passed different forms of labor legislation. In 1946, the Allied Control Council passed a works council legislation; the United States had introduced this legislation because it appeared as the lesser evil compared to a potential socialization of industries.[7] The Soviet Union was expected to socialize industries in its own occupation zone. But in the late 1940s, demands for socialization were also widespread among Germans in the British and American zones of occupation. Influenced by Catholic social ethics, even the Christian Democratic Union (CDU)—the future "American Party" of Chancellor Adenauer—called for the socialization of the heavy coal and steel industries in its 1947 political program, named the Ahlener Program after the small Westphalian town where the party delegates had met.[8] To the US occupation forces, codetermination may have seemed like a welcome alternative to socialization because it protected private ownership.

Britain and the United States each implemented a different form of corporate governance in their own occupation zones. The British occupation forces relied heavily on works councils in the administration of the coal and steel industry in the Ruhrgebiet, Germany's major industrial region. The large Ruhr cartels, like Krupp and Thyssen, were delegitimized for having supported the Nazis, while the labor movement was less tainted because many members had fought in the resistance, had been interned in Nazi labor camps, or went into exile, typically in Britain, the United States, or Scandinavia. While the British also pursued de-cartelization and deconcentration efforts, works councils—established by individual corporate contracts—helped revive the Ruhr coal and steel industries, which then served as an economic engine for German reconstruction.[9] The American occupation forces acted more hesitantly. Recognizing that codetermination implied a departure from the established system, they suspended codetermination legislation temporarily in two German states, Hesse and Württemberg-Baden, and effectively delayed the recognition of works councils in the American occupation zone until 1950.[10] With this provisional system in place, one of the first items on the legislative agenda in the young republic in 1950 and 1951 was the uniform regulation of corporate governance.

These labor relations were to be determined within the framework of the new German economic order, the social market economy—developed by a few liberal economists in response to the shortcomings of corporatist and planning elements in the Weimar and particularly the Nazi economic regime.[11] The liberal economists thought of their social market economy as a third way, different from socialist planning and nineteenth-century laissez-faire economics in which the state had only the role of a night watchman. While today the social market economy includes corporatist labor policies and social insurance, for example, this was not the case originally when it was meant as a free market economy.[12] The state only provided a framework to protect free competition, especially legislation to prevent cartels. The state thus had a more active role than in nineteenth-century laissez-faire capitalism, but the role was strictly limited and was not meant to interfere with the economic process. What was "social" about the social market economy? It was embedded in a legal framework that would allow all economic actors to fulfill their potential. None of the social policies were envisioned to be part of the social market economy; to the contrary, the creators of the social market economy reviled social policies that might invoke elements of planning. With its emphasis on free market competition, the original German conception of the free market economy was freer than the US economic practice, with its New Deal planning elements that continued during World War II.

A small group of ordoliberal—later also referred to as neoliberal— economists developed the tenets of the social market economy during the 1920s and 1930s. Among them were Walter Eucken, a professor of economics at Freiburg; Wilhelm Röpke, a rising economist at Jena University; Alexander Rüstow, the director of economic and political research at the Association of German Engineering Manufacturers; Alfred Müller-Armack, a university lecturer in economics at Cologne, called to a chair in Münster in 1938; and Erhard, a research associate at an institute for market research at the Nuremberg Commercial College.

Eucken, the best established in the group, came from a notable academic family; his father was a Nobel laureate in philosophy, and his brother held a chair in chemistry. A proponent of the deductive method, Eucken in the

1920s defended the receipt of loans from the United States against, among others, Hjalmar Schacht, a Weimar economist, banker, center-right politician, and strong critic of German reparations payments, and who would later serve as economic minister under the Nazi regime. Eucken argued that technological improvements would help Germany pay its way, and that thrift and self-denial were not enough. Eucken formed the center of a circle in Freiburg that gave liberal economists a forum. During the Third Reich, he kept his distance from the Nazi regime, including the Nazi-friendly leadership of his own university under philosopher Martin Heidegger.[13]

Röpke and Rüstow both had left-leaning sympathies during the Weimar years. In the 1920s, Röpke approved of works councils and supported high wages—a rare position for a liberal economist. While he argued against socialization, he was open to moderate government investments to counter economic slumps. Rüstow even identified as a socialist at the end of the First World War and briefly worked for the nationalization of the coal industry, but soon became disillusioned with socialist planning. Both Röpke and Rüstow emigrated to Turkey in the early 1930s, and a few years later, Röpke moved to Geneva for a position as a research professor with Rockefeller funding.[14]

Müller-Armack and Erhard were the less established outliers in this group.[15] Both came from lower-bourgeois backgrounds: Müller-Armack's father was a Krupp factory manager in Essen, and Erhard's father a small shopkeeper in Fürth close to Nuremberg. Neither of them was in a secure university position by the late Weimar years. Müller-Armack sympathized with Nazi ideology in the early 1930s and even joined the Nazi Party. He later headed two research institutes in Münster—one on the textile industry in Germany and abroad, and one on housing policy; his own research interests focused on religious denominations and economic development in Europe. Erhard, by contrast, refused to join any Nazi organizations, although until 1942 he continued to work at a research institute that significantly benefited from Nazi patronage. When he left the institute—at least partly because of an acrimonious conflict about the institute's leadership—he founded his own institute on industrial research, basically a one-man enterprise with corporate funding, and began to develop proposals for a postwar economy, still a sensitive topic in the last war years.[16]

These ordoliberal economists were only loosely connected before 1945, with stronger ties formed by correspondence between Eucken, Röpke, and

Rüstow, but they shared certain convictions. First, they all supported the absence of any parochial nationalism; all favored open markets over protectionism. Second, they were convinced that the interests of the people were important, and third, they believed that spreading wealth among hardworking people could have positive effects. They also thought of taxation as an aid to enable dynamic parts of the population to help themselves. Fifth, they shared concerns over cartelization. And sixth, they viewed the state as above the economy as well as interest groups.[17] Their ideas were domestically grown; none of them engaged in transatlantic exchanges during the Weimar years. Indeed, Röpke studied in the United States for a year on a Rockefeller fellowship in 1927, but was not impressed by what he saw. When Erhard had a chance to put these economic ideas into practice in the postwar years, his ties with other ordoliberals from this group were strengthened, and he received administrative and academic consulting support.

Erhard's career took off with American patronage after the Second World War. He was one of a few competent economists in Germany whose reputation was untarnished even by membership in the Nazi Party and proposed free market concepts similar to free enterprise. Erhard was appointed economics minister of Bavaria, and quickly rose to heading the Special Bureau for Monetary and Currency Matters (Sonderstelle Geld und Kredit), and then the economics administration of the British and American zones of occupation. While the American occupation authorities largely ignored his plan for a currency reform, Erhard, in winter 1948, assembled an academic board to debate the future economic order in Germany. One major question was the role of the state: Should the state steer the economy, or should it just set the framework? After extensive and heated debate, the board made recommendations that suited Erhard's own convictions: immediately limit the role of the state, and free price controls in all areas except for housing and food. Erhard pushed legislative reforms in this regard through the Bizonal Economic Council, the first parliamentary body in the American and British occupation zones, only two days before the currency reform.[18]

Over the next year, Adenauer embraced the social market economy for the CDU's 1949 federal election campaign. Adenauer was no purist about economic concepts; for him, promoting free markets positioned the CDU against its main rival, the Social Democratic Party (SPD), and united Protestants, the bourgeoisie, and social reformers within the CDU, increasing the party's reach from the limited Catholic milieu of the Weimar Center Party. In response to reform critics who said that the "social" in "social market economy" was only cosmetic, Adenauer commented mockingly that it was better than a "bureaucratic planned economy"—marking his own pragmatic as well as cynical approach to economic concepts.[19] Purportedly, Adenaur only proposed the social market economy as the CDU's economic platform for the 1949 campaign, while he officially supported the party's first political program, the Ahlener Programm, which called for socializing core industries, in obvious contradiction with a free market economy. Effectively, the election platform, called *Düsseldorfer Leitsätze*, soon became the party's economic program, although it never passed a party vote.

To Adenauer, embracing Erhard and his economic concept meant forestalling possible coalitions with the SPD at the federal and state levels. With the *Düsseldorfer Leitsätze*, Adenauer positioned the CDU to diverge from the SPD, which still called for socialization.[20] This was a continuation of the early Weimar years, when industrialists had also agreed to codetermination in an effort to prevent more radical forms of economic and political governance. For Erhard, working with the CDU meant siding with a winner and establishing the social market economy, which was still a hotly debated concept in Germany, particularly as prices soared in the months following currency reform.[21] Although Erhard did not join the CDU until 1963—he previously leaned toward the liberal Free Democratic Party—he vigorously campaigned for the party and likely contributed to its slim win over the SPD (31 to 29.2 percent). Erhard thus became the first economic minister of West Germany.

With the social market economy, the young republic gained an economic order that was built on homegrown ideas: it embraced free market competition with state supervision to prevent market failures such as the emergence of cartels. Free competition, the reasoning went, would settle wages and prices. The republic aimed for a more competitive economy than the US postwar economy with its remnants of New Deal planning as well as wartime wage and price controls. And the social market economy was more competitive than the postwar socialization experiments of the British Labour Party. Indeed, Erhard's ideas and personality irked the Allies. American occupation forces, which had allowed Erhard to rise to positions of influence, resented his urge to eliminate market controls—including Allied controls over the German economy. And British diplomats attacked his personality, comparing his "small eyes and diminutive nose" surrounded by an "expanse of gently undulating rosy flesh" to the facial appearance of the Nazi Hermann Göring.[22]

Codetermination Debate in West Germany

In this larger economic context, the newly formed umbrella organization of German trade unions, the DGB, fought for codetermination to protect workers' interests against industrialists, and improve their social and economic status. In Germany, codetermination eventually involved the formation of works councils that participated in business decisions, such as job arrangements, personnel planning, hiring guidelines, social services, time registration, and performance assessments. Works councils also needed to be informed about production changes, such as the introduction of new technologies, although they only had consultative rights in this area. In large companies, codetermination meant that labor representatives were added to the company's board of trustees—separate from the managing board of directors who controlled daily operations—which consisted of bankers and other professionals who helped secure financing and knowledge transfer.[23] The British occupation forces had implemented a form of codetermination in the Ruhr coal and steel industries, and the DGB hoped to expand this to other industries. But codetermination also had roots in German labor traditions. In the German Empire, in the late nineteenth and early twentieth centuries, 10 percent of private companies with over twenty employees and 50 percent of machine tool companies—a new industrial sector—had already introduced voluntary works councils. Codetermination first became law in the aftermath of World War I, when spontaneously formed councils of soldiers and workers threatened more radical forms of government, such as the people's councils of the Russian Revolution. In this situation, giving workers a say in corporate decisions was a concession that helped prevent broader anticapitalist social and economic policies. Weimar legislation implemented three levels of codetermination: works councils at the company level, labor and employer organizations through a Central Working Group (Zentralarbeitsgemeinschaft) at the industry level, and a preliminary economic council at the national level.[24] Although companies often usurped their works councils, and the national economic council was never permanent, the post–World War II German labor movement returned to the Weimar precedent, and in late 1950 and early 1951, a debate about codetermination ensued in the coal and steel industries.

The word "codetermination" was new in English. To translate the German term *Mitbestimmung* into English, "codetermination" eventually won out over "comanagement," "codirection," "joint management," and "workers' participation in management."[25] While codetermination was a literal translation, it did not capture the meaning of *Mitbestimmung* with its reference to the Weimar era dispute over governance through people's councils. Indeed, codetermination, together with planning and socialization, was one of the main goals of Fritz Naphtali's economic democracy. A journalist and union economist, Naphtali envisioned economic democracy, which was embraced by a 1928 union congress in Hamburg, as a step toward a socialist society. In addition to giving workers a say in corporate decisions, codetermination was a step toward ending entrepreneurial autarky.[26] By the post–World War II period, *Mitbestimmung* in German had taken on various connotations including an association with social democrat and syndicalist ideas as well as with social Catholic notions of self-governance. American Marshall Plan administrators, however, primarily saw codetermination as a means of limiting private industrial power; they ignored the fact that codetermination also included, within the German system, the objective of improving workers' social and political status.[27]

Talks between labor and employer representatives about the form of corporate government in Germany, under guidance of the department of labor, proceeded slowly through 1950, with no agreement in sight by the end of the year. In early 1951, negotiations came to a head. Over 80 percent of workers in the Ruhr coal and steel industries voted to walk out on February 1, 1951, if the new West German parliament had not passed a codetermination law by then. Chancellor Adenauer now took charge, presiding over the discussions between labor and employer representatives. While Adenauer was said to personally reject codetermination, he also sought to settle the governance question in the coal and steel industries to enable Germany to participate in the European Coal and Steel Community—the so-called Schuman Plan—for which his government was then negotiating. In late January, an agreement was produced that formed the basis for legislation before the West German parliament.

This prospect alarmed American observers. Throughout January 1951, the *New York Times* reported on the German strike votes and increasingly tense negotiations, and *Business Week* noted that Germany was "headed for a Ruhr showdown."[28] By February 5, the *New York Times* compared code-termination to two landmark achievements in German labor history: Kaiser Wilhelm I's proclamation of labor's full right to strike and the Weimar establishment of works councils.[29] On the same day, the *New York Times* reported on a letter by Earl Bunting, the NAM's managing director, to the German consul general in New York City, Heinz L. Krekeler, stating code-termination "places political goals above economic accomplishments and destroys the fabric of efficient industrial production."[30] Back in August 1950, Bunting had already expressed the NAM's concerns, claiming that under the conditions of codetermination, "collective bargaining becomes a struggle for power, a form of bloodless civil war, in which economic power rather than the ballot box is utilized to create a new form of government," and warned that codetermination would lead to "dictatorship whether it be of the right or of the left" because management would be hindered by the "veto power exercised by specific interest groups," leading to an inability to produce.[31] By 1951, Bunting and the NAM had honed their message, asserting that codetermination would pose a "serious roadblock" to foreign capital investment in Germany, particularly from the United States.

"Two authorities heading any operation," Bunting explained, "have always meant delayed action and compromise decisions resulting in wasted time, effort and material."[32] Combining the rhetoric of productivity with Cold War concerns, Bunting claimed that it was "generally agreed that increased production is one of the most urgent requirements in all democratic nations if we are to attain the high standard of living desired by all of the people, as well as to gird ourselves against the aggressor." Analysis of successful business, he continued, indicated "the need for managerial freedom to steer a business through the complex problems of industrial production" that required management to be "free to exercise flexibility and speed of decision as well as have the authority necessary to accomplish results."[33] Bunting and the NAM were immediately publicly accused of interfering in international affairs by Lewis Galantiere, a writer and member of the Council on Foreign Relations. Galantiere called Bunting's correspondence "brutish" because it gave the impression that the United States was dealing with its Allies in the same way that Russians were commanding their satellites—with a kind of economic imperialism—a charge led by both Communists and non-Communist anti-Americans. Galantiere wrote, "American business does itself a disservice when it comes roaring, bull-fashion, into the diplomatic china shop."[34]

American Business Intervention in the German Debate

Despite accusations of interfering in international affairs, conservative US business organizations remained involved in the codetermination debate. In late March 1951, when the West German parliament debated the legislation, the NAM even sent a delegate to Europe: Haynes, the chair of the NAM's Committee on Europe. Haynes worked closely with Michler, a Standard Oil executive and chair of the NFTC's Committee on Germany. After conversations with the Allied occupying authorities, German bankers, government officials, employers, and officials from the DGB, Haynes realized that some form of codetermination would pass. So Haynes and Michler focused their efforts on two goals: first, to ensure that the eleventh member of a company's board of trustees (who would break a tie) be determined by the shareholder representatives—in their words, "save the eleventh man for the shareholders"; and second, to have all the labor representatives on the board be elected from and by the plant workers, not by outside organizations, such as trade unions.

On April 3, 1951, the day before the second of three debates in the parliament, Haynes and Michler held a press conference in Bonn. In a joint statement of the NAM and the NFTC, they said that they did not wish to "impose" their views on the German people but instead believed that "exchanging facts and opinions between free German citizens and free American citizens is mutually beneficial." They warned that "reducing stockholder representatives on the Board to less than 50 percent is a step to Socialism." If the number of shareholder representatives on the board of trustees was reduced to five out of eleven, private capital would be "deprived of its management rights." Capital investors, from Germany and elsewhere, would seek investment opportunities outside Germany, forcing industries to pursue needed capital from the state, which would lead to state ownership. They also warned that union officials serving as members of boards would interfere with their capacity to represent employees, and would lead to a labor monopoly in that industry and even across industries. They concluded that there were "other tested and proved methods" to achieve "mutually beneficial high productivity, lower prices, higher wages, and a constant flow of capital needed for expansion and modernization."[35]

Haynes had traveled to Europe as the representative of an organization that championed free enterprise. Because the NAM represented mostly small and medium-size businesses, where labor often represented a significant cost, it was more conservative than other US business associations; it also was a main player behind a free enterprise campaign in the United States.[36] At the same time, the NAM encouraged a realistic approach to collective bargaining, recognizing that it was here to stay. The association publicly endorsed collective bargaining, calling for negotiations to be "conducted in a manner which promotes mutual understanding between management and workers and which fully protects the public interest."[37] The NAM accepted collective bargaining as better than government-determined wages and working conditions, but also wanted to limit collective bargaining to wages, hours, and conditions of employment, preferably on the basis of an incentive pay system. Moreover, the association opposed any expansion of collective bargaining to other issues (figure 6.1).

Although not a typical NAM member, General Motors practiced the association's recommendations on plant-level collective bargaining. Rather than reacting defensively to unions, the company took initiative to demand that the union act responsibly, discontinue attacks on the company, and agree to other changes such as the introduction of piecework and pay differentials as well as promotions based on a merit versus seniority.[38] In 1948, General Motors gained additional stability through the first multiyear contract with the United Automobile Workers that tied automatic wage adjustments to the standard of living and productivity.

With plant-level collective bargaining established in the United States by the late 1940s, the NAM now railed against industry-wide collective bargaining, calling it a labor monopoly and charging that it would lead to socialism.[39] In May 1949, the NAM board of directors demanded that employers and unions that engaged in bargaining practices involving more than one plant "be subject to statutory regulations" because such monopolies were "inimical to the public welfare."[40] The NAM argued that because labor was a significant cost factor, companies were to compete on their labor costs as they did on other manufacturing costs; any attempt at fixing labor costs would be reflected in the price of a product and therefore was illicitly interfering with free competition.

Labor Relations, NAM Style

Mr. Ruffin and His Cotton Mill

Figure 6.1
This cartoon by the Textile Workers Union of America accused William H. Ruffin, the NAM president, of a hypocritical stance toward collective bargaining. While Ruffin publicly praised the high American standard of living, the cartoonist accused his textile company of paying below-average wages and installing fountains on street corners rather than running water in company housing.
Source: Textile Workers Union of America, "Labor Relations, NAM Style." Courtesy of Hagley Museum and Library.

Because industry-wide collective bargaining implied a threat of national strikes, and the federal government was the only agency powerful enough to counter industry-wide unions and employers, the NAM contended that it would lead to "control and direction by government of production, prices and profits."[41] The NAM interpreted federal fact-finding commissions during the 1946 strike wave as evidence of this trend. Leo Teplow, one of the NAM industrial relations experts, argued that industry-wide collective bargaining would introduce class conflict to the classless US society. Once collective bargaining reached the national level and, because of its national proportion, the federal government got involved, the importance of the labor vote would introduce political sensitivities to the negotiations. Instead of resolving economic disputes, this would "bring them to the ballot box," and "divide people politically and economically into classes having diverse interests." Teplow charged that industry-wide collective bargaining would turn an economic question into a political one, alleging that "there are some labor leaders who endorse industry-wide bargaining because it is bound to result in government intervention and possible government nationalization of industry"—or in other words, a socialist society. Teplow warned particularly of this "long-range sociological influence."[42]

In the United States, unions negotiated contracts for one in six unionized workers in collective bargaining sessions with more than one employer— that is, "multiemployer collective bargaining," the term preferred by economic analysts because these bargaining constellations did not necessarily involve a whole industry. Only the coal and railroad industries practiced national industry-wide collective bargaining that covered more than 90 percent of unionized workers. The pottery and glass industries also practiced national industry-wide collective bargaining, but contracts covered only 60 to 80 percent of unionized workers and also excluded nonunion workers. Industries concentrated in a geographic region, due to natural resources or access to water, such as fishing, longshoring, metal mining, and lumber, sometimes practiced regional collective bargaining with multiple employers. And frequently, smaller companies engaged in local or citywide multiemployer bargaining because they found it easier for a business association to conduct contract negotiations than doing so themselves. In San Francisco, for example, multiple industries conducted citywide collective bargaining to prevent the strong local unions from gaining concessions from one enterprise and then extending them to others.[43]

Haynes thus came to West Germany as a representative of an organization that championed free enterprise and fought regional collective bargaining practices. To him, codetermination posed the same problems as industry-wide collective bargaining because it extended the influence of labor unions directly into managerial decision making. Not surprisingly, German labor unions were incensed by Haynes and Michler's intervention.

Figure 6.2
This cartoon equated the NAM intervention in the German codetermination debate with a torpedo targeting the debate.
Source: Welt der Arbeit, April 6, 1951. Courtesy of Friedrich-Ebert-Stiftung.

The DGB released an indignant press statement, calling it "unnecessary to take a factual position to a tendentious"—that is, biased—"release which every decent German and every democrat must reject as a stupid interference of private foreign circles in the legislative measures of a foreign country." A few days later, the DGB press organ followed up with a cartoon showing a torpedo labeled, "Appointed interference of foreign enterprises," heading for a ship called "Codetermination."[44] Haynes and Michler had struck a hornet's nest, and it immediately backfired on them, as they had been warned. In addition to possibly helping pass the codetermination law, they made the climate more difficult for US Productivity Program officers. As a labor officer reported, Haynes and the NAM were "poison to the German Federation of Trade Unions," and for a while the officer found it impossible to lead conversations with German unionists without being asked whether Haynes's actions had been approved by the US administration.[45]

US Labor Unions and the US Government on Codetermination

US trade unions, which had fought for plant-level collective bargaining, did not immediately embrace the German drive for codetermination. Rather than pursue legislation that protected workers' rights, US trade unions sought to make improvements by negotiating with employers; their focus was mostly on wages and grievances, and they shunned any participation in managerial responsibilities, maintaining a clear separation between management and labor. As we saw in chapter 4, AFL and CIO officers served as labor advisers in Europe for the Productivity Program in an effort to convince European workers of the merits of the American model and lure them away from Communist promises. US trade unions thus supported the Productivity Program's model of collaborative labor relations, and helped implement it in West Germany and other European Marshall Plan countries.

Still, in late January, when the debates between West German employers and unions were coming to a head, and a strike appeared imminent, US trade unions only cautiously backed the German unions' call for codetermination. On January 25, 1951, the CIO issued a press release stating that CIO president Murray had urged US high commissioner John McCloy to use his "good offices to avoid the threatened industrial conflict over the issue of co-determination." The demands for codetermination were a legitimate aspiration of West German unions with which the CIO could "sympathize," although "just as the industrial problems of Western Germany are different from our own, so are the objectives of our trade unions." Murray warned that a strike would "weaken the West German free trade union movement and . . . strengthen the Soviet-Controlled Communists." He concluded that "the objective of the free German trade union movement is to prevent the return to power of the old German industrialists with Nazi sympathies," and that this objective was "one which the American people share." Therefore, the CIO called for "friendly mediation" from the US high commissioner.[46] Unlike the NAM and the NFTC, the CIO thus worked through diplomatic channels, rather than making public statements directed at Germans and parliamentarians, or lobbying them in private. But German unionists appear to have been disappointed by the US unions' "lack of understanding and the lukewarm attitude."[47]

Clinton Golden, one of the main proponents of collaborative collective bargaining in the United States and the former chief labor adviser for the Marshall administration, visited Germany during the codetermination debate.[48] Golden concluded that "co-determination stemmed less from a desire to participate in management than a distrust in management" because German unionists felt that industrialists were primarily responsible for Hitler's rise to power, and their influence therefore needed to be checked by labor unions to prevent a repetition of history.[49] Notably, Golden situated the demand for codetermination in the context of the Nazi regime rather than the larger history of class conflict in Germany. Like Golden, other labor advisers claimed that codetermination was born out of "fear, uncertainty and suspicion," not a "lust for power." "Fear of managerial actions that may adversely affect the workers' livelihoods" and the "need for protection from an economically stronger employer" led to demands such as veto power in issues of fundamental interest to labor along with the right to see balance sheets and profit-and-loss statements.[50]

US unions supported the right of German unions to demand codetermination, and asked the US government to back these demands. At the same time, Murray stopped short of endorsing codetermination in the press release, making it clear that the CIO had a different position on corporate governance.[51] And beyond press statements, US unions did nothing to support the DGB demands. The CIO and the AFL both sought to achieve labor goals through collective bargaining rather than legislation. Despite the CIO's rapprochement with the federal government since the New Deal, US unions sought legislative support only to secure their right to bargain collectively, not for the government to be involved in determining labor conditions. In addition, they sought their independence in collective bargaining and therefore shunned any participation in corporate decision making.[52]

In the end, the US government remained neutral with regard to codetermination. Officers of the Productivity Program had advocated for the US model of collaborative labor relations. Although codetermination posed a rejection of this US model and presented a return to German corporatist cooperation, the US government did not seek to influence the codetermination debate. Even before receiving Murray's cable, McCloy had already "issued a clear-cut statement that the US would follow a policy of strict neutrality and that the American Authorities would not take any action which would prejudice the case for or against co-determination since this was viewed as strictly a local German matter."[53] This neutral position was consistent with previous policies and actions on codetermination by US occupation authorities. While the British occupation forces had relied on works councils in the coal and steel industries—a de facto form of codetermination that the DGB hoped to preserve through legislation—McCloy had temporarily suspended codetermination legislation and effectively delayed the recognition of works councils in the American occupation zone until 1950.[54] This official neutrality may have spurred the NAM and the NFTC to take action rather than rely on diplomatic channels. In his response to Galantiere's criticism of interfering in foreign affairs, Bunting declared that the NAM need not limit its statements on matters beyond the United States because it could make remarks that the US Department of State, "as an official representative of government, might not be able to make," and because it could hold views that were contrary to the State Department ones.

Outcome: Codetermination and Corporatism

The German parliament eventually passed codetermination in the German coal, steel, and iron industries on April 10, 1951. As Meyer Bernstein, a US labor adviser in Germany, reported, the final debate was raucous. Around six in the evening, the Social Democratic Party delegates realized that they had voted for a part of the law that they didn't agree with. More debate followed, interrupted by several recesses, and the erroneously passed passage was retracted; shortly before midnight, delegates voted. The final version of the law was read aloud to delegates instead of being distributed in written form.[55] In the coal and steel industries, a company's board of trustees now was to be composed of eleven members: two to be selected by and from the company workers, two selected by the respective trade union, one public person related to labor (usually a labor-friendly government official), four shareholder representatives, and one public person related to employers. Thus labor achieved parity representation, with five out of ten board members related to labor. The eleventh member, whose vote would break a tie, was to be elected by the previous ten. Should there be a deadlock, an industry-wide senate or court was to propose three additional candidates, and should these not find a vote within thirty days, the shareholders could vote for the eleventh person. By the following summer, the NAM called these regulations "window dressing" and rejoiced that the eleventh man was "saved for the shareholders"—meaning that shareholders only had to hold out long enough to get their candidate. The NAM, however, lost its second cause: to have all labor representatives on the board of trustees be elected from and by the plant workers, which assumes that they would be motivated to advance a company's competitive position rather than advance labor solidarity beyond the company. Instead, representatives from outside a company, including union members, became part of the board of trustees.

By early December 1951, the DGB announced its goal to expand codetermination to other industries; this was at a time when many major West German industrialists were out of the country visiting the United States through the Productivity Program's Project Impact discussed in chapter 4—a fact that was not lost on some observers. In July 1952, the parliament extended codetermination to other industries through the Corporate Governance Law (Betriebsverfassungsgesetz). Yet the DGB was forced to agree to a weaker form of codetermination than in the coal and steel industries: only a third of a company's board of trustees were to be labor delegates. And only companies with over two thousand employees—not one thousand employees, as in the coal and steel industries—had to have works councils.

In the strong or weak form, codetermination differed significantly from the US model of labor relations. The DGB implemented workers' rights through federal legislation, while US trade unions shunned government regulation, focusing instead on improvements through collective bargaining. German workers expected that participation in managerial decisions meant that their interests would be included in corporate decisions, even as US trade unions feared that it would compromise their bargaining position. German trade unions sent external representatives to corporate boards, while in the United States, collective bargaining procedures were mostly internal to a company. West Germans resisted adopting the US model of labor relations, regardless of the Productivity Program officers' attempts at bringing collaborative labor relations to Europe, and regardless of the NAM and the NFTC's intervention.

Some historians have argued that the codetermination debate was the first step in reintroducing corporatist elements into the West German economy by allowing corporate entities like labor unions and business associations—versus individual actors or the state—to make economic decisions, sometimes at the expense of consumers.[56] Corporatism had a long tradition in Germany, stretching from the German Empire through the Weimar Republic to the Nazi regime. In the West German Republic, having external labor representatives on the board of directors—if not on the works council—was to avoid labor representatives from being co-opted by management and serving company goals rather than acting as an independent interest group, as had happened with many weak works councils during the Weimar years, yet it also introduced a corporatist element. Even if the forms of labor relations in West Germany had, at the most, weak corporatist elements, they still deviated from Erhard's original vision of the social market economy and US productivity ideas. Erhard wanted free individual or corporate actors, not labor unions or business associations, to make decisions. Similarly, free enterprise and collaborative labor relations foresaw plant-level collective bargaining, not works councils. The "real existing" capitalism in West Germany thus differed from both the pure social market economy and free enterprise. If not a full-scale corporatism, it at least introduced corporatist elements in lieu of free competition.[57]

Not only may Haynes have inadvertently helped pass codetermination, but his actions threatened the goals of the Productivity Program. Meeting with German union officials, he had presented himself as an editor of a NAM magazine gathering information. While not incorrect, labor officers contended that this pretense would make German union officers more suspicious of the ulterior motives of American officers. Even worse, Haynes had come to Germany in a second function: he, together with Miles Standish from the National Management Council, was tasked by the Marshall administration in Paris with organizing a productivity seminar for coal and steel industrialists. Labor officers immediately protested Haynes's dual mission as "impossible" and demanded that he withdraw from the second one. While Haynes may not have stepped down, his singular focus on codetermination ensured that this second task did not become public knowledge in Germany. US and German officials were still negotiating the German Productivity Program, and it was not until April 1952 that the DGB conditionally agreed to participate in the program.[58]

After the codetermination vote had passed, US officers realized that they needed to let things cool off before taking any further steps. Based on the Weimar experience, the DGB had concluded that the success of codetermination depended on the "personalities, abilities, and the knowledge" of labor representatives to prevent management from undermining codetermination, and therefore it planned to train its officers as well as the labor representatives elected by the works councils.[59] It was in this area that US officers saw an opportunity to help. Golden had originally recommended that the Productivity Program send a US consultant team to Germany to help train labor representatives for their codetermination function. Yet labor officers in Germany protested that such a step would be seen as too interventionist because "the blood is still running hot"; this was late March, even before the public statement by Haynes and Michler. US labor officers in Germany argued that the German side needed to request their help because US offers of assistance would be misconstrued as an "attempt to fashion co-determination after the American model."[60] They also maintained that rather than sending a US consulting team to Germany, German officials should be sent to the United States to "let them see for themselves what we [Americans] have."[61]

Conclusion

In the end, codetermination proved not to be the revolutionary change in labor relations that German and US observers hoped for or feared during the codetermination debate in 1951. "Codetermination has not brought about any revolution," Frederick Harbison, the director of the Princeton Industrial Relations Section, concluded in 1956 on the basis of a two-year study of the practices of codetermination in Germany, "but it has given workers a means of making their grievances and desires felt at the plant level." In practice, labor and management divided their spheres of influence, with labor given a say in questions of wages and working conditions, and management retaining its authority over operational and other decisions. As such, the study showed, codetermination did not follow the course expected by either labor or management. Rather, the study's author, sociologist Michael Blumenthal, stated, "Codetermination did not solve as many problems as predicted, nor did it create as many as were prophesized."[62] But codetermination created the institutional framework for workers and labor unions to respond to new technologies, including computing and automation technologies in the 1950s.

For this reason, understanding codetermination is important for comprehending German responses to technological change and productivity technology. If labor representatives on the board of trustees of a coal or steel company wanted to oppose investments in new technologies, they would only have to convince one employee representative on the board. In other industries, where labor only had a third of the members of the board, such a vote would be more difficult. In addition, the works council had to be consulted on changes in production technologies. As we saw in chapter 2, labor unions already embraced rationalization technologies in the 1920s because they had hoped to ease the labor burden and spread the wealth of higher productivity; we will see in chapter 8 that the DGB continued this position on technological change after World War II. Thus the union representatives on corporate boards of trustees—one of the aspects of codetermination that Haynes and Michler had fought so vigorously—proved not to be an obstacle, at least not with regard to new computer technology. Resistance to new technologies, including electronic computers, came from individual workers who were afraid of losing their jobs, and the German press vigorously added to their concerns.

The codetermination legislation also revealed the limits of shaping German economic and labor relations after the US model. Among West European Marshall Plan countries, Germany, with its delegitimized elites and destroyed infrastructure, promised the best chance of implementing US productivity ideas, including mass production and distribution, free enterprise, and collaborative labor relations. Yet rather than following the US model, Germans turned to their own homegrown ordoliberal ideas of free competition along with its tradition of works councils and corporatist economic decision making. In his analysis of productivity politics, historian Charles Maier considered America's policy a "resounding success" that convinced the DGB to accept a more circumscribed codetermination law than it had originally called for.[63] Any form of codetermination, however, be it strong or weak, represented a departure from the model of US labor relations, based on plant-level collective bargaining, which Productivity Program officers brought to Western Europe. The resulting system of corporate governance in Germany thus presented a complex amalgam that resembled neither the US model nor the German precedent.

The US concept of free enterprise and German theory of the social market economy are both based on competition. Yet each system accorded different roles to the state. The German fathers of the social market economy acceded an ordering role to the state to create a framework that protected free competition. US free enterprise proponents, by contrast, sought liberty from state intervention, at least in principle. In practice, however, each economic system came to incorporate tensions between contradictory forces not foreseen in theory. In Germany, codetermination legislation acceded roles to corporate entities such as labor unions and business associations in the free market economy, introducing corporatist elements that favored cooperative agreements over competition. In the United States, the 1890 Sherman Act had already created a de facto role for the US government in protecting free competition. In addition, New Deal policies as well as wartime wage and price controls—revived during the Korean crisis—introduced elements of planning to the US economy that contradicted the idea of free enterprise. The notion of the capitalist "West" commonly serves as an identifier of countries unified by geopolitical interests. But this chapter suggests that the "West" denotes an alliance of countries that heatedly debated their economic models and implemented different forms of industrial relations. The variety of economic systems persisted rather than European economies conforming to the emerging US model.

Regardless of the complexities and contradictions within the US model, Haynes and Michler sought to introduce free enterprise in West Germany. While Marshall officers created demonstration projects to expose Germans and Europeans to the benefits of nonrestrictive business practices and collaborative labor relations, Haynes and Michler looked to other actors to highlight free enterprise: US corporations operating in Germany and Europe. Haynes and Michler declared that the maxim "When in Rome, do as the Romans do" was "not the way to spread the American gospel of competitive, free enterprise." They said that many US corporations were "known to be leaders toward lower prices, higher wages, and improved living standards," but they conveniently left out the Productivity Program's idea of collaborative labor relations. "Every US corporation in Europe," Haynes and Michler continued, "has an opportunity to adopt the efficient methods and equipment common in America and thus demonstrate the soundness of the system which we cherish and for the defense of which we are willing to fight."[64] In the next chapter, I will look at one of these companies, IBM, and investigate how far it implemented American productivity methods and values in its West Germany operations.

7 IBM: An American Corporation in Germany

In April 1950, IBM's longtime chief executive Thomas J. Watson Sr. had a heated phone conference with his company's main executive in Europe, Harrison Chauncey, about a newly formed works council in the company's West German subsidiary. While Chauncey responded matter of factly, Watson was indignant. He demanded that management and employees in Germany get "on a basis like before," and abolish the works council, while Chauncey merely tried "to get the agitators out of the Works Council." Watson wanted it to be made clear that IBM was "not going to back Communism under any conditions"—revealing what he thought of workers committees. Chauncey, closer to the European situation, said that he did not believe IBM "had many Communists in the organization" and pointed out that members of the works council were by law protected from being released, which made Watson ask, "What's the use of going on with the business?"[1]

IBM was among the US multinational companies in Europe that Eldridge Haynes and Gordon Michler wanted to hold to US productivity ideas. In 1950, IBM was not yet the global behemoth that it would become in the 1970s, when the company dominated computer markets, holding more than 70 percent of the US computer market share by value and over 60 percent in every Western European country except the United Kingdom.[2] In the early 1950s, when Watson talked to Chauncey, the company was still only carefully exploring electronic technologies. But it had sold office machinery, including punch card machines, in Germany since the early 1900s under licensing agreements with the Deutsche Hollerith-Maschinengesellschaft (Dehomag). And the company had reorganized its foreign operations, including its operations in West Germany, under a wholly owned subsidiary, IBM World Trade Corporation, only one year earlier, in 1949. Within the next few years, IBM moved into electronic computing, built electronics laboratories in Europe, including in West Germany, and grew to be the main manufacturer of the new productivity technology, electronic computers. IBM thus became an increasingly visible poster child for US productivity and business methods.

Yet as this brief excerpt from Watson's phone conference with Chauncey indicates, the company's labor relations were far from the Marshall Plan's model of collaboration based on plant-level collective bargaining; instead, the company practiced welfare capitalist labor relations. Considered progressive compared to the violent suppression of labor unions, welfare capitalism met employee needs by offering various corporate benefits in return for foregoing union organization or accepting a company union.[3] IBM was a latecomer to welfare capitalist policies and implemented measures only in the 1930s. As David Stebenne has shown, the company introduced welfare capitalist measures during the Great Depression, when many other companies turned away from this approach. Like other welfare capitalist companies, IBM now paid above-average salaries, built bright and air-conditioned production facilities, provided training and recreation facilities, and supplied generous insurance programs—all with the goal of avoiding union organization.[4] Similar to other welfare capitalist companies, IBM's corporate culture was highly focused on the chief executive who posed as the benevolent originator of these often-paternalistic measures.

In his company's international operations, Watson Sr. sought to implement the same corporate culture and labor relations in the foreign subsidiaries—an attempt that frequently sat uneasily with local practices and cultures. For example, in Latin America, not offering alcohol at corporate functions was considered offensive, and IBM's dress code—dark suits, button-down shirts, and ties—evoked associations with Mormon missionaries.[5] When West German IBM workers formed the works council in 1950, such bodies were not yet protected by law. Works councils, however, had been provisionally recognized in the US occupation zone where IBM was located, and American military authorities advocated for an expansion of works councils' rights, although their eventual status still awaited the passing of codetermination legislation in 1951 by the German federal government, as we saw in chapter 6.

This chapter looks at Watson's attempts at implementing a homogeneous corporate culture and labor relations across his company's international operations. It argues that Watson's welfare capitalist approach favored decisions by managerial fiat rather than by engaging in democratic decision making with workers in cooperative industrial relations. With his wife, Jeanette Kittredge Watson, by his side, Watson Sr. used the gendered rhetoric of an "IBM family" to make such welfare capitalist labor relations palatable to West German workers. While the socialist labor movement shaped the larger national context in Germany, a closer look at labor relations in the West German subsidiary reveals that Watson's family rhetoric fit nicely with the local tradition of Christian labor organizations in Swabia, where IBM's main manufacturing facilities were located. German workers and executives brought their local perception of class relations as well as their expressions in everyday choices, habits, and consumer practices to IBM's welfare capitalist relations, reforming them according to local traditions. An American productivity company, IBM thus offers a close focus on transatlantic labor relations.[6]

This chapter first looks at IBM's corporate background and the organization of its international operations. It then analyzes Watson's rhetoric of the IBM family with its gendered implications, and how it shaped labor relations, and made paternalist labor relations palatable in Germany within German labor traditions and class relations. Then it examines the IBM works council and its relation to Watson. The chapter concludes with an analysis of IBM's exceptional step of abolishing distinctions between blue- and white-collar workers within the company, while also making these distinctions symbolically visible through the introduction of uniforms. The chapter explores the complex labor and class relations at IBM, and how they fit in the German context. It also shows that IBM veered far off from the goals and ideals of the Marshall Plan.

IBM's Progressive Corporate Culture in the United States

While it is impossible to reduce a corporate culture to a single person, this chapter will nevertheless concentrate on Watson because of the central position of corporate owners in welfare capitalist firms, and because of his charismatic personality and how he used it to shape IBM's corporate culture. One of the signs of the paternalistic culture at IBM was its songbook. At company events, for example, employees regularly sang "Ever Onward," the IBM song that proudly boasts of Watson as the "man of men," the "sterling president" whose name meant "courage none can stem." One of the songs dedicated to Watson describes him as an inspiring leader as well as a fatherly figure, a "friend so true," who increases the joy of his "boys"— emasculating grown men as corporate sons— and made them "smile, smile, smile."[7] Visits to foreign subsidiaries and other celebrations heightened this cult of personality.

Watson came from simple roots. Born in 1874 on the family farm in upstate New York, he grew up as the youngest of five, and the only son in a family facing many setbacks in the farming and lumbering business. He failed at several jobs as a peddler and salesman before joining the National Cash Register Company (NCR), where he learned salesmanship and then turned around the struggling Rochester, New York, branch office. He ran an (illegitimate) undercover used cash register business to drive out the competition and quickly moved up the corporate ladder.[8] At NCR, he experienced welfare capitalism as practiced by the company's head, John H. Patterson. In 1912, Watson was one of thirty top managers at NCR, along with Patterson, found guilty in an antitrust suit.[9] Within a year, Watson left NCR and soon became the chief executive of Computing-Tabulating-Recording Company in New York, a newly formed public company, which he later renamed the International Business Machine Company, or IBM. Watson led the company for more than forty years, until passing the chairmanship to his oldest son a few months before his death, at age eighty-two, in 1956. He transformed the struggling and disjointed company into a global corporation focused on office machinery. Along the way, through his charismatic, somewhat-eccentric, and abrasive personality, he shaped the culture of the entire corporation.

During the interwar years, Watson pursued progressive corporate policies. The late nineteenth-century global economy, dominated by the colonial operations of British imperial firms, collapsed in the period between the two world wars, with companies turning to cartel and price-setting agreements instead of free market interactions.[10] During the 1920s, a time of American political isolationism and corporate protectionism, Watson extended IBM's operations beyond the United States to build networks in South America and Europe, and then reach into Africa and Asia after World War II. A staunch internationalist, Watson took his first trip to Latin America in 1919 and expanded the company's foreign operations in the following two decades. During the 1930s, he also served as the president of the International Chamber of Commerce (ICC), an international association of businessmen, financiers, and traders to promote free trade and eliminate existing trade barriers, which Watson encapsulated in the slogan "World Peace through World Trade."[11]

At IBM, Watson implemented the kind of welfare capitalist labor relations that he had experienced at NCR. In the company's early years, he was unable to implement many measures, but he reportedly refused to lay off employees, even against advice. Although he was not technically the founder or owner of IBM, a publicly held company, his role as longtime chief executive resembled that of many owners who implemented welfare capitalist measures in their own companies.[12] During the Great Depression, when many companies let employees go in large numbers, IBM fared comparatively well. After an initial wave of layoffs in 1932, IBM garnered enormous data processing contracts for the Social Security program, providing a windfall that allowed for the implementation of welfare capitalist measures such as air-conditioned production facilities, replacing piecework pay with hourly wages, paid holidays for all employees, and life, accident, and family hospitalization insurance.[13] In return, the company expected employees, during a decade of union-friendly politics and organization drives, to refrain from organizing into unions.

Anecdotal sources have described Watson as a gender progressive who promoted women's professional careers within IBM during the interwar years, when large companies usually relegated women to secretarial positions. In 1935, Anne van Vechten, an acquaintance of Watson's oldest daughter, Jane, at Bryn Mawr, reportedly asked for a personal interview with Watson and challenged him to explain why large companies did not offer professional opportunities to women. Watson hired her and twenty-four other young women for the first all-women class at the IBM school, where new recruits, usually hired directly out of college, were introduced to IBM's operations and culture. Watson's decision did open career paths to women, but these paths were limited to positions such as customer training and human relations, and IBM still expected women to resign from salaried employment on marriage. Apparently Watson took interest in the women, and frequently attended their almost-weekly parties, which included the men at IBM's school.[14] In the following years, Watson often surrounded himself with young professional women, taking them out for luncheons or on trips, and at least once buying a group new shoes after a rainy outing in New York City.[15] But this progressive image was not what the Watsons presented to the employees of the West German subsidiary and elsewhere around the globe.

IBM's International Operations

After World War II, IBM reorganized its global operations. Previously, IBM had mainly granted licenses to foreign-owned and foreign-operated companies. The company's operations in Europe dated back to its predecessor companies in the early twentieth century.[16] One of them, Hermann Hollerith's Tabulating Machine Company, granted a license in 1908 to Ralegh Phillpotts in Great Britain for the exclusive right to sell Hollerith machines in Britain and the British Empire, and a license to Willy Heidinger, who founded Dehomag for sales in Germany.[17]

By the 1920s, IBM had operations in all major European countries, including France, Italy, Austria, Hungary, and Switzerland. IBM opened manufacturing facilities in Sindelfingen, Germany, in 1924, in Vincennes near Paris in 1925, in Berlin in 1934, and in Milan in 1935. The foreign licensee companies were independently owned, funded through local capital, and paid licensing fees and patent royalties to IBM. Cash flow problems frequently meant that they were unable to meet their royalty obligations, however, and constantly struggled to maintain their independence. During one such crisis in 1922, Watson acquired 90 percent of Dehomag shares for IBM and thus controlled the German company. Still, the foreign companies mostly operated independently; some of them even undertook their own research and held their own patents. For example, in the 1930s, Dehomag obtained a patent for the D11, a tabulating machine with a printing function.

In 1949, IBM reorganized its foreign operations; company lore has it that this was to create a position for Watson's younger son, Arthur K. Watson.[18] Yet given Watson Sr.'s business acumen, the significance of IBM's foreign operations, and the need to justify such major changes to the board, one has to assume that other factors played a role in the corporate reorganization, such as that foreign operations now were subject to US tax laws, a parent company could now claim tax benefits from foreign operations, and foreign operations would now be held accountable to US courts.[19] In addition, the US government's free trade policies may have influenced IBM's decision.[20] The reorganization meant that first, IBM was no longer tied to individually negotiated licensing agreements and could deal with any foreign subsidiary on the same basis, and second, that IBM had more direct control over its foreign operations. Instead of granting licenses to foreign corporations, IBM now organized its foreign operations in a wholly owned, independently operated subsidiary: the IBM World Trade Corporation. With this step, IBM recognized the significance of its foreign operations.[21] Watson Sr. announced the creation of the World Trade Corporation in a global telephone conference, the first of its kind, connecting IBM employees from facilities worldwide.

True to the company's pompous style, the conference began with a performance of the IBM song "Ever Onward" by the Endicott IBM choir. Watson Sr. then addressed the "IBM family," emphasizing the internationalist motives guiding the creation of the IBM World Trade Corporation.[22] Watson urged his employees to continue contributing to their local communities because "we cannot be good business people unless we are first good citizens." Watson extended "good citizenship" to mean supporting the United Nations, and urged his employees not to "sit back and criticize" but to "think instead of the accomplishments of the United Nations." This announcement took place on the day of the cornerstone-laying ceremony for the UN building in New York, which Watson attended in the afternoon. Watson expressed his full support of the United Nations, and expected it to "create a world in which each country and its people can live in freedom and independence, with respect for their rights and for the rights of other countries and peoples." He tied the business of IBM to the goals of the United Nations when he argued that the World Trade Corporation "furnishes us an agency through which we can all work, not only to advance our own interests, but to further the great objectives of the United Nations and help make 'World Peace through World Trade' a living, lasting thing." He closed by asserting that the "United Nations means more to me than hope for peace. I have the conviction that it will develop policies that will bring peace, happiness, prosperity and security to people in every country." In other words, the Marshall Plan's goal, extended to the whole world.[23]

The board of trustees of the newly formed company included only one woman: Watson's wife. Born in 1883 to a prominent railroad family in Dayton, Ohio, Jeanette Watson graduated from Wheaton College in 1902, and in 1913 married Watson, nine years her senior. Apparently the two connected over not drinking alcohol at a social club function.[24] Within a year, the couple moved to New York for Watson's new position, and between

1914 and 1919, they had four children. Mrs. Watson ran the rapidly growing household with the help of one or two servants, making it a place of ref-
uge and regeneration as well as family fun. The Watsons often entertained
business guests, with many of them brought home unannounced by Watson Sr.—apparently a strain on their relationship and the reason why Mrs.
Watson asked for a divorce in 1929, although she stayed in the marriage.[25]
The Watsons liberally mixed private and corporate life, with Mrs. Watson
increasingly accompanying her husband on business trips, and the whole
family coming along for extended visits to Europe every few years.[26]

Mrs. Watson frequently traveled with her husband—over thirty thousand miles in the year preceding the 1949 phone conference—who often
emphasized that he could not have built IBM without her support and
advice. Her most concrete technological contribution may have been the
suggestion for a remote control keyboard for typewriters, inspired by visits
with injured veterans and in hospitals.[27] More generally, she is recognized
as a corporate mother who shared in the concerns of employees and their
families. While she usually remained silent in public, the phone conference
was one of the rare occasions when she extended a brief address to IBM
employees worldwide. She said that she was excited to speak and felt hon-
ored to be a member of the board of trustees; she offered her greetings to all
employees and their families.

Mrs. Watson's presence within IBM allowed her husband to present himself as the "pater familias" who provided good care for the members of his
own as well as the IBM family. A photograph, published in IBM's new global
employee magazine, of the board during the 1949 phone conference shows
Mrs. Watson seated next to her husband at the head of the table in a circle
of fifteen men. While the men looked directly at the camera, she looked to
the side and bowed her head, indicating her largely supportive role. Unlike
women office workers or the young professional women in IBM's training
class, Mrs. Watson wore a dark blazer that blended in with the men's suits,
yet her hat, the only head covering worn in the room, marked her as an
accessory (figure 7.1).[28]

Watson's IBM Family in Germany

During their visits to corporate subsidiaries abroad, the Watsons appealed
to conservative family values, and Mrs. Watson became an important part

Figure 7.1
Board of Trustees of IBM's World Trade Corporation during a 1949 global phone conference announcing the reorganization of the company's international operations. *Source: IBM World Trade News* (October 1949): 2. Courtesy of International Business Machines Corporation, © International Business Machines Corporation.

of the presentation of the company's gendered culture, providing an emotional and caring complement to her charismatic yet sometimes-brusque husband who, for example, did not shy away from publicly criticizing IBM executives.[29] In summer 1953, she accompanied her husband on a European trip that included a multiday stop at IBM's factories in Böblingen and Sindelfingen, Germany, during which Watson met with the German works council and sought to reinforce welfare capitalist labor relations. Mrs. Watson was present at the meetings, although she usually stayed in the background. Rather than going shopping or visiting museums and sights, she chose to spend considerable time with IBM employees. In the afternoon, she and her husband individually greeted about thirty-six hundred employees and their spouses. Watson stressed that he and his wife always enjoyed meeting employees personally—although pictures reveal that Mrs. Watson wore long white gloves on the greeting lines, suggesting that she may have loathed the close physical contact. Watson also acknowledged his wife's work for the company, and in her name expressed thanks for the set of mocha coffee dishes that the works council had presented on behalf of all IBM employees.[30] Clearly the employees related to her as a motherly figure, giving her gifts as a sign of appreciation.

Mrs. Watson presented a gendered image customary for American women coming of age after World War II, although she herself was of an older generation. During the 1950s, more American women (and men) married than in the previous generation, and they did so at a younger age and had more children than their parents' generation. As the newlyweds moved to the fast-growing suburbs, young women devoted themselves to homemaking and child rearing. Some feminist historians argue that the widespread turn to family values and domestic ideals was a result of the individual decisions of a generation that had experienced women's work as an unfortunate consequence of economic hardship during the Great Depression; others contend that welfare programs helped legitimize the idea of the "family wage"—where men earn enough to support their families on a single income—and that this system cemented and expanded the division between male and female jobs.[31]

For Watson Sr., his wife provided the female foil against which he could present himself as a strict yet caring executive. To do so, he often used the rhetoric of the IBM family. During the 1953 visit to Germany, for example, he said that it had "always been his objective for all employees to be dealt with like family members." He tied these relations to a typical feature of welfare capitalism: the open-door policy that was meant to avoid grievance procedures. He said, "It may happen that, here or there, something is not right, or it may happen that you are disappointed and feel you have been treated wrongly. In such cases, turn trustingly to your group or department head or, if necessary, talk about your concern with Mr. Borsdorf or Mr. Hörrmann," the two executives of the German IBM subsidiary. Watson later described his delight in being able to personally greet all IBM employees because wherever he and his wife "have been, in each country of the IBM-Family, [we] have always met the same kind of people. People who think right and people who have always let us feel their friendship." He assured German employees that he had always tried to "return this friendship to the full extent" because friendship was above everything else for him and his wife.[32] Watson often talked about the IBM family, with himself as the father and his wife as the mother; together, they deeply cared and provided justly for the members of the large IBM family.

Watson also underscored the importance of the individual employee as a member of the corporate family and a thinking person making suggestions for improvements. He wanted to sharpen this direct relationship of employees with their company, unmediated by labor unions. Watson frequently said that IBM was all about individuals and their ability to think—thus the ubiquitous "THINK" signs posted in various languages in IBM buildings around the world. IBM employees were to "hone their cognitive abilities and avoid impulsive actions—whether from lack of consideration or habit."[33] It remains an open question whether forming a labor union would have counted as such an impulsive, irrational action; instead of dealing with a democratically elected representation of IBM employees, Watson said that regardless of the job, IBM always cared about the same: the human being. At seminars, for example, Watson stressed that every "man" was important. He sometimes listed corporate roles on a board: the manufacturer, general manager, sales manager, salesman, service man, factory manager, factory man, office manager, and office man. Then he crossed out the beginning and/or ending of each phrase, highlighting what was crucial in every single employee: being a "man," a human being. While highly gendered—women were obviously excluded—presentations like these underlined the significance of humanity and personality within IBM's corporate culture (figure 7.2).

Family Rhetoric in West Germany

Watson's family rhetoric found easy acceptance in West Germany's conservative political climate. During the 1950s, many West Germans longed to return to the "normalcy" of the nuclear family, with its wage-earning father, homemaking mother, and two to three children. Throughout the decade, half the women of eligible age in Germany worked, fueling growing public concern about "women's work."[34] Representatives across the political spectrum passed legislation that promoted women's family roles. For example, the 1949 German Basic Law—the German constitution—granted special protection of marriage, motherhood, and families as well as equal rights to women for the first time. But guided by conservative family values, the law also stated that housework constituted women's primary contribution to their families, and women could hold gainful employment only when it did not interfere with their duties as housewives and mothers.[35] These policies reflect the dominant social values that provided fruitful ground for the notion of the IBM family.[36]

In Germany, executives and employees embraced Watson's family rhetoric and appreciated the company's welfare capitalist programs, albeit sometimes cautiously. At the same time, employees pursued local forms of labor organization such as joining national unions, forming legally required

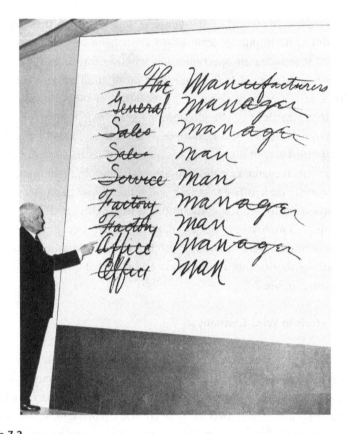

Figure 7.2
Thomas J. Watson Sr., IBM's chairman, during a presentation on "The Human as a Task" (*Der Mensch als Aufgabe*).
Source: Combined issue of *IBM World Trade News* and *THINK* (July 1956): 8. Courtesy of International Business Machines Corporation, © International Business Machines Corporation.

works councils, and electing labor representatives to the board of directors. Watson thus was unable to achieve a main welfare capitalist goal: to avoid labor organization. Yet Watson's family rhetoric translated his American welfare capitalist approach into the ideas of Catholic social ethics such as personality and personal dignity that formed the basis of the Christian labor movement prevalent in Swabia in southern Germany, where IBM's main manufacturing sites were located; this made Watson's rhetoric acceptable to IBM's workforce in Germany, and ensured the company a well-intentioned and cooperative workforce.

Although German executives employed the family rhetoric more spar-
ingly, they copied the Watsons' gendered presentations at company events.
For instance, to celebrate Watson's forty years at IBM in 1954, the German
subsidiary invited all employees and their spouses to an evening reception,
similar to the festive events marking the Watsons' visit to Germany the previous summer. The Watsons were not present at this event, but following
their example of the previous summer, the German executives Johannes H.
Borsdorf and Oskar E. Hörrmann along with their wives personally greeted
over thirty-six hundred guests by individually shaking their hands. The
event was held after the usual work time, signaling IBM's reach into family time and the private sphere. And again, the attending women, both
employees and wives of employees, received a properly gendered gift—this
time a bottle of Eau de Cologne.

Similar to other IBM anniversary celebrations in the United States and
around the world, the evening opened with a prayer and speeches celebrat-
ing Watson's life and success. There was a dinner, presentations by IBM
Clubs, and music and dancing. In their speeches, Borsdorf and Hörrmann
referred to Watson's habit of thanking his wife for her involvement with
the company, and said that Watson knew that his wife always influenced
him positively and that the wives of IBM employees likewise influenced
their spouses positively. Therefore, they explained, it had been Watson's
wish that the wives be invited to his anniversary celebration.[37] Addressing
women in their role as a spouse only rendered invisible the many women
employees working at IBM in manufacturing and office positions.[38] Still,
through its hospitality, the company extended its reach into IBM families,
seeking to enlist spouses to support the company's bidding.

For German employees, the anniversary celebration must have been an
unusual way to mark the evening of April 30, the eve of the May holiday.
On this day in many regions of Germany, including Swabia, people erect
maypoles on the village green, accompanied by processions and May dance
celebrations. Also, May 1 is the traditional holiday of the international
labor movement as well as a bank holiday in West Germany. At the least,
the May holiday provided employees with a chance to sleep in the day after
the anniversary celebrations.

In contrast to the German IBM executives, the editors of IBM's German
employee magazine, *IBM Deutschland*, embraced and even heightened
Watson's family rhetoric. Reporting on the visit of Watson and his wife in
summer 1953, they wrote that IBM was more than a mere place to work;

it was "a chair at the table of a large and faithfully providing family." At IBM, every employee could find "more than a mere basis for existence, he could also find moral support and meaning of life. He knows that a large organization stood behind him. He knows that IBM will provide him with support, a secure place in a unique community, he knows that he will find help and consolation as well as work and bread. And where would he find that outside of IBM?"[39]

The magazine quoted a young employee who, after participating in the festivities for Watson's visit, considered Watson a "simple and humble man who expressed logical and almost trivial thoughts, but whose speech had strength, an indescribable atmosphere, a deep wisdom and sincere love that was more than convincing." The young man had read Watson's biography and stated that he "no longer had the feeling to just arrive at his workplace when he arrived at IBM in the morning. He now knew that IBM was more for him, thanks to the ingenuity of only a single man"—Watson.[40]

Besides embracing Watson's family rhetoric, German IBM employees enjoyed the company's generous remuneration, vacation policies, social services, and benefits. But other aspects of IBM's welfare capitalist culture transferred less easily to the German context, and needed to be explained and promoted to find acceptance. For example, suggesting improvements was a long-established structure in IBM's domestic operations; German employees, however, hesitated to submit recommendations for fear that management may perceive them as overstepping boundaries because culturally, supervisors organized work processes. The employee magazine campaigned for ideas for improvements, explaining how to submit successful proposals. The drive, which began in 1953, resulted in many submissions, which increased each month, reaching a peak of 560 suggestions in April 1954, the month of the anniversary celebration. This was more suggestions than the total number in the first half of 1953.[41]

IBM's German Works Council

During their trip to Germany in summer 1953, the Watsons met with members of the Böblingen works council and Arnold Stender, the labor representative on IBM's West German board of trustees. By then, the members of the original works council (that Watson and Chauncey had so heatedly discussed in their phone conversation) had resigned because they were unable to garner the support of the majority of IBM employees. Chauncey had then hoped that "several of the old members [would] not be elected"—presumably anticipating that a different set of worker representatives might be more pliable to management wishes.[42] Addressing the members of the works council as "members of the IBM family," Watson immediately clarified his view of labor relations. He talked about how he and his wife greeted over three thousand German employees and their spouses by shaking their hands "to keep the IBM family spirit going." He also invited council members and their wives to the luncheon for sales managers the next day. He said that he and his wife "like to meet with the members of the family as often as possible," and emphasized that "there is no one important man in the company—everybody is important."

Then W. Berger, the head of the works council, thanked Watson for the care packets sent in the immediate postwar years and life insurance. Stender, the newly elected labor representative on the German IBM board of directors—also a provision of the codetermination legislation—was the only one to request something. He began by identifying himself "as a devoted Christian and as a democrat"—presumably to indicate that he was a member of the conservative Christian Democratic Party, which was closer to the Christian wing of the labor movement than the Social Democrat wing—and said that even as an elected labor representative on the board of directors, it was a little scary to bring up an issue.[43] He asked that IBM employees may "in the near future [have] an old age pension so that older employees may not no longer [sic] have the fear of old age." This was the only (slightly) adversarial moment in an overly friendly meeting, and Watson brushed it aside, remarking that management had been "at it." Watson said that IBM employees had job security and IBM provided employment for three thousand people in Germany; the unemployment rate in West Germany in 1950 was 11 percent and was still over 8 percent in 1953.[44] Watson liked to grant corporate benefits by managerial fiat rather than by negotiations with labor representatives, and did not want anyone to think that forming a labor organization was the way to achieve benefits.

After the visit, Berger carefully embraced Watson's family rhetoric, and like Watson, linked it to the company's labor relations. In 1954, at the celebration of Watson's fortieth anniversary, Berger sent the "warmest thank you" across the Atlantic to the "head of our so often praised IBM family in the name of all IBM employees in Germany and their family members."

Appreciating the company's welfare capitalist measures, he praised how Watson's outstanding personality showed itself in his direct personal relationships with his employees, as expressed in the company's open-door policy as well as its social benefits, such as above-tariff salaries and vacation days, Christmas bonuses, insurance contributions, emergency funding, the pension system, and recreation homes, all of which, in the end, were due to Watson's goodwill. Berger acknowledged the wives of employees, explaining that their invitation to the festivities was an expression of gratitude and honor for their work as "guardians of domesticity and family life and source of all happiness and creative power"; wives, in this way, contributed to IBM's undisturbed operation.[45]

During contract negotiations the following summer, Berger used the IBM family rhetoric in an editorial explaining his view of labor relations. He noted that IBM employees often talked about the IBM family, where all members belonged and felt responsible for each other, and encouraged each other to do their best. At the same time, he cautioned, the notion of a family needed to be used carefully because there were a number of differences between the corporate family and one's own family: one could leave one's company, while one was part of one's family forever; companies were organized rationally to make a profit, while the family transcended this world, and was the place of birth and death; and one had a material-economic task in one's company, while one belonged to one's family with one's soul. Nevertheless, Berger continued, the company was not independent from an employee's family life because employment determined his or her standard of living as well as his or her satisfaction with work. After a brief historical overview on industrialization and Marxist class conflict, he asserted that by the mid-twentieth century, "partnership" had grown between employers and employees; "what used to be a fight against each other has turned into a struggle about cooperation."[46] In other words, Berger acknowledged the forming of IBM's welfare capitalist family culture as a way of overcoming labor conflict.

Berger's remarks were clarified in an earlier editorial; there, he had asked whether workers were "proletarians" or "personalities." Berger stated that the Marxist labor movement initially saw workers as proletarians who, in their class fight, had to eliminate other social classes, especially the capitalist class. By the mid-twentieth century, according to Berger, it was clear that the labor movement was a force shaping societal relations to allow everyone to work and live according to his or her abilities. Berger maintained that in Swabia, a Marxist proletarian movement never took shape. By contrast, the local precision industry promoted workers who were "smart in their heads and clever with their hands, and worked in a precise, conscientious, reliable and loyal way," or who, in Berger's words, became proud craftsmen and citizens—or personalities (*Persönlichkeiten*)—rather than proletarians.[47] Berger's view of workers as personalities matched well with Watson's emphasis on the individual "man."

While little is known about Berger's background and union affiliation, he and his works council connected Watson's family rhetoric to Catholic social ethics. Historically, the labor movement in Germany had splintered into competing unions that worked with political parties across the ideological spectrum, as we saw in chapter 6, including Christian unions, which were dominant in Catholic areas such as Swabia.[48] Based on the 1891 papal encyclical *Rerum Novarum*—an open letter written by Pope Leo XIII on the rights and duties of capital and labor—Christian unions were built on the principles of Catholic social ethics, including human dignity and subsidiarity—the idea of responsibility and self-help in small social groups, starting with the family. Berger connected Watson's focus on the "man" and "human" with the principle of human dignity (*Personalität*), and linked Watson's emphasis on the family with the principle of subsidiarity.

At quarterly employee meetings, Berger advocated for union membership and shared statistics on the number and percentage of employees who were union members; for instance, in 1953 in Böblingen and Sindelfingen, of the 2,141 employees, 75 percent were union members—93.8 percent of the 1,425 blue-collar employees and 38.4 percent of the 716 white-collar employees.[49] Berger urged 100 percent union organization.[50] German IBM employees thus continued their German tradition of organizing in national labor unions.

These kinds of conflicts with local labor traditions in IBM's international operations were not unique to West Germany. In 1955, for example, a union representative visited the new IBM factory in Greenock, Scotland, and told the management that the union "did not want us [IBM] to do anything outside the payroll"—that is, the union did not want benefits other than wage increases. The union seemed to fear that nonmonetary welfare capitalist benefits like a country club—which IBM planned to create at the time—would assuage IBM employees and divide the labor movement.

Watson considered this a "shock," given what he considered his "friend-ship" with "outstanding labor leaders in the United States for many years," and immediately instructed his son Arthur to "move the planned export business from Greenock to the Netherlands."[51] Although this instruction was never implemented, Watson rejected any interference with his busi-ness.[52] While IBM operated within the realm of local labor legislation, it used the broadest interpretation of the law to shape labor relations in its European subsidiaries, and Watson sought to avoid any impression that labor representation would benefit IBM employees.

Elimination of Class in IBM Germany?

German employees welcomed IBM's welfare capitalist measures, but daily interactions and habits exhibited the typical German class distinctions. For example, when three young workers completed their three-year appren-ticeship program—which marked the end of their formal education and training—IBM Germany did not celebrate this with a formal ceremony, where the young men could dress up. Instead they were recognized as part of a quarterly employee meeting. A photograph shows Borsdorf, the exec-utive, dressed in a suit and tie, shaking hands with the three men, who are wearing their apprentice uniform—a short, buttoned-down dark jacket with rolled-up sleeves and matching pants—and handing them each a gift of a book. This event was nothing like graduates proudly walking across a stage to receive their diplomas; indeed, the body language of the three men, standing with sunken shoulders and torsos, exudes unease, if not embarrassment, with the unusual attention. Outside the ritualistic greet-ing line, this event probably marked one of the rare occasions for these men to personally interact with top management, shaking hands across class borders.[53]

Images in the employee magazine confirm that work clothing distin-guished different employee groups at IBM and that the different groups rarely interacted. Only executives and sales operatives wore dark suits, dress shirts, and ties; scientists and engineers in the laboratories wore white over-coats; technicians often wore long dark overcoats, which protected their clothing from dirt; some men in manufacturing also wore protective cloth-ing; the large number of women working in light manufacturing, such as the assembly of typewriters, typically wore their own street clothing;

and women in office positions—who were rarely depicted in employee magazines—appear to have worn more fashionable, feminine clothing. At employee assemblies, the different status groups frequently sat together, identifiable by their different outfits.

The employee magazine reveals that different clothing at work also reflected different habits and practices outside work. One article addressed questions of "prestige," a common German term for regard given to higher social groups.[54] A sales assistant recommended a dress to a customer not because it fit her nicely or had other fine features but rather because the higher price of the dress would preserve the prestige of the customer's husband. Similarly, the article talked about a man working in a position that required him to often visit large companies. He reluctantly bought a new Mercedes to replace his Volkswagen not because he liked or cared for the car but instead because it was more representative of his status. The article also described a family that grudgingly decided against purchasing a house in a subdivision it liked because the husband held a prestigious administrative position, and a clerk in a lower position in the same office lived in the subdivision. Living near each other would have required interactions that the family considered socially inappropriate. While the article bemoaned the concern for prestige in each case and urged readers to overcome it, it shows the importance of class distinctions in everyday life.[55]

In 1958, IBM took the unprecedented step of creating the first all-salaried workforce in an industrial company, eliminating the differences between blue- and white-collar employees in the United States. Thomas J. Watson Jr.—who had succeeded his father—announced in January that hourly employees would now receive annual salaries. The new policy affected about a third of over sixty thousand domestic employees, who also received more generous health benefits, compensation for sick days, and full pay for authorized personal absences, and they no longer lost pay for occasional late arrivals.[56] He described the new policy as an opportunity "for every single employee to develop within his or her responsibilities and to grow with the company."[57]

IBM's German executives Borsdorf and Hörrmann followed suit in August 1958, and unilaterally gave all workers the status of salaried employees, yet this policy had different consequences and meanings than in the United States. Blue-collar workers gained better benefits, such as the continuation of pay in case of illness or death, longer vacations, and a longer period of notice for layoffs—benefits that not even white-collar employees in other German companies enjoyed. The policy, however, affected employees only within the company. It did not change their social status beyond the company, and legally, with regard to social insurance, they continued to be considered workers.[58]

In welfare capitalist fashion, the executives in the United States and Germany granted the new labor policy by managerial fiat, and due to its sensitive nature, it required careful negotiations within the company and did not go without criticism outside IBM. While little is known about the reasoning behind the decision, Borsdorf and Hörrmann may have felt compelled to follow directions from the United States.[59] Cognizant of the policy's explosive nature in the German context, they announced it in carefully worded statements at the IBM employee assemblies in Sindelfingen and Berlin. Positioning the policy in the tradition of welfare capitalist labor relations, the statement declared it a long-established goal of IBM to moderate the difference between blue- and white-collar employees because the company "expected from every employee that he [sic] performed at his best for the corporation regardless of the kind of his role."[60] The policy focused on intrinsic motivation and personal responsibility in a culture of respect, rather than one of supervisory and managerial control; IBM was to become a classless corporation for the purpose of higher productivity.

Notably, the employee magazine, *IBM Deutschland*, accompanied the announcement with the image of a young woman at an IBM employee assembly. Her fashionable blouse suggests that she worked in an office position. She is seated next to another young woman, and is surrounded by men in suits (executives or salesmen) and lab coats (laboratory employees). The woman looks pensively skeptical. She is seemingly depicted as an employee benefiting from the new policy, but neither she nor the other white-collar employees in the image would benefit from it, and they would no longer distinguish themselves through their rank. In Germany, salaries for white-collar employees had decreased since the 1920s, frequently to levels below blue-collar employees. While white-collar employees continued to rank higher in social respect, they often found it difficult to afford the lifestyle that they and their families aspired to.[61] Now they had even lost the distinction as white-collar employees—a source of social prestige in their company.

Not surprisingly, the unilateral announcement raised criticism. Indeed, neither employer associations nor labor unions condoned the decision; employer associations worried that unions would seek the same social benefits that IBM's former blue-collar employees now enjoyed, and unions feared for their membership if companies abolished workers by turning them into salaried employees. Possibly anticipating such reactions, IBM limited its public relations work, and the new policy only found mention in short notices in the economic section of newspapers. Internally, the IBM works council celebrated the company's pioneering role and claimed that the new policy would guarantee IBM a place in German social history.[62]

Still, even within IBM, the new policy remained controversial, albeit buffered by long-established cooperative labor relations and the willingness of the works council to work with management. In Scotland at the IBM Greenock plant, unions demanded that IBM refrain from nonmonetary compensation, which reveals the precarious nature of such benefits. In Germany, the works council had not asked for workers to become salaried employees; instead, in 1958, it fought for other social benefits and monetary remunerations, such as a 42.5-hour workweek, forty-year anniversary recognition, vacation homes, salary increases for team leaders and difficult work conditions, payments to bridge the gap before pension payments, support for older employees by moving them from a night to day shift, support for early disabled persons, and employee shares in capital gains from large profit increases. IBM management refused to grant these in order to instead reduce the prices of IBM products in view of a global recession.[63] Although Borsdorf and Hörrmann emphasized that their announcement was "in agreement with the board of trustees and the works council," the new labor policy surprised the works council, which was pursuing many other goals. It took half a year for management and the works council to negotiate a new contract, indicating the preliminary nature of the initial agreements. Indeed, in December 1958, Berger, the head of IBM's works council, stated that the transition to the new labor policy had been handled in a "smooth and satisfactory manner despite the short announcement," making the new salaried policy sound more like a burden than an achievement.[64]

In an editorial for the employee magazine, the works council interpreted the new labor policy in the context of IBM's welfare capitalist culture.

Rather than calling it a decision that was unexpectedly bestowed on IBM's employees—which, of course, it was—the works council stressed IBM's goal to provide a "meaningful place for humans and humanity," in addition to manufacturing and selling office machinery as well as making a profit. The editorial said that at IBM, employees were not to be "estranged, or even enslaved, burdened by worries about their own future and the future of their families"—a rejection of Marxist interpretations of the situation of workers. Instead, the works council presented IBM's welfare capitalism as a way for workers and employees to "ban their worries, eliminate their distrust and grow their trust in each other" to begin an "era of humanity" that allowed workers to express and realize their personality and potential.[65] The works council editorial ended with an allusion to a popular Marian song, revealing its Catholic affiliation: it stated that their new status provided a safe roof under which all employees would find "protection and shield against the vicissitudes of life."[66]

Notably, the German IBM subsidiary announced its new labor policy just half a year after requiring all employees at manufacturing sites to wear uniform work clothes. In March 1958, IBM's German management introduced the new clothing policy, and the company offered financial support for the purchase of the first set of work clothes. Of course, practical reasons—the protection from dirt—could explain this new requirement. Yet the clothes were made of a color that easily showed stains and dirt, and the works council asked whether the uniforms could be made in a darker color, or whether workers in dirty positions could receive an extra allowance to pay for the increased need to wash their work clothes. In addition, the clothes were not of sufficient quality, and many workers complained that they no longer fit after the first wash.[67] Aside from such practical issues, the new work uniforms were symbolic: they visibly distinguished manufacturing employees from office and other employees just at the time when the new labor policy was put in place. Management had replaced a legal distinction between blue- and white-collar employees with a symbolic one that marked daily interactions.

A remark by one of the German executives, Hörrmann, revealed that at least he did not intend the new labor policy to eliminate class distinctions; rather, he emphasized the belief that everyone was born into his or her social station, and should be content with it. After reading the carefully worded announcement at the Sindelfingen employee assembly, Hörrmann counseled employees not to continually burden themselves with new worries that prevented them from enjoying what they had already achieved: "When we have achieved the vehicle, the refrigerator, the television set or the new house, we should be glad about the success and enjoy it—we should not hastily immediately hunt for the next material things and thus overstep the boundaries that are set for everyone."[68] As Hörrmann announced the elimination of class distinctions within IBM, he upheld the German understanding of the same, rejecting US ideas about increasing the standard of living.

Conclusion

IBM pursued its own welfare capitalist labor relations in its company's West German subsidiaries. This approach differed distinctly from the collaborative labor relations propagated by the Marshall Plan's Productivity Program: while the Productivity Program promoted plant-based collective bargaining to ensure that productivity gains were equally shared between management, workers, and consumers, IBM shunned organized forms of labor representation and instead bestowed—admittedly generous—corporate benefits on its employees by managerial fiat. IBM's labor relations were couched in a corporate culture of gendered family rhetoric that disguised the hierarchical relations between the charismatic patriarchal figure of Watson Sr. and his corporate sons and daughters. The company avoided democratic decision making and union affiliations, seeking instead to build individual relationships with employees based on employees' loyalty and desire to perform at their highest capabilities. Despite the company's growth abroad in the two decades following World War II, IBM was not a representative of the American economic model along with its form of labor and class relations in the way that Haynes and Michler recommended that US companies serve.

IBM's careful navigation of labor relations and traditions in West Germany may have contributed to the company's corporate success.[69] IBM followed, although reluctantly, local labor laws such as the codetermination regulations. This helped IBM attract a highly qualified workforce when, for example, it expanded its expertise into novel electronic technologies in the 1950s. With higher wages and access to the latest technological developments in the United States, IBM also promised the same employee rights and security as a local German company.[70]

At the same time, the company's labor policies had different implications in Germany than in the United States. Watson's welfare capitalist family rhetoric transferred easily into the Christian social teaching prevalent in IBM's works council and among the staff, and German IBM employees appreciated welfare capitalist benefits such as above-average pay, vacation time, and social insurance benefits. But some labor policies played out differently in Germany. For example, in the United States, turning hourly employees into salaried ones gave blue-collar employees the security of a regular income—a demand of industrial unions like the CIO-UAW in the mid-1950s in hopes that workers would be protected from technological change, as we will see in the next chapter. In Germany, the new labor policy not only guaranteed annual pay to workers but also affected their social status: inside IBM, the new labor policy legally abolished the distinction between blue- and white-collar employees. But outside IBM, blue-collar employees continued to be regarded as working class, for example, for the purposes of social and pension insurance. At the same time, new uniforms for manufacturing workers were introduced to symbolically distinguish them from white-collar workers in daily interactions.

In the early 1950s, IBM was but one competitor in manufacturing electronic computers; two decades later, it dominated computing markets in the United States and Europe. The introduction of this new technology, particularly changes in the manufacturing process through computer automation, raised questions about technology, labor, and markets, and how to shape these relations. In the next chapter, I will look at how European and American engineers, politicians, labor representatives, and others negotiated these questions.

8 Computing Technology: Productivity Promise or Automation Threat?

In June 1956, union officer Ted F. Silvey sent ballpoint pen sets to a dozen correspondents in Europe. In an accompanying leaflet, Silvey hailed automated mass production technologies, which had allowed the price of ballpoint pens to be slashed from $15 for a single pen in the early postwar years—when ballpoint pens first became available—to 98¢ for a set of four pens in 1956. This was a *sixtieth* of the price, within less than a decade. And the ink came in four different colors. Over the next two years, Silvey distributed more than one hundred ballpoint pen sets around the globe, including to European union officers and administrators of the Organization for European Economic Cooperation (OEEC), explaining, with notable nonchalance for a union officer, that "the number of workers needed for this fabulous production" of over seven hundred million pens a year in the United States "is insignificant." He emphasized, however, that "the number of people needed with money in their pocket to buy this output is highly significant—in fact gives full meaning to the point that with automation the machines and instruments can do almost everything except buy what they make!" Within two years, the newly founded journal *Computers and Automation* printed the text from Silvey's leaflet, suggesting that computer-automated production enabled the cheap mass production of the pens, and that electronic products like computers would be among the next automatically produced items.[1]

Silvey was one of the first in the Congress of Industrial Organizations (CIO) to sound the alarm about automation technologies and the looming threat of overproduction. In his leaflet, Silvey proposed the "traditional trade union answer to increased productivity from science and technology": more income for workers and fewer hours of work—a four-day workweek instead of a five-day workweek or a ten-month work year instead of twelve-month work year—with full wage or salary income. Silvey's background as a journeyman printer may have been the reason for this interest in factory and office automation, which was rare among union officers. By the mid-1950s, Silvey had worked for two decades as a CIO union organizer and officer, joining the CIO in its founding days.[2] In the early 1950s, apparently on his own initiative, Silvey began collecting information on automation technologies, perhaps because he was alert to its particular threat to the printing profession. Silvey urged CIO president Walter Reuther in 1953 that the unions needed to find new answers to automation technologies.

Silvey's ballpoint pen campaign shows how nongovernmental parties initiated transatlantic dialogues after a surprising breakdown of government-established channels in the face of automation. In 1955, European officers started asking their US counterparts for guidance on automation technology and requested a tour to study automation in the United States—requests to which US government officers, who had always been so eager to engage in conversations about productivity, responded with delay tactics. Exchanges eventually happened outside government channels, such as through labor unions, and Silvey became one of the point persons for these exchanges. He was predestined for this role through his early interest in automation technology as well as his experience in transatlantic productivity exchanges; in 1946 and 1947, Silvey had first traveled to Germany to conduct a study of occupational disease control, industrial accident prevention, and workmen's compensation for the US military government, and later served as a labor officer for the Marshall Plan's Productivity Program in Washington, DC. With many transatlantic connections, but no longer in government employ, Silvey traveled to Europe in 1956 and 1958 for conferences and lectures on factory and office automation.

Automation is typically defined as machine control of mass production through feedback mechanisms or electronic computers; it can also mean the computerization of routine office work. Raising questions about the effects of new production technologies on workers and working conditions, automation became a counterconcept to productivity. Some automation proponents like John Diebold—who claimed in his 1952 book to have invented the term, although Ford engineers already used it in 1947—promised that automation technology would take over repetitive work, end "the subordination of the worker to the machine," and release the worker "for work permitting development of his inherent human capacities." Diebold suggested that those performing simple repetitive tasks could be trained to carry out more engaging maintenance and repair jobs.[3] The lesser-known automation proponent Hunt Brown, whose prolific publications include loose-leaf handbooks on office automation, stated more bluntly that automation would replace "drudgery, not people," and create more jobs for more employees who would "not be doing the same dull, drab, routine work" but instead would "have more interesting assignments."[4]

But many workers, journalists, politicians, and some scientists on both sides of the Atlantic did not expect that computer-mediated productivity technologies would help increase standards of living; they feared that technological improvements would lead to de-skilling and even unemployment. While journalists published sensational articles and politicians held public hearings, setting a critical tone in the public debate, union rank-and-file members sought protection against technological change from their organization leaders. The US CIO and German Deutscher Gewerkschaftsbund (DGB) both argued that new technologies would allow workers to produce more in the same time—in other words, to increase productivity—which would allow businesses to raise wages without raising prices. Thus, both organizations embraced technological change, pursuing a proconsumerist, macroeconomic approach to technological innovation, including computer automation, the most controversial form of technological change in the postwar period. Unions in both countries, however, also used the prospect of automation technology to push for reforms and more worker protection: a guaranteed annual wage in the United States, and codetermination and control of corporate concentration in Germany. In the United States, union leaders primarily expected short-term consequences and did not seek government protection for affected workers until the early 1960s. German union leaders reasoned that the need for costly investments for automation technologies would favor industrial concentration—a problematic tendency because large industrial companies had previously supported the Nazi government—so they sought to control automation through stronger codetermination rights.

This chapter compares US and West German perceptions and attitudes toward automation technology, particularly computing technologies, in the postwar years. The biggest difference between the two countries is that the public debate in the United States was less critical than it was in West

Germany.[5] The chapter first looks at the debate in postwar West Germany before discussing the interruption of transatlantic exchanges over the question of automation technology. Next, it examines the US public debate and moderate union response, dominated by Reuther and his United Automobile Workers union (UAW). Finally, the chapter returns to the West German side to explore the more alarmist debate, why the union umbrella organization DGB embraced rationalization technology, and how corporations in Germany legitimized computer automation.

Debate about Automation Technology in West Germany

"Next year you'll have it easier, Daddy," a little boy in a cartoon promises his father in a distinctly Frankfurt dialect. "We'll have an electronic brain in Frankfurt that you can rent by the hour. That way you no longer have to always rack your brain over my math homework" (figure 8.1). The little boy was referring to plans to open a computing center in Frankfurt in 1956. Remington Rand, IBM's main competitor in the 1950s, had built the Univac computer, the first commercially available electronic computer that sold for over $250,000 and rented for about $12,000 a month; only forty-six machines were ever built. One of the last was flown to Germany—the heaviest airlift to that date—and installed in a computing center, available for companies to rent by the hour for their data processing needs.[6]

In Germany, the publicity for the computing center raised many questions about the new technology, as this cartoonist pointed out. Could these "giant brains"—as Germans often called computers—think and solve intellectual problems? If so, how would computerization affect employees doing intellectual work in offices and on shop floors? Would those employees suffer the same dire fate as manual laborers had when machines took over many of their physical tasks during the first Industrial Revolution? With unemployment rates already high in Germany and other European countries, many Germans saw computer automation as a threat to their jobs.

In the German context, the threat of unemployment from technological improvement was a major part of the public debate. Weimar German union analysts in the 1920s had already linked technological improvement to rising unemployment rates. With increasingly pessimistic tones, Germans moved from seeing technological change as causing temporary shifts in employment to considering it the reason for long-term, structural

DER GEZEICHNETE WITZ

Figure 8.1
This cartoon from a Frankfurt newspaper commented on the upcoming opening of a computing center in Frankfurt—"Next year you'll have it easier, Daddy. We'll have an electronic brain in Frankfurt that you can rent by the hour. That way you no longer have to always rack your brain over my math homework"—indicating that electronic computers would take over intellectual work.
Source: "Der gezeichnete Witz," *Frankfurter Neue Presse,* n.d., press review, Kuratorensammlung, Deutsches Museum.

unemployment.[7] The American economist Robert Brady was among the first scholars to link rationalization technologies to the high unemployment levels at the end of the Weimar period—soaring from 15.7 percent in 1930 to 30.8 percent in 1932—and during the rise of the Nazi regime.[8] Later historical analysis indeed showed that rationalization measures contributed to late Weimar unemployment. For example, the Ruhr mines hired large numbers of workers even during the hyperinflation of 1923–1924, and acted like a sponge soaking up returning soldiers and young men in the immediate war years. But they released workers in the second half of the 1920s when they introduced new machinery. Based on the productivity

levels of 1910–1915, the mines in 1929 would have had to hire over 107,000
additional workers to achieve their output had they not acquired improved
machinery; this means that about half of the more than 208,000 unemployed miners that year were out of work due to technological improvement, not because of the Great Depression.[9]

By the early to mid-1950s, press coverage of computer technology often
took on sensationalist tones, with German magazines, like the left-leaning
Der Spiegel, painting a daunting picture of the automation age and asking
whether "electronic brains" could think.[10] Norbert Wiener, a professor of
mathematics at MIT, soon became the go-to person for European journalists looking for critical sound bites on computer automation, and his
book on cybernetics, the science of automatic control systems, found wide
reception in European countries.[11] "Since machines had already developed
superhuman physical powers," Wiener declared, "people had to get used
to machines accomplishing superhuman brain work."[12] Wiener, who had
become an activist scientist in the wake of the nuclear bomb, complained
that not even union officials in the United States would heed his warnings.[13] In Germany, however, Wiener's critique of automation fell on fruitful soil.

The German press warned that computer technology would affect not
only factory shop floors—as it had in the first Industrial Revolution—but
would also affect corporate offices by replacing employees who conducted
routine clerical and mental tasks. For example, one of the first German-
built computers, the G1 at the Göttingen Max-Planck Institute, conducted
mathematical calculations ten times as quickly as the women doing them;
thus "the electronic 'thinking' machine . . . seemed to compete today with
office clerks."[14] Similarly, Alwin Walther of Darmstadt University, the head
of a leading center in electronic computing that included a large punch
card installation, warned at the opening of the Univac Computing Center
in Frankfurt that because of computers, "accountants and similar middle
clerks who perform principally simple but long and monotonous calculations, are threatened by unemployment."[15] Routine office work appeared
even more ready for computerization than factory work, rendering super-
fluous the "(female) secretaries who pursue their occupation as a side line
before marriage"; only office machinery operators would be needed in the
future.[16] While the German press recognized this threat to office work, the
national debate focused mainly on factory automation.

Politicians, especially of conservative persuasion, often used more measured, positive tones when discussing automation technology. Economic minister Ludwig Erhard, for example, argued that automation—which promoted mass production and consumption—would need to be balanced with the "personal, individual and typically German variety of products"; the German way of life would resist any uniformity or mass dimension in the cultural, intellectual, and economic spheres.[17] Similarly, Fritz Berg, the president of the German Industrial Association (Bundesverband der Deutschen Industrie, BDI), turned West German conditions into arguments for automation technology. He asserted that unlike in the Soviet Union, German companies would not pursue technological progress on the backs of workers; rather, German industry, including small and medium-size companies, would help achieve a better future without disturbance and destruction. Berg thus suggested that German industry would proceed with automation in a way that would alleviate the potential negative effects. Berg and Erhard both cautiously admitted that automation technology could have detrimental results, and both were convinced that German companies would remediate such effects and benefit from an automated future.

Likewise Siegfried Balke, the Christian Democratic minister for nuclear research, said in his 1956 address at the inauguration ceremony for the Univac Computing Center that automation allowed "human beings to fulfill their destiny as human beings" because "the electronic brain does not think for me; it leaves me time to think." An industrial revolution was not imminent, Balke contended; rather, technological change would be evolutionary or incremental, because organizational tasks consisted of psychological components that could not be automated. Balke claimed that with the initial introduction of computers, "less people will perform work which can be done by machines," including routine mental activities, "and more people will perform work which can only be done by human beings," such as programming computers.[18] While generally supportive of automation technology, Balke nevertheless questioned how suited American computers were for postwar German corporations, which he thought should first improve their workflow and implement other rationalization methods before investing in expensive electronic machinery. Here Balke was in agreement with other German economists who warned that surplus labor, low salaries, and few available bank credits rendered computing technology too expensive, compared to manual or electromechanical data processing.[19]

In a feisty comment—which was excluded from the English translation of his speech—Balke asked that Remington Rand not accuse him of injuring the company's interests with his emphasis on organization over computerization.[20] Overall, Balke was convinced, automation technology would humanize work.

One of the primary issues at stake in the German debate was whether computer technology was revolutionary or evolutionary. Automation critics, such as the Social Democrat Carlo Schmid, claimed that computer technology was revolutionary like the new technologies of the first Industrial Revolution and would cause a major upheaval that would have detrimental consequences for replaced workers. Thus Schmid charged at his party's 1956 convention that "omissions are deadly in revolutionary situations." If labor neglected to oppose computer automation, Schmid argued, it would be overrun by the revolution.[21] Supporters of automation, by contrast, emphasized the evolutionary character of computing technology. In their eyes, electronic computers were just a continuation of ongoing mechanization and technological improvements, and therefore did not warrant any special concerns. Automation critics occasionally allowed that computing technology did build on the technological advancements of earlier times, but they maintained that like steam engines and power looms before them, computers would have far-reaching and negative ramifications for labor, and that labor interests required continued protection.

Interrupted Transatlantic Dialogue

Because of these heated public debates, Europeans, especially Germans, who had no experience in automation technology, looked across the Atlantic for guidance on the new technologies. Communication channels established through the Marshall Plan's Productivity Program were among the natural ways for them to inquire about automation technology. German administrators did so through the European Productivity Agency (EPA), an agency that had been established in 1953 within the European administration for the Marshall Plan in Paris to continue and coordinate the activities initiated through the Productivity Program.[22] The EPA administrators repeatedly asked their counterparts in Washington, DC, in 1955 for guidance on automation technology, requesting information on the status of automation in sectors such as banking, insurance, and air transportation as well as in the automobile, chemical, and petroleum industries. The EPA also requested information on university courses on automation in the United States, such as a yearlong graduate course at Harvard on control systems engineering. Finally, the EPA proposed a productivity study group where Europeans could travel to the United States and investigate automation technology.[23] All these suggestions were activities that the EPA routinely conducted in other areas of technical exchange.

But American administrators did not respond to the EPA's questions concerning automation technology. After months of requests by the EPA, which finally culminated in a desperate appeal for policy guidance from Washington, the American administrators furnished the EPA with some brochures produced by the AFL and the CIO, and in October 1955, sent Jack Conway, a leader of the AFL's automobile union, to address the EPA staff. US administrators often delayed answering questions and claimed that a response would require considerable research effort—although union officers like Silvey had already gathered some of the information that Europeans were requesting—and that the EPA would have to pay for the information—an embarrassing demand since European countries were furnishing the same information for free within European cooperation. Worse yet, the study group on automation was flatly rejected. In a remarkable shift within only two or three years, the US administration turned from a proselytizer of productivity into an unresponsive partner; this was not just a failure to convince Germans of the US model, as in the case of collaborative labor relations versus codetermination, but an interruption of transatlantic dialogues too.

One of the reasons for this drastic change may have been US perceptions of the European economic and labor situation. Some US officers, such as the later Marshall administration head Richard Bissell, thought that productivity measurements could have adverse labor effects under the European postwar conditions of shortages in machinery and capital as well as high unemployment, but they chose not to emphasize these potential problems. Indeed, the Marshall Plan administration's Labor Information Department instructed its staff "not to mention in their publicity cases where capital investment was known to have reduced job opportunities."[24] This instruction may have applied to changes in the United States, where, for example, Metropolitan Life Insurance replaced 135 punch card operators when installing a Univac computer, Ford went from 117 workers to

41 on an engine block line after automation, and a chemical plant had 700 production workers and 300 maintenance men before automation, and 550 and 450, respectively, afterward.[25] The Marshall Plan instruction certainly applied to European cases, where, for instance, the French automobile manufacturer Citroen discharged 3,000 workers after the installation of American equipment through Marshall Plan aid, and Renault, also a French automobile maker, saved 38 percent on labor costs after installing automatic machinery.[26]

US labor administrators were convinced that a strong free labor movement was needed to prevent such negative effects. They believed that while European labor leadership condemned low standards of living, European unionists also were reluctant to change the "equilibrium of scarcity"—that is, the assumption that there was only a limited amount of resources to be distributed—because of the labor movement's weakness. "Unless a labor movement is strong enough politically and economically to insist that the necessary changes be carried through in a socially enlightened fashion with adequate safeguards for the welfare of workers during the transition," labor administrators understood, "it might be considered irresponsible to demand such a change." Consequently, US labor administrators sought to strengthen the free labor movement in European countries, and relied on the help of the AFL, the CIO, and the ICFTU— the international non-Communist labor organization associated with the Marshall Plan—to do so.[27]

US administrators sought to present productivity technology and related labor relations ideas in a nonprovocative way, if at all. For example, John G. Harlan Jr., the assistant director for the Program Office of Industrial Resources, objected to the proposal to hold a seminar on automation technology in Europe after meeting with Diebold, the independent consultant who popularized the term automation, particularly if Diebold himself were to run the seminar.[28] As Harlan reported, Diebold "shrugged off lightly" the question of technological unemployment with an "abbreviated explanation of an expanding economic situation." Diebold struck Harlan as "one of those intellectuals who doesn't want to be bothered by the facts-of-life and, therefore, has given no serious thought on the question of technological unemployment."[29] If the US administration was to facilitate automation seminars in Europe at all, Harlan strongly recommended that they be conducted by someone who gave more credence to labor concerns.

But concern and anticipation about the domestic situation in the United States may have been another reason for the US administrator's unresponsiveness with regard to automation questions. Indeed, when US administrators rejected the European study tour on automation technology, they ostensibly did so because "the question of automation had become highly controversial between labor and industry in the US," which is why US officers advised their European counterparts to first investigate automation technology in Europe.[30] US officers thus found themselves in a similar situation as with concerns about exposing European work-study exchange students to the full reality of US labor relations. In both cases, they were not sure that the realities in the United States lived up to the image that they wanted to convey to Europeans, and in both cases, relations between management and labor were the reason for their hesitation. With exchange students, the US officers eventually resolved to show them everything, "the good, the bad, and the ugly"; with automation technology, they chose to interrupt transatlantic exchanges.

Automation and Labor in the United States

By fall 1955, when Americans interrupted transatlantic exchanges, automation had become controversial in the United States. Concerns about industrial technology in general and technological unemployment in particular went back to the early nineteenth century, and increased during the Great Depression.[31] Yet in the early 1950s, when the German magazine *Der Spiegel* already sounded the alarm and eagerly trotted out Wiener's sound bites, Wiener was still a lone critical voice in the United States; neither his book on the *Human Use of Human Beings* nor his talks appear to have generated much resonance. Instead, American newspapers like the *New York Times* quietly published brief, technical announcements about new automation equipment.[32] The topic of technological unemployment mainly found expression in US popular culture. One example was Kurt Vonnegut's 1952 science fiction novel, *Player Piano*, which describes a society fractured by automation technology: a third of the population had no satisfying work options, and could only participate in useless works projects or join the military.

Americans initially disagreed about the labor effects of the new technology. Some automation proponents, like Diebold and Hunt, promised that automation technology would free workers from routine tasks and create more intellectually challenging, skilled work. But others claimed the opposite: automation technology would solve the shortage of skilled labor in the United States, especially if the "pool of available manpower may diminish under the impact of unsettled world affairs."[33] In the future, these proponents claimed, only unskilled workers would operate manufacturing sites, with automatic technology performing all operations requiring skills such as testing and inspection (figure 8.2). Claims like these need to be seen in the context of the Cold War competition and nuclear arms race that led to a debate over a "manpower"—or labor—shortage in the United States, exacerbated by mobilization for the Korean War. The main concern was the

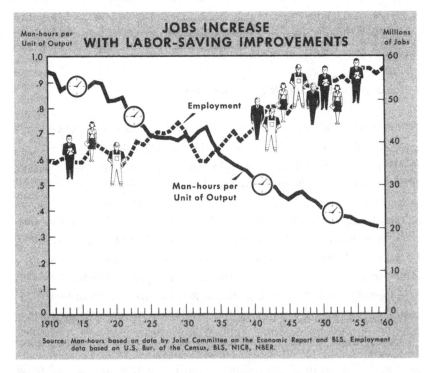

Figure 8.2
This NAM brochure claimed that higher productivity as a result of technological improvements was correlated with increases in industrial employment.
Source: Edward I. Maher, *Automation: A Background Memorandum* (New York: National Association of Manufacturers, 1960). Courtesy of Hagley Museum and Library.

supply of men for the military and defense production, and the need for highly qualified professionals—scientists and engineers.[34] In this situation, automation technology promised to free the skilled "manpower" of white, able-bodied men for military service and research.

Not surprisingly, the US military generously funded automatic manufacturing technologies, particularly for the manufacture of electronics. For example, the American Bosch company produced instrumentation for automatic control for military purposes worth more than $1 billion from 1949 to 1952, and the US Navy, the US Bureau of Standards, and a half-dozen corporate partners collaborated on a method to automatically edge ceramic wafers for electronics manufacturing, called project "Tinkertoy."[35] Eliminating the slow, labor-intensive manual wiring of electronics components, the project had already produced electronics worth $6 billion by 1952. While it was expected that the technology would be needed in a "possible future military conflict and ensuing manpower shortage for electronics production," the federal agencies also hoped to disseminate the technology for the civilian production of electronics, such as radios and TV sets, declaring that there was "nothing secret about the Tinkertoy pilot plant," and inviting industry to learn about the process. Indeed, electronics manufacturing would become one of the prime areas of automatic control over the next few years.[36]

In view of tremendous uncertainty about the overall labor effects from automation technology, Americans distinguished between long- and short-term effects. In the long run, some people believed, automation technology would spur the need for more workers through economic growth.[37] In the meantime, automation technology would likely "put workers out of jobs," even some manufacturers admitted. They saw the need for training and education, and professed their "increasing awareness of their responsibilities to themselves and to their co-workers as human beings." Earl Bunting, the managing director of the National Association of Manufacturers (NAM), also called on manufacturers to meet the challenges from automation technology with the "highest order of industrial statesmanship in the public interest."[38] But such balanced responses were not universal. L. T. Rader, for example, the general manager of General Electric's specialty control department, bluntly labeled the "hopes—or fears—that workers will be replaced universally in the automatic factory tomorrow" as "unadulterated rubbish."[39]

Reuther, finally, the president of the UAW union and the CIO, embraced technological change with little hesitation. In January 1954, in his first reported public statement on automation technology, Reuther expressed his "qualified approval of 'automation' of the [automobile] industry," because apparently, the writer speculated, "better and cheaper production methods increased the general welfare and therefore were good for worker as well as management."[40] Public commentators acknowledged Reuther's openness toward automation, congratulating his "anti-featherbedding, progressive approach to 'automation,'" and emphasizing that he wanted "supervision . . . and some method to keep men at work and to maintain and even improve their purchasing power."[41]

In late 1954, it still remained unclear whether automation technology would free skilled workers or create more skilled positions. Aside from scholarly and literary critics such as Wiener and Vonnegut, most manufacturers cautiously acknowledged that automation technology would require workers to adjust to new production technologies, while others dismissed the effects on labor.[42] And labor leaders generally embraced technological change. Yet the public debate dramatically heated up in spring 1955, just before Europeans sought guidance on automation technology from their American partners. The tone of the debate quickly turned acerbic, possibly spurred by anticipation of the collective bargaining season in the automobile industry. Automobile companies and unions had not met at the bargaining table since 1950, when they agreed on five-year contracts that tied salaries to trends in productivity and the standard of living, giving management the labor security to plan for model changes. Another factor may have been a concern about a spike in the US unemployment rate at 6 percent in 1954.

Automation technology was not a major bargaining issue in 1955; the UAW's major objective was achieving a guaranteed annual wage. Although Reuther generally endorsed automation technology, he declared that the UAW now would only accept a two-year contract because of the rapid changes from automation. In a climate of unemployment, this may have been a concession to the concerns of rank-and-file union members who feared for their jobs. Then, in February 1955, Reuther publicly criticized President Dwight Eisenhower's economic report for not providing guidance on automation technology. While he charged that the administration was "selling the American people short" with "petty, half-hearted economic programs," he nevertheless continued to embrace automation technology, claiming that the report was "based on an hypnotic preoccupation with statistical indices of the long-run past rather than with the tremendous strides toward economic abundance."[43] Although Reuther's positive views on automation had previously been acknowledged, now his position was misconstrued.

A day after Reuther's provocative statements, US Steel chairman Benjamin F. Fairless called the fear of automation "a miserable fraud" and "just plain silly," and complained that automation technology had become "a kind of bogeyman with which to frighten people."[44] While Fairless may not have intended his statements as a reply to Reuther, public commentators quickly made the connection, and Fairless's notion of the "bogeyman" stuck. Barely a month later, US secretary of commerce Sinclair Weeks even went so far as to accuse "some scaremongers"—presumably Reuther and the UAW union leadership—of "frighten[ing] automobile workers with the bogeyman of automation" through claims that automation was "a vicious Frankenstein devouring their jobs." Reuther publicly protested against these accusations, appealing to Eisenhower to prevent such "an attempt to divide and disunite our people."[45] Reuther's own provocative tone as well as the conflation of his position with the more critical union statements may have heated up the public debate.

There were now many formal inquiries into automation technology. The CIO organized a congress on automation technology in Washington, DC, in April 1955.[46] The next month scholars held academic conferences at Michigan State University and Cornell University, and later at Yale University, Pennsylvania State University, and elsewhere.[47] The Bureau of Labor Statistics (BLS) conducted five case studies on the introduction of electronic computers and their labor effects: in an electronic equipment manufacturer, a large insurance company, a mechanized bakery, a petroleum refinery, and an airline reservations system.[48] In October 1955, the US Congress conducted hearings on automation and technological improvements.[49]

The Moderate Response of US Unions

Although Reuther was occasionally accused of opposing automation, he and other labor leaders repeatedly declared their support for technological progress. "The UAW-CIO and its one and one-half million members welcome automation," said Reuther at the 1955 CIO convention, and we "offer our cooperation to men and women of good will in all walks of life in a common search for policies and programs within the structure of our free society that will insure that greater technological progress will result in greater human progress."[50] James Carey, the CIO's secretary treasurer, emphasized that "organized labor did not oppose labor saving devices," although he asserted that the benefits from new technology "must be distributed promptly." Similarly, I. W. Abel, the international secretary treasurer of the United Steelworkers of America, CIO, argued that automation was "here to stay and if it brings progress we're for it," presumably hoping for advancements in the form of better work conditions and a higher standard of living.[51] This positive attitude toward automation technology was a position that Reuther and his union had developed over several years.

In 1949 already, individual voices had confronted Reuther with critical views of automation; Wiener, the MIT mathematician, wrote Reuther, warning about the impending problem of technological unemployment. Wiener had recently turned down an offer to serve as a consultant for the construction of an automatic production line because he thought automation would "undoubtedly lead to the factory without employees," and in the US capitalist system, "the unemployment produced by such a plant can only be disastrous." He cautioned Reuther that organized labor in an automated industry would be competing with "slave labor whether the slaves are human or mechanical"; Wiener urged that labor unions develop a policy on automation to ensure that the profits benefit labor.[52]

Four years later, in 1953, Ted Silvey, the union officer who mailed the ballpoint pen sets to Europe, added his warning voice; he advised Reuther, who had just been elected CIO president, to have unions devote funding to industry-wide research on automation and develop a staff to work on the issue. Unless labor knew "how this problem of a lowered total employment load is to be met," Silvey wrote, unions would have a "super job on [their] hands when it comes to snapping out of the next depression." The answer would need to be "of a type quite different from that which saved us in the middle thirties."[53] In other words, Silvey doubted that legislation, public works programs, and demand-side economics could successfully address automation technology.

While Reuther, who proved central in the union response to automation, may have acknowledged these warnings, he took a different approach toward technological change. Reuther himself saw technological progress as a way to solve economic problems through economic growth. Having grown up in a family with strong socialist traditions and worked in a Soviet automobile plant for two years in the 1930s, he sympathized with Communist ideas, although he later enforced the anti-Communist pledges for union officers.[54] Still, Reuther thought of himself as a radical because he viewed strikes as a legitimate means to achieve improvements for workers. In late 1945, Reuther called a strike against General Motors, asking for higher wages without a price increase, to avoid inflation immediately eating up workers' new purchasing power. Although the strike did not achieve its goals, it carried Reuther to the UAW presidency. Under his leadership, the union developed a strategy of attacking individual companies and setting model contracts that created patterns for each collective bargaining session—effectively achieving the same results as industry-wide collective bargaining. In 1948, Reuther initially opposed General Motor's proposal for a two-year sliding scale contract tied to cost-of-living and productivity improvements because he feared wages could go down if the cost of living decreased. But later he hailed the contract as a union achievement, and in 1950, General Motors and the UAW agreed on a five-year contract on the same terms.[55]

Reuther and the UAW quickly acclimated to plant-level collective bargaining without state participation encoded by the Taft-Hartley Act—a departure from the tripartite wartime collective bargaining procedures. But their contracts also meant accepting the distribution of income between profits and wages as normal, if not fair, as even contemporary socialists noted, because they consented to "the objective economic facts—cost of living and productivity—as determining wages, thus throwing overboard all theories of wages as determined by political power of profits as 'surplus value.'"[56] While Reuther may not have called himself a socialist, he did aim for larger social changes, such as an expanded welfare state with public health insurance and more generous unemployment benefits. Unions, however, were unable to turn these issues into federal responsibilities, so Reuther hoped to use collective bargaining to induce federal programs through the backdoor. Therefore, he never relinquished his belief in the government's redistributive role.

Reuther's rationale was to turn large companies from opponents into supporters of public spending by increasing the level of fringe benefits provided by companies to pick up the slack from the government. If companies found themselves assuming the high costs of social benefits, Reuther reasoned, they would no longer object to the state taking over these tasks. This reasoning seemed to pan out when Congress increased social security benefits by more than two-thirds after a union breakthrough on pensions.[57] The UAW's 1955 collective bargaining campaign for the guaranteed annual wage followed the same rationale. The guaranteed annual wage would raise the costs for releasing workers during seasonal slumps, with the intent to incentivize corporations to plan work more evenly over the course of the year. But the guaranteed annual wage also raised the costs of fringe benefits overall, and after the union's success, several state legislatures increased unemployment compensation. In other words, industrial unions used the collective bargaining process to create social and economic conditions conducive to the introduction of social welfare programs that they could no longer directly demand through the political process.

Yet in the 1955 bargaining season, the guaranteed annual wage soon became tied to automation. The year before, at a collective bargaining conference, Reuther argued that the guaranteed annual wage would serve as an incentive against layoffs, in addition to giving workers wage security. If companies had to pay a guaranteed annual wage, management would have an interest in assuring that the transition went without disruption, and that workers would be retained and retrained for new positions; doing so would be cheaper than paying workers without getting anything in return. Thus, the guaranteed annual wage would incentivize companies to introduce automation technology in times of economic expansion, when new jobs would be available for displaced workers, rather than in times when layoffs would be the result.[58] In the public perception, guaranteed annual wage and automation technology soon became closely entangled, as the public attacks on Reuther showed.

Reuther and the UAW led the CIO in a moderate response toward automation technology, focusing on macroeconomic demand-side economics, and emphasizing that employers and workers both had the same interest in ensuring stable demand for their increasing productivity, based on the New Deal economic model. An often-repeated anecdote about Reuther encapsulates this position: "CIO President Walter Reuther was being shown through the Ford Motor plant in Cleveland recently. A company official proudly pointed to some new automatically controlled machines and asked Reuther: 'How are you going to collect union dues from these guys?' Reuther replied: 'How are you going to get them to buy Fords?'"[59]

Reuther's policy was consistent with Clinton Golden's ideal of collaborative labor relations, discussed in chapter 4; both supported technological improvements because of their overall positive effects for the corporation, despite possible negative outcomes for individual workers. Technological progress, enabling economic growth, was essential to Reuther's socialist program of improving society by distributing ever-increasing profits.[60] Technological progress allowed individual workers to be more productive—without working harder or faster—and generate greater wealth in which workers could participate. Productivity increases meant that workers could earn more without companies needing to raise prices, which preserved the purchasing power of the increased wages. It also meant that workers created demand that prevented overproduction.

The US labor movement, though, was not united behind Reuther. AFL economists warned in early 1955 that unemployment would likely increase by 750,000 to 1 million workers over the coming year, partly due to the introduction of labor-saving machinery and more efficient production methods. Within the CIO, the railroad, machinist, foundry worker, and musician unions initially took rather critical stances toward automation.[61] But Reuther and the UAW's mellow response to automation soon set the tone. The UAW thus relegated the response to collective bargaining, without making demands for a political response except for wanting a congressional investigation. Anticipating short-term disruptions from technological changes, the UAW called on collective bargaining committees to broaden seniority rights, strengthen the rights of workers to be transferred to other jobs, work toward the preferential hiring of displaced workers from other plants of the same company and other companies in the same industry and area, negotiate job classifications that properly reflected increased worker responsibility for costly equipment and enlarged output volume, protect the integrity of skilled trades, and establish joint management and union programs for training as well as retraining at company expense and without loss of wages.[62] As critics have claimed, US unions focused on issues that they could win and that had remunerative value for workers; in doing so, they lost the chance to influence how automation was introduced on shop floors and in offices, such as whether technology was to be controlled by practitioners or management.[63] Otherwise, the UAW and particularly

Reuther hailed the long-term macroeconomic benefits of automation through increased productivity along with a higher standard of living.

As quickly as the automation debate in the United States heated up, so it died down again. The BLS took a middling position in its series of case studies on the effects of automation. It argued that while "some individual workers inevitably suffer losses as a result of displacement; others are benefited, as a result of up-grading." The bureau also warned that firms would have to adapt new technologies to avoid becoming uncompetitive, and eventually go out of business and release all their employees. The Department of Labor also expressed hope that a shorter workweek, industrial expansion, the leisure industry, new products, and public programs such as highway and school constructions would alleviate the effects of automation technology.[64] The congressional hearings on automation in October 1955 noted that none of the witnesses opposed automation, and the committee refrained from recommending legislation on the topic.[65] By 1957, the Hollywood film *Desk Set*, starring Katharine Hepburn and Spencer Tracy, would similarly assuage Americans' fear about the prospect of unemployment from office automation. The movie is set in a television station where computers have just been installed in its research and payroll departments. Hepburn's character, afraid of being replaced, sarcastically remarks how the computer will surely "provide more leisure for more people." Finally, all the research staff members receives pink slips, notifying them that they have been let go, but it turns out that the pink slips were printed by mistake by the computer in the payroll department. In the end, the technical—not the managerial—staff save the day.

The German Union Response to Automation Technology

Like US unions, German labor unions reacted to automation with moderation. Contrary to the sensation-grabbing public debate in Germany, the DGB, the union umbrella organization, embraced technological progress, albeit somewhat less enthusiastically than the CIO, with more concern about technological unemployment. Also, the DGB tied automation to economic concentration—a concern in Germany because of the collusion of large German corporations with the Nazi government. In addition, the DGB called for an expansion of codetermination rights to decisions about technological change, thus tying automation to one of its pet projects, similar to the CIO linking automation to the guaranteed annual wage.

Since the Weimar union trip to the United States in 1926, German labor unions held positive views of rationalization technology. At its 1949 founding congress in Munich, the DGB declared that "macroeconomic rationalization is to be pursued systematically and with full energy as an encompassing problem," including technological improvements in industrial manufacturing, distribution and transportation, construction and farming, and research and development.[66] The union association conceded that in a capitalist economy, replacing humans with machines would lead to persistent unemployment. But throughout the 1950s, the DGB still fought for a planned, socialistic economy with nationalized core industries. In such an economy, the DGB argued, rationalization would lead to an optimal overall economic performance with the highest-possible productivity, increasing purchasing power and raising standards of living through full employment. At its federal congress in 1952, the DGB espoused rationalization through the Marshall Plan's Productivity Program, under the condition that automation benefits be fairly divided between workers, consumers, and investors, based on transparent information on costs and profits.[67]

A highly orchestrated conference, Automation—Benefit or Danger?, was held in January 1958 in Essen, the center of the Ruhr mining and steel industry (figure 8.3).[68] Fritz Sternberg, a Jewish Marxist public intellectual who, during his wartime exile in the United States, had contacts with Victor Reuther, Walter's younger brother, gave the opening speech for the conference. Contrasting twentieth-century automation with the mechanization of the first Industrial Revolution, Sternberg assured the audience that workers would not again turn into Luddites destroying machines, but demanded that this time, the increase in productivity benefit workers by an increase in the standard of living, not after generations of suffering, hardship, and agony.[69] At the same time, Sternberg warned of technological unemployment. In the chemical industry, one of the prime automation sectors alongside the electronic sector, he said that productivity had increased by 50 percent from 1947 to 1954, while the overall number of employees had increased by only 15 percent, and the number of production workers a meager 1 percent. So far, economic expansion had absorbed the workers released through automation improvements, but Sternberg remarked that

Figure 8.3
Fritz Sternberg (at lectern) giving the keynote address at the DGB conference, Automation—Benefit or Danger?, in Essen in January 1958.
Source: Deutscher Gewerkschaftsbund, *Automation—Gewinn oder Gefahr? Arbeitstagung des Deutschen Gewerkschaftsbundes am 23. und 24. Januar 1958 in Essen* (Düsseldorf: Deutscher Gewerkschaftsbund, 1958), 3. Archiv der Sozialen Demokratie, Friedrich-Ebert-Stiftung.

economic growth was already in decline in the United States and Germany, rendering future unemployment likely.[70]

Unlike Reuther and the CIO, Sternberg and other German unionists assumed that automation technology would create short-term dislocation and unemployment during a transition phase, and structural long-term unemployment. This difference persisted even when, in 1957, US officers finally accommodated European wishes for a study group on automation; not even the experience in the United States swayed one of the German participants, Herbert Vogel, in his views of technological unemployment. Vogel, an official of the German white-collar union Deutsche Angestelltengewerkschaft, was part of a group of labor delegates from eight European nations visiting the United States to study automation technology and its social consequences in spring 1957.[71] Vogel reported about the strong team spirit that soon developed across different nationalities, and he was convinced that only an economically integrated Europe could face automation, hoping the political unification would be supplemented by technological unification.[72] But Vogel dismissed the US union position, suggesting that unions were supporting the management position.

By 1957, Vogel found that the automation debate in the United States had cooled down—possibly a reason why the US administration had agreed to the study tour. Most Americans with whom Vogel talked dismissed automation as a buzzword of the immediate postwar years, thereby revealing a distorted memory since the automation debate had peaked less than two years earlier. The Americans were satisfied that the sensationalist "hype" about automation had died down since it had not brought any unsolvable problems.[73] When the European group asked about technological unemployment, it was often told that automation technology had, by contrast, solved the problem of a shortage of skilled workers; obviously this argument continued to have credence in private conversations even though it was no longer used in public discourse in the United States. The group found that automation technology had not caused significant unemployment in the United States, but this was due to early retirements, higher output or higher quality, and retraining and transfers within a company, all motivated partly by managerial social responsibility and concern about a company's public reputation, and partly by guaranteed annual wage provisions. Vogel contended that while such measures protected the existing workforce, hiring freezes transferred the employment effects to those seeking work, reducing their ability to find a job.[74] In addition, the protection of the organized workforce and its seniority rights also breached the solidarity within the working population, although Vogel did not explicitly make this charge. Vogel expected that the concealed employment effects would lead to technological unemployment once economic conditions changed, such as through an economic downturn, the demobilization of the US military, or cheaper electronic computers permitting more automation. When US local unions expressed their pride in having prevented layoffs, Vogel saw the union assurances as a sham.[75]

In addition to expecting structural unemployment from automation, German unions thought that automation would mean companies would merge to form larger corporations with more capital. This argument was absent from the US automation debate and carried a particular meaning in the German context: unionists accused larger corporations of supporting the Nazi regime. Thus Otto Brenner, the militant head of IG Metall, the metal workers union that also organized workers in the automobile and electronics industries, including IBM, and who had gone to jail for being a member of the socialist resistance, reminded everyone that industrialists had supported Hitler. At the Essen conference, Brenner led the working group on economic power. He reported that under the economic system in West Germany, automation would lead to higher corporate concentration because only large companies, with a high production volume, could profit from the large investments required for automation technology and electronic computers. To prevent the accumulation of economic power in the hands of a few industrialists and managers, which could again result in "catastrophic political relations," Brenner demanded comprehensive democratic control of corporate decisions. "Industrialists cannot and must not decide on their own about the volume and targets of production, investments, prices, marketing and sales, and the make-up of the enterprise," Brenner insisted.[76] Brenner said that in the immediate postwar years, unions—along with many other social and political forces—had fought for the nationalization of core industries in a "democratically controlled social economy."[77] There were still open questions for the German Left in 1958—a year before the Social Democratic Party would distance itself from its economic reform program in order to appeal to a larger demographic. Brenner demanded that codetermination be fully implemented and expanded to ensure such democratic control.

The DGB also demanded that codetermination rights be expanded in view of automation.[78] Codetermination granted works councils a say in hires and dismissals, but only consultative rights on questions of work organization such as the introduction of new technology.[79] The DGB pursued a strategy similar to the CIO, tying the response to broader union goals: the guaranteed annual wage in the case of the CIO, and codetermination in the case of the DGB. Well into the 1960s, the DGB and Brenner along with his IG Metall continued to link automation to calls for a broadening of codetermination within and across corporations.[80]

In response to automation technology, German unions demanded that companies retrain their employees for positions of equal qualification and remuneration, pay for the transition period and, if needed, severance, and reimburse moving costs for relocation. They also considered demanding a reduced workload in the form of a shorter workweek, more paid breaks, more paid vacation, and a lower retirement age. Unions demanded that labor representatives be informed early about automation decisions so that they could evaluate the potential social effects of new technologies.[81] Ludwig Rosenberg, a member of the DGB managing board, declared in his closing statements at the Essen conference that technology could lead to "barbarism if it did not serve human beings." Following the generally balanced union view, however, he emphasized that whether automation led to "benefits or dangers" depended on union members and their actions; "labor unions see it as their task to ensure that the second industrial revolution brings workers economic and cultural benefits and thus serves social progress."[82]

The First Automation Machines in Germany

In the United States, companies had installed the first electronic computers on shop floors and in corporate offices *before* the automation debate ensued. By contrast, in Germany, the first electronic computers—a Univac computer by Remington Rand, installed in a European computing center in Frankfurt, and a smaller IBM 650, rented by the Allianz insurance company in Munich—arrived in 1956, just as the heated automation debate on both continents had interrupted exchanges.[83] But how exactly did the automation debate shape the technology and its use in Germany?

The companies manufacturing and acquiring computers all promoted the view that computers created more skilled positions for workers. IBM not only marketed its computers as productivity machines, as we saw in chapter 4; its top management insisted that the machines not be called "computers" because at the time "computers" referred to the women who manually performed mathematical calculations, and executives wanted to limit concern about the machines replacing employees.[84] Instead, IBM called its machines "electronic calculators" and fought the widespread tendency to anthropomorphize computers. For example, the mathematician John von Neumann used the term "memory" to refer to stored programs—rather than the more neutral "storage"—and compared computer memory to the "associative neurons of the human nervous system." He also talked about input and output "organs," comparing them to sensory and

motor neurons. Likewise, the insurance executive and computer prose-
lytizers Edmund Berkeley titled his book *Giant Brains*, and journalists on
both sides of the Atlantic liberally used similar terms.[85] But Thomas Watson Jr., addressing the annual meeting of the International Chamber of
Commerce in Paris 1955, emphasized that it was not correct to call IBM
machines giant brains because they could do nothing that a human had
not programmed them to do.[86] Computers would not replace human labor
because they lacked the capacity to think creatively.

In Germany, IBM reinforced this message in its customer magazine,
IBM Nachrichten. Max Woitschach, later the head of IBM Germany's basic
research department, argued that if thinking meant calculating, then
machines had been able to think for decades; if thinking meant routine
decision making, then machines had been able to think for many years;
yet if thinking meant the creative power of the mind, machines could not
think—they could merely alleviate humans' mental efforts by serving as an
aid for thinking.[87] Therefore, computers would not replace human intellectual work. While Woitschach admitted that automation would have
effects on labor, he expected it to require higher skills and lead to more
leisure time for workers. An illustration accompanying the article showed
how Woitschach may have imagined future workers returning to school to
receive more training and then exchanging their blue jumpsuits for white-
collared shirts, signifying their elevation from working to middle class.[88]
Woitschach assured German workers that automation would not lead to
poverty and unemployment, such as that caused by nineteenth-century
factory mechanization. There would always be a need for qualified workers who manually handled exceptions that could not be automated, and
qualified staff would be needed for workflow analysis and computer programming in newly created positions.[89] Employment effects, Woitschach
maintained, would be positive, leading to higher productivity and standards of living rather than to technological unemployment.

Allianz installed its first electronic computer in an effort to rationalize
the company's office operations, together with other measures from new
filing cabinets, typewriters, and dictation machines to lighting systems and
wall paint.[90] Heinz-Leo Müller-Lutz, the head of the department responsible for the company's computerization strategy, addressed automation in
an internal memorandum in late 1956, several months after the installation of the IBM 650 and when the department had begun planning for the

next computer. Looking to the United States for the latest technological developments, Müller-Lutz embraced the term "automation" and assured his superiors that the primary purpose of automation was not to exclude workers from economic life. Instead, computers were to take over the "repetitive manual and also intellectual" activities, and thus free humans beings from the monotony of work. The number of unskilled employees would decline, and the numbers of qualified skilled workers and technicians would increase. Müller-Lutz blamed the press for the negative public perception of automation technology, and hoped that the positive attitude of labor unions and the SPD would influence employees.[91]

He also warned not to overestimate the employment effects from automation. At Allianz, Müller-Lutz expected the computer to free a number of routine employees—mostly women—including keypunch operators. He predicted their number to be low since the previous transition to punch card technology had already resulted in personnel reductions. Because the releases would occur over a longer time period, Müller-Lutz expected all employees to be relocated to other positions within Allianz.[92] He may also have counted on some attrition because these routine positions had a high turnover, with women occupying them for a few years between graduating from school and marrying. In the end, however, Allianz found itself employing more—rather than less—women in routine keypunch positions because computerization required transferring all information onto punch cards before a computer could process it.[93] Allianz thus introduced its electronic computer as a German rationalization technology that would increase both the output of its office staff and the company's profits by reducing its operation costs; it did not introduce the computer as a productivity machine in the sense of the Marshall Plan with share-out agreements that would allow employees and consumers to participate in the benefits from higher productivity.

Conclusion

The debate about computer technology along with its social and economic implications was urgent enough to interrupt transatlantic exchanges. US officers had been intent on communicating their productivity ideas to West Germans and Europeans, and they had set up official channels and institutions to do so, yet when confronted with questions about automation, they chose to stop these exchanges. Of course, the automation debate was not merely about computers as technological objects; it raised larger issues about the economic system, relations between employers and employees, and the role of the state in economic questions. Automation in this way became a counterconcept to productivity.

Practically, the automation debate appears to have had little effect on deterring or even delaying the introduction of computing machinery—certainly not in the United States or West Germany. The debate, though, did affect the social and cultural meaning of computers. In the United States, computers were productivity machines that promised to raise the standard of living for everyone, including workers and consumers. Computers were technological wonders that companies acquired almost like prestige objects more than utilitarian tools and proudly displayed behind glass in entrance halls with palm tree decorations.[94] In West Germany, by contrast, computers meant the threat of technological unemployment. In a postwar climate with scarce capital resources and cheap labor, computer installations needed to be tightly justified and prove their financial benefits through meticulous accounting.[95]

Larger economic trends further determined the perception of computer technology: German worries about unemployment came true in the United States, and American expectations of higher productivity were realized in Germany (figure 8.4). In Germany, rapid economic recovery—the so-called economic miracle—led to precipitous drops in unemployment: from 11 percent in 1950 to 1.3 percent in 1960. During the 1960s, unemployment was so low—around 0.8 percent, except for a "record high" of 2.1 percent in the recession year 1967—that the German government recruited Italian and Turkish guest workers for heavy industry.[96] Because foreign workers were considered unqualified for clerical jobs and other tasks that required linguistic skills, computers came to be seen as the solution for labor scarcity in these occupations. Germans therefore enjoyed the benefits of computer technology that had been so prevalent in the minds of Americans.

In the United States, while unemployment rates initially sank from a postwar high due to the release of military forces to around 3 percent in the early 1950s, it peaked in the mid-1950s because of repeated recessions, and reached 6.8 percent in 1958 and 6.7 percent in 1961 without receding to less than 5.5 percent in the interim years. These developments changed the views of many. By the early 1960s, CIO officers like Silvey, who had

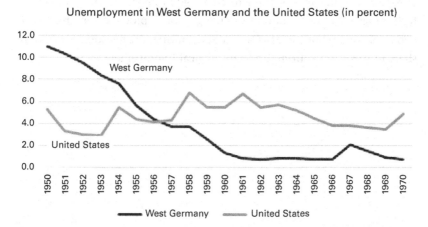

Figure 8.4
Unemployment in West Germany and the United States followed opposing trends. Sources: Bundesagentur für Arbeit, *Arbeitsmarkt in Zahlen: Arbeitslosigkeit im Zeitverlauf* (Nuremberg: Bundesagentur für Arbeit, 2017); US Bureau of Labor Statistics, *Employment Status of the Civilian Noninstitutional Population, 1940s to Date* (Washington, DC, 2017).

followed automation technology from the beginning, and promoted the view that automation would spur economic growth and higher standards of living, for the first time called for government intervention in technological improvement—normally an absolute no-no for US unions. During the 1962 presidential campaign, even John F. Kennedy identified the "manpower question"—that is, the problem of "maintain[ing] full employment at a time when automation is replacing men"—as the major domestic challenge of the 1960s.[97] Computing technology continued to soar regardless of manpower and automation questions. In the mid- to late 1950s, two IBM computer models had turned electronic computers into mass-produced items, with the IBM 650—the so-called Model T of computing—selling about two thousand copies, and the IBM 1401 selling over ten thousand copies. In the mid-1960s, IBM's System/360 promised companies a fully compatible series of computer models that would allow switching to a larger model without needing to reprogram applications, and by 1971, IBM dominated global computer markets, with more than 70 percent of the US computer market share, and more than 60 percent in every Western

European country except the United Kingdom.[98] The iconic productivity technology, electronic computers changed work processes on shop floors and in offices on both sides of the Atlantic, raising the question, who would share in the benefits from the technological improvement?

Conclusion

When Henry Luce, in 1941, declared the twentieth century to be the American Century, the astute editor and commentator noticed the trend of Europeans increasingly looking to the United States as well as studying US technological, economic, and cultural developments. The growing European interest in all things American marked the gradual ascendancy of the United States as the vanguard of modernity; the country rose to global leadership despite US political isolationism during the first decades of the twentieth century and in the face of military conflict in Europe again threatening the global order. Luce, born the son of a Presbyterian missionary in China, became a secular "missionary" promoting US global leadership for the good of the world; he also anticipated the missionary zeal that would characterize the officers of the postwar Marshall Plan program who brought American technologies along with economic and political values abroad.[1] *Productivity Machines* has provided a detailed historical study that fleshed out Luce's notion of the American Century, and its analysis of transatlantic transfers of productivity technology and culture has revealed the complex ways that technological and economic factors shaped the United States' ascendancy to global leadership, and how these factors, in turn, were shaped by it.

More specifically, *Productivity Machines* has offered important insights into the interpretative flexibility of productivity technologies as well as the persistent social and economic differences between the United States and European countries—even during the Cold War, when they belonged to the presumably homogeneous bloc of Western capitalist countries.[2] First, the book examined the interpretative flexibility of productivity culture and technology. Productivity technologies, such as Fordist mass production methods and electronic computers, carried social, economic, and political values. These values were not inherent in the technologies. As Andrew Feenberg and historians of technology have shown, values are inscribed in technologies during the process of creating them; in other words, technological values are socially constructed.[3] This inscription of values into productivity technologies did not end once the technologies were created; their values continued to be negotiated as the technologies moved across the Atlantic. Americans linked productivity technologies to high wages and low prices in a dynamically growing economy that promised a higher standard of living. By contrast, many European business owners sought to continue their economic system of low wages and high prices protected by cartel agreements. Productivity technologies thus became embedded in larger sets of economic and social ideas.

Also, a broad cast of actors—not just the small corporate elite that Feenberg identified—was involved in the continued negotiation of values. The actors included top executives, middle managers, engineers, government officers, journalists, politicians, and union officials. Yet a divide existed between labor union officials and rank-and-file union members. The former welcomed productivity technologies because they saw these technologies as a means to achieve higher economic profits that would allow higher pay without price increases, thereby protecting the purchasing power of salaries. Labor union officials willingly gave up education and relocation benefits for workers whose jobs were affected by technological change from automation to gain this macroeconomic promise. In contrast, individual workers feared losing their jobs from automation; lofty macroeconomic promises did not convince them.

In addition, the meaning of productivity along with its social and economic values changed over time. In the 1920s, officers at the Bureau of Labor Statistics initially denied any direct relationship between a worker's productivity and the wages received. At the same time, welfare capitalist entrepreneurs like Henry Ford linked productivity to lower prices and higher wages; Ford granted workers higher wages in the interest of creating markets for his mass-produced products, while he shunned any worker representation. By the late 1940s, Marshall Plan officers continued the emphasis on high wages and low prices, promising an equal sharing of productivity between industrialists, workers, and consumers. Marshall Plan officers, moreover, replaced welfare capitalism in Europe with collaborative labor relations that ensured workers a say in share-out agreements. US politicians agreed with this and made their continued support of the Marshall Plan dependent on the success of share-out agreements. The German DGB made formal share-out agreements a precondition for its participation in the US Productivity Program. Together, the continued inscription of values, wide range of social groups involved in the process, and changes of values over time revealed the interpretative flexibility of productivity technologies.

Second, differences in class relations and economic systems persisted on both sides of the Atlantic. Thus, productivity culture and technology became embedded in different conceptions and realities of class relations. As Charles Maier argued, the American economy of abundance replaced the need to (re)distribute scarce economic resources; instead, everyone would participate in dynamically growing resources.[4] Officers of the US Productivity Program sought to use this promise of higher living standards for everyone in order to change class relations in European societies—unsuccessfully. In the United States, the working class benefited from higher wages and lower prices and identified as middle class. In West Germany, many industrialists proved hesitant, if not unwilling, to share productivity gains in the form of higher wages or lower prices, averting any distributive effects. Even when workers received financial gains from productivity, social class differences persisted. While these differences were no longer necessarily marked by income disparities, they continued to be expressed in distinct consumption patterns—from clothing to housing and vehicles—that were deemed socially appropriate for different classes and marked daily interactions.

Also, differences continued between the US and West German capitalist economies—mostly in line with political economists' arguments about "varieties of capitalism" identified as liberal market economies and coordinated market economies.[5] By not taking these notions as heuristic concepts and instead investigating the historical evolution of the economic system in each country, *Productivity Machines* has exposed the messy comingling of different economic characteristics in both countries. In the supposedly liberal economy of the United States, we have discovered elements of planning originating from New Deal programs as well as wartime price and wage controls; these elements continued into the postwar period and even increased through military mobilization during the Korean crisis. In West Germany, economic minister Ludwig Erhard sought to implement a social market economy based on free competition protected by the

government—an economic model diametrically opposed to the mechanisms of a coordinated market economy that supposedly characterized
West Germany. But over time, codetermination and other changes introduced corporatist elements, giving labor unions and business organizations
more influence in economic decisions, in lieu of free competition. West
Germany's coordinated market economy thus was revealed as a complex
amalgam of free competition and German corporatist traditions. One could
say that the "real existing forms of capitalism" on both sides of the Atlantic
proved more complex than any clearly distinguished economic models or
varieties of capitalism would have us believe.

Notably, these social and economic differences persisted despite intense
transatlantic interactions, and even despite American efforts at reshaping
European social and economic relations after the US model. Before World
War II, individual American businessmen such as Ford communicated ideas
about labor and economic relations to audiences abroad, and Weimar Germans and other Europeans eagerly studied the production technologies
behind the US economic prowess. After World War II, the Marshall Plan
administration—with support from US corporations and labor unions—
undertook a concerted effort through its Productivity Program to reshape
class and labor relations in Western Europe. Yet even West Germans—
whose political and economic institutions had been delegitimated through
Nazi excesses—opted for their own tradition of codetermination over, for
example, the US model of collaborative labor relations. These persistent differences are particularly remarkable during the emerging Cold War competition, which many think of as the confrontation between a homogeneous,
capitalist West and the communist East. By contrast, *Productivity Machines*
has cast new light on the notion of the "West," revealing continuing transatlantic differences between capitalist economies even during the height of
the Cold War.

Finally, technology was an important factor in the ascendance of the
United States to global power; *Productivity Machines* thus has substantiated
the calls by Walter LaFeber and others that scholars pay attention to the
role of technology in foreign relations. Different from military or nuclear
technologies, the productivity technologies investigated here—mass production methods and electronic computers—are not among the technologies employed to achieve government objectives in foreign relations
through the threat of armed conflict. Rather, they allowed Americans to

buttress their country's economic prowess and claim superiority through the means of soft versus hard power. By investigating the transition from production lines (the productivity technology of the second Industrial Revolution) to electronic computers (the productivity technology of the third Industrial Revolution), *Productivity Machines* has also shown that electronic computers began to replace production lines as iconic productivity technologies in the 1950s and 1960s. Two decades before information technologies started to capture the imagination of diplomats as well as political and academic analysts in the 1970s, information technologies already shaped foreign relations and were shaped by them.[6] *Productivity Machines* has, as such, revealed the deep historical roots of information in our globalized network society—roots that got their start decades before computers and computer networks became pervasive parts of corporate and daily lives.[7]

Teasing out the persistent differences between capitalist economies and their labor and class relations, *Productivity Machines* has also related to the work of scholars exploring the exceptionalism of the US economy. At the turn of the twentieth century, the German sociologist and economic expert Werner Sombart opened a century-long debate about American exceptionalism when he asked why there was no socialism in the United States. Sombart was then one of the leading German experts on socialist and Marxist theory, even though some of his later writings proved conducive to the Nazis. Following a visit to the United States for the 1904 World's Fair in Saint Louis with a group of German intellectuals (including Max Weber), Sombart noted that the United States, arguably the most capitalist country in the world, did not develop a working-class consciousness and working-class political party.[8] He laid out some of the observations that German visitors would continue to grapple with over the next five decades: the lack of feudalism and its impact on social relations; the higher standard of living among working people; more friendly relations between classes; universal male suffrage giving workers a sense of political citizenship; the electoral college favoring two dominant parties; the dominant parties incorporating working people's interests; higher social mobility; and higher geographic mobility because of the open western frontier.[9] While Sombart died in 1941, the year that Luce declared the twentieth century to be the American Century, *Productivity Machines* has followed the questions that Sombart raised at the beginning of the twentieth century into midcentury transatlantic relations, providing a history of capitalist technologies and thought.

By midcentury, the perception of exceptionalism led American businessmen, union officers, engineers, and government officials to bring
American technologies and economic ideas across the Atlantic, couched
in the terms of US productivity. From Ford and Edward Filene, to the Marshall Plan's Productivity Program officers, to business and labor delegates,
Americans viewed their economy and labor relations as a model to bring
to other industrialized countries on the European continent. German businessmen, unionists, engineers, journalists, and politicians—from Weimar
travelers like Köttgen and the General German Union Association union
group, to the Marshall Plan's productivity study groups, exchange students,
and Project Impact participants—perceived US technology and economic
relations as worthwhile to study, but they did not unthinkingly accept the
US model. While there certainly were technological and economic differences between the United States and Europe, I do not claim that America
was indeed exceptional; instead, I have written the history of transatlantic exchanges that were grounded on the mutual *perception* of exceptionalism, where historical players acted on what they made out as American
exceptionalism.

At the same time, *Productivity Machines* has revealed implicit assumptions
in the American economic system that served as the basis for these perceptions of American exceptionalism. Most important, the US economic model
was based on economic growth in a dynamically expanding economy. Economic growth and expansion may generally be inscribed in capitalist economic relations, with the use of resources for investment in higher gains in
the future.[10] But the American view of *growth* as the engine of the economic
system was diametrically opposed to European assumptions of *stable economic resources* that needed to be redistributed; Americans and Europeans
differed in the role that they accorded to economic growth. *Productivity
Machines* has shown that American expectations of economic growth and
higher standards of living through increasing productivity from technological improvements were unique to the American economic model, and
not inherent in capitalist economic relations. Even critics of the history of
capitalism and its implicit assumptions of teleological development have
acceded that capitalism had "long-term temporal dynamics that possess a
powerful directionality," including sustained economic growth, allowing
for population growth and income per capita growth since the early eigh-
teenth century.[11] Without doubting this general dynamic, this book has

illuminated the details of why and how different capitalist systems have different conceptions of economic growth, revealing historical contingency in what could be seen as temporal directionality. It has explained why different varieties of capitalism existed, and persist.

While *Productivity Machines* has focused on transatlantic relations, it connects to the history of global economic development. As Marshall aid began flowing to Europe, President Truman announced the Point Four Program, so named because it was the fourth point in his inaugural address in January 1949. The program, which provided technical assistance to develop and modernize countries in the Global South, had roots in US domestic economic development, such as the hydroelectric dams and electrification projects of the Tennessee Valley Authority, and built on US engagement in East Asia and the Western hemisphere. Like the Productivity Program, one of the goals of the Point Four Program was to raise the standard of living by the application of technology.[12] Still, power plants, dams, roads, and bridges played a more important role in raising agricultural economies into industrial modernity than the mass production and electronic automation technologies used to manufacture the consumer goods, such as cars, refrigerators, television sets, and ballpoint pens, that flooded markets in the United States and Western European countries. Despite many differences between the Marshall Plan and the Point Four Program, their connections raise questions about the relations between transatlantic and global technological and cultural exchanges in the postcolonial world; at the very least, both programs aimed at reshaping other countries around the globe according to the US model and strengthening US global leadership during the Cold War. It remains to be studied whether and how they held similar values with regard to free enterprise, assumptions about the role of the government and economic planning, and labor and social relations.

While proponents of productivity technology and culture have historically promised higher standards of living, analysts have in recent years diagnosed increasing economic inequality within and between countries around the world. In the last four decades, income inequality has increased in many countries, albeit more significantly in the United States than in Europe. In the United States, the top 10 percent of earners have increased their share of national income from 35 percent in 1980 to 47 percent in 2016, while in Europe, the share of the top earners increased only from 33 percent in 1980 to 37 percent in 2016. In the United States since 1996, the top 1 percent earn more than the bottom 50 percent, with the top earners capturing 20 percent of the national income in 2016, and the bottom earners capturing just 13 percent.[13] This provokes the following questions. Does productivity technology and culture realize different things based on local social, economic, and political values? Does productivity technology and culture have an inherent logic that causes (or is one cause of) increasing inequality? Through this book I hope to raise awareness about the gap between productivity promises of dynamic growth and actual growing inequality.

In addition, the promise of dynamic economic growth assumes that it is acceptable to use ever-increasing amounts of natural resources, yet we are becoming aware that environmental and climate effects are the hidden costs of capitalism.[14] There are also environmental consequences of our information technologies. For example, information devices are built from rare earth metals that are often mined in China and other countries of the Global South under minimal environmental standards, and issues of materials sourcing may soon make questions of energy independence mute. The energy use of our computers, including cloud storage and Bitcoin mining, at the global scale exceeds the energy use and carbon dioxide emission of our transportation technologies. And the short life cycles of electronic devices—which are dumped in the Global South and dismantled by young laborers wearing no protection from dangerous ingredients—pose questions about electronic waste.[15] *Productivity Machines* has raised the question about how far these environmental consequences are embedded in the logic of the productivity promises of a dynamically growing economy. In the end, *Productivity Machines* asks, What kind of capitalism do we want in our globalized world?

Closer to home, the expectation of ever-increasing economic growth—based on the productivity technology and culture studied here—has pervaded lives on both sides of the Atlantic. For example, in my own university, the faculty are encouraged to teach more students every year, publish more papers and books, and win more grant dollars than the year before; in other areas, employees may be expected to serve more customers, write more lines of codes, or raise more donations. Politicians and the media lead us to expect ever-increasing gross domestic product and stock market indexes. Sometimes, we are asked to achieve more with the same number of people and financial resources, and without any technological improvements. In other words, we are urged to increase our output by working harder or working more hours—just as Köttgen urged Germans to do after World War I—not by improving work processes. In an effort to combat this expectation of continuous growth, the slow movement has emerged—beginning with the slow food protest against industrialized production of fast food in the 1980s. Today, even faculty are counseled to slow down and set their goals deliberately rather than seeking to simply increase the quantitative measures of their output.[16] As we become aware of the inherent forces of productivity technology and culture, we can debate their values and make conscious choices about what we allow in our own lives.[17]

Notes

Introduction

1. Luce, "The American Century."

2. For the US ascendancy to global power, see White, *The American Century*; Nolan, *The Transatlantic Century*.

3. Adas, *Dominance by Design*; Adas, *Machines as the Measure of Man*; Rosenberg, *Spreading the American Dream*; Ekbladh, *The Great American Mission*.

4. I developed this project in the context of the Tensions of Europe network and the participants' understanding of circulation and appropriation as a process that involved agency on both sides of the Atlantic. See Misa and Schot, "Inventing Europe." For results of the network, see Schot and Scranton, *Making Europe*.

5. For interpretative flexibility, see Pinch and Bijker, "The Social Construction of Facts and Artifacts."

6. Feenberg, *Critical Theory of Technology*. A decade earlier, political theorist Langdon Winner asserted that the invention, design, and arrangement of technological devices could settle political issues, and that some technological systems were compatible with and could reinforce certain political relationships. Despite criticism of his historical examples, his arguments remain important. Winner, "Do Artifacts Have Politics?" Similarly, David Noble has shown how management used automation technology to shape labor relations and increase its workplace control with the choice of numerical control over record-replay technologies. Noble, *The Forces of Production*. Gabrielle Hecht developed the concept of "technopolitics" to describe the strategic design and use of technologies to pursue political goals. Hecht, *The Radiance of France*, 15.

7. In their studies of technology transfers in the colonial period, Daniel Headrick and Michael Adas have examined the mutual constitution of technologies and political values in transnational contexts, showing that technologies were not neutral means but instead structured power relations between colonizers and colonized, and

how they perceived each other. Headrick, *The Tools of Empire*; Adas, *Machines as the Measure of Man*; Adas, *Dominance by Design*. For the technopolitics of postcolonial relations, see Hecht, *Entangled Geographies*.

8. Feenberg, *Alternative Modernity*.

9. For excellent surveys of the history of computing in the United States, see Campbell-Kelly et al., *Computer*; Ceruzzi, *A History of Modern Computing*. For computing in countries other than the United States, see Gerovitch, *From Newspeak to Cyberspeak*; Medina, *Cybernetic Revolutionaries*; Mullaney, *The Chinese Typewriter*; Peters, *How Not to Network a Nation*; Hicks, *Programmed Inequality*. On the need for an international history of computing, see Schlombs, "Toward International Computing History."

10. Cortada, *The Digital Flood*, 17. In addition, historians, economists, and political scientists have provided comparative studies of national innovation systems and technological development policies that were colored by concerns about national competitiveness in the computer industry. See, for example, Flamm, *Targeting the Computer*; Aspray, *Technological Competitiveness*; Mowery, *The International Computer Software Industry*; Coopey, *Information Technology Policy*. More recently, see Campbell-Kelly and Garcia-Swartz, *From Mainframes to Smartphones*.

11. LaFeber, "Technology and U.S. Foreign Relations." In the same year, Odd Arne Westad suggested technology history as a new approach to studying Cold War history, and he explicitly called for work on the birth of the electronics industry. Westad, "Bernath Lecture."

12. The US developed its military prowess over the first half of the twentieth century, culminating in its command over nuclear weapons by mid-century. Epstein, *Torpedo*; Abraham, "The Ambivalence of Nuclear Histories." For a recent survey of technology in US international relations, see Winkler, "Technology and the Environment in the Global Economy." See also the 2006 OSIRIS issue on technology in international affairs, Krige and Barth, eds., *Global Power Knowledge*.

13. Van Vleck, *Empire of the Air*; Muir-Harmony, "Project Apollo"; Ekbladh, *The Great American Mission*. Of course, military and civilian technologies cannot be strictly separated—for example, military procurement shaped corporate manufacturing processes.

14. For the original argument, see Hall and Soskice, *Varieties of Capitalism*. For a later discussion, see Coates, *Varieties of Capitalism, Varieties of Approaches*; Hancké, Rhodes, and Thatcher, *Beyond Varieties of Capitalism*. The varieties of capitalism can be seen as the empirical validation of Feenberg's argument that there is not *one* industrial civilization that is shaped by its technologies. Instead, different industrial societies can be imagined, based on how their technologies are shaped.

15. Werner Abelshauser pursued the same goal, at a higher level of generality. Abelshauser, *Kulturkampf.* The literature on Americanization—that is, the American cultural, economic, and political influence on other countries—similarly focused on interaction between American and European economies. See chapters 5 and 6. While varieties of capitalism proponents emphasized differences between the US and European economies, scholars of Americanization tend to highlight assimilation processes. *Productivity Machines* agrees with the more recent research on Americanization that transatlantic cultural and economic differences persisted.

16. For a discussion of the "Stunde Null" hypothesis, see, for example, Geppert, *The Postwar Challenge*, 5.

17. Abelshauser, "The First Post-Liberal Nation," 304.

18. Waring, *Taylorism Transformed.* See also Devinat, *Scientific Management in Europe*, 3, 41, 45; Spender and Kijne, *Scientific Management*, vii; Bloemen, "The Movement for Scientific Management in Europe between the Wars"; Guillén, *Models of Management*, 8, 100–109.

19. By contrast, in social policy and social science, ideas primarily flowed westward, initially through US students and thinkers visiting Germany, and later through European and particularly German visitors and émigrés to the United States, such as the members of the Frankfurt school. Rodgers, *Atlantic Crossings*; Logemann and Nolan, *More Atlantic Crossings?*

20. For the domestic German computer industry, see Petzold, *Rechnende Maschinen.*

21. Milward, *The Reconstruction of Western Europe*; Abelshauser, *Wirtschaft in Westdeutschland.*

22. Histories of capitalism have seen a resurgence. For overviews, see Rockman, "What Makes the History of Capitalism Newsworth?"; Beckert, "History of American Capitalism." Yet this literature focuses mostly on the nineteenth century and preceding centuries.

23. Sombart, *Why Is There No Socialism in the United States?* Historians have since addressed Sombart's question. See, for example, Foner, "Why Is There No Socialism in the United States?"; Lipset and Marks, *It Didn't Happen Here*; Archer, *Why Is There No Labor Party in the United States?*; Beilharz, *Socialism and Modernity*, 189–200.

Chapter 1

1. Clague, "Productivity and Wages in the United States," 292.

2. "Ewan Clague, Ex-Labor Data Official, Dies."

3. Clague, *Reminiscences of Ewan Clague: Oral History, 1958*, 50–51.

4. In the late nineteenth century, workers began to see themselves as consumers, asking for a living wage that would guarantee an "American standard of living." Glickman, *A Living Wage*, 78–83.

5. Montgomery, *Workers' Control in America*.

6. Stapleford, *The Cost of Living in America*; Clague, *The Bureau of Labor Statistics*.

7. Draft report, "Labor Efficiency in the Manufacture of Common Brick," n.d., 1, 11, 13, box 10, entry 31, RG 257, NARA; "The Development of the Baking Industry," August 30, 1923, 1, box 10, entry 31, RG 257, NARA; "Relative Productivity of Workers under Hand and Machine Methods of Manufacture in the Canning of Tomatoes," n.d., box 10, entry 31, RG 257, NARA; "Relative Productivity of Workers under Hand and Machine Methods of Printing Newspapers," n.d., box 10, entry 31, RG 257, NARA.

8. Stapleford, *The Cost of Living in America*, 184–252.

9. Memo from James Silberman to Ewan Clague, "Assignment as OSR Consultant on Productivity, May–June 1950," July 12, 1950, Silberman Papers, Truman Library.

10. Porter, *Trust in Numbers*.

11. "Ewan Clague, Former Chief of Labor Statistics, Dies." For interpretative flexibility, see Pinch and Bijker, "The Social Construction of Facts and Artifacts," 40–44.

12. Clague, *The Bureau of Labor Statistics*, 3, 8–15.

13. Stapleford, *The Cost of Living in America*, 22–58.

14. Following World War I, conservative federal governments cut the BLS budget, forcing the bureau to go from quarterly to half-yearly publication of the cost-of-living index, which was also decoupled from wage negotiations. Work on the cost-of-living index again picked up during the New Deal years. Ibid., 78–95, 149–183.

15. Wright, *Hand and Machine Labor*, 5.

16. Ibid., 11.

17. Ibid., 11.

18. The report covered 672 product "units," which typically referred to aggregates such as 12 half hoses or 1,000 bricks rather than single items. It excluded products from new industries for which no comparable hand production methods existed, such as sewing machines or bicycles, but included products that had changed from hand to machine production, although doing so resulted in comparing unequal counterparts such as handmade plows with wooden moldboards compared to machine-made cast iron plows.

19. Wright, *Hand and Machine Labor*, 5–6.

20. Ibid., 20.

21. Notably, both hand and machine methods were in use in 1895. It appears that BLS officers chose facilities that worked with different clay qualities in each method: one where the bricks required drying before being molded, and one where the bricks were directly molded and stacked.

22. Wright, *Hand and Machine Labor*, 614–617.

23. The salary differences do not seem to reflect differences in age or experience. Ibid., 614–617.

24. Ibid., 30–31.

25. Montgomery, *Workers' Control in America*; Braverman, *Labor and Monopoly Capital.*

26. Montgomery, *Workers' Control in America*, 140.

27. Glickman, *A Living Wage.* While labor leaders resisted quantifying this standard of living, it generally included the costs for food, clothing, taxes, schoolbooks, furniture, newspapers, medical bills, and contributions to religious causes. White, male wage earners were to earn this American standard of living for themselves and their families; different standards were to apply to female, African American, immigrant, and unskilled workers.

28. While the 1898 study had examined different product units, in the 1920s, the BLS officers bundled different products in industry reports.

29. Desrosières, *The Politics of Large Numbers*, 204–206, 210–235. See also Igo, *The Averaged American.*

30. Letter from Ethelbert Stewart to H. C. Schranck, H. C. Schranck Co., May 29, 1923, box 10, entry 31, RG 257, NARA; letter from Ethelbert Stewart to A. J. Todd, labor manager, B. Kuppenheimer and Co., October 25, 1923, box 10, entry 31, RG 257, NARA.

31. See, for example, reports on the relative productivity of the canning of tomatoes, peas, and corn, the manufacture of cans, the manufacture of men's collars, negligee shirts, and half hoses, and the printing and publishing of newspapers and magazines. Box 10, entry 31, RG 257, NARA.

32. The archival records do not indicate whether these reports were ever published.

33. These kinds of reports were not untypical for the statistical practices of the day, although British statisticians followed more arithmetic methods than German statisticians. Desrosières, *The Politics of Large Numbers*, 19–23.

34. The four plants for which BLS raised more detailed numerical data are not identified in terms of the processes that they used or their location. "Year 1921. Number

of Brick per One-Man Hour Produced in Plant," n.d., box 10, entry 31, RG 257, NARA.

35. The archival records do not include any information on the provenience of these data. Thus, it is also unclear how these plants were selected, and whether they provide a representative sample or even exhaustive coverage of the brick manufacturing industry.

36. Draft report, "Labor Efficiency in the Manufacture of Common Brick," n.d., 3, 6, 7, box 10, entry 31, RG 257, NARA. Wright's *Hand and Machine Labor* presented the dry-clay process as the hand method.

37. Draft report, "Labor Efficiency in the Manufacture of Common Brick," 11.

38. Ibid., 5.

39. Ibid., 3–5.

40. Letter from Ralph P. Stoddard, secretary-manager of the Common Brick Manufacturers' Association, to Ethelbert Stewart, BLS commissioner, May 16, 1922, box 10, entry 31, RG 257, NARA.

41. Ibid.

42. The report appears to suggest a possible correlation between plant size and productivity, with the bigger plants using more machinery. "Year 1921. Number of Bricks per One-Man Hour Produced in Plant."

43. Clague, *Reminiscences of Ewan Clague: Oral History, 1966*, 2, 16.

44. Stapleford, *The Cost of Living in America*, 149–169. See also Cortada, *All the Facts*, 136–159.

45. Clague, *Reminiscences of Ewan Clague: Oral History, 1958*, 48, 51.

46. Although the shift from a handful of voluntarily reporting companies to the large corpus of census data may appear like a huge methodological step, it would have been a more radical shift had BLS officers used data from a random sample—a methodology that national market surveys and election polls would popularize in the United States in the 1930s. Desrosières, *The Politics of Large Numbers*, 19–23, 204–206, 210–235.

47. Clague, "Index of Productivity of Labor," 1.

48. Clague published these articles between July 1926 and January 1927 as follows: Clague, "Index of Productivity of Labor"; Anonymous, "Productivity of Labor in the Cement, Leather, Flour, and Sugar-Refining Industries, 1914 to 1925"; Anonymous, "Productivity of Labor in Slaughtering and Meat Packing and in Petroleum Refining"; Anonymous, "Productivity of Labor in the Rubber Tire and the Iron and Steel (Revised) Industries"; Anonymous, "Productivity of Labor in Eleven Industries."

49. Clague, "Index of Productivity of Labor," 2.

50. "Man-hours," a term used by the historical actors—and in some industries, such as computing, was common until at least the 1970s—of course implies a gendered conception of the (appropriate) workforce. It has been used here as a historical actors' category rather than replacing it with a more neutrally gendered term such as "worker hours."

51. Clague, "Index of Productivity of Labor," 2.

52. Using 1914—a year of economic depression—as a base year resulted in an over-estimation of productivity increases, particularly in industries that immediately reacted to economic decline such as the iron and steel industry. For this industry, Clague thus calculated an average productivity for 1914 and 1916, a year with stronger overall economic performance. Anonymous, "Productivity of Labor in Eleven Industries," 37.

53. Ibid., 35–36.

54. Clague, "Index of Productivity of Labor," 5.

55. Ibid., 6.

56. For the employment index, Clague relied on the Census of Manufactures and collated the information with BLS data on month-to-month employment to account for problems from occasional changes in the classification of employees, the exclusion of smaller companies from the census, and data compilation errors. Ibid., 2–4.

57. Ibid., 2–4. The employment index or index of the total man-hours, however, did not take into account the differences in skill and ability among different classes of workers. For example, a new machine may have replaced five unskilled (lower-paid) workers but required two highly trained (higher-paid) employees to tend to it, saving physical labor without necessarily translating into financial savings. Ibid., 17–18.

58. Anonymous, "Productivity of Labor in Eleven Industries," 48.

59. Productivity was set at 100 for the base year 1914. A productivity index of 100 in a succeeding year would indicate flat productivity compared to 1914; an index of 106 indicates a 6 percent increase in productivity compared to 1914.

60. Anonymous, "Productivity of Labor in Eleven Industries," 45.

61. Productivity in the automobile industry was measured in the number of cars produced, and the industry manufactured other products in addition to cars during World War I, explaining the seeming decline in 1918.

62. Anonymous, "Productivity of Labor in Eleven Industries," 38.

63. For both the cane sugar refining and slaughtering and meatpacking industries, labor costs played less of a role than the cost of the raw material; productivity improvements thus played a less vital factor for the overall profits in these industries than in other ones. Ibid., 39–40, 45.

64. Ibid., 49.

65. Clague, "Productivity and Wages in the United States."

66. Clague, *Reminiscences of Ewan Clague: Oral History, 1958*, 1–5, 96.

67. Ewan Clague, "Notes on Iron and Steel Productivity Data," November 16, 1927, 4, box 4, RG 200, NARA.

68. In his planning, Clague sought to make the best use of the data while limiting the amount of calculation. Ibid.

69. Stewart, "Standardization of Output by Agreement," 263. See also Clague, *The Bureau of Labor Statistics*, viii, 17.

70. "A Proposal for Improving Railway Shop Production," n.d., 4, box 10, entry 31, RG 257, NARA.

71. Letter of O. S. Beyer Jr., consulting engineer for the Labor Bureau, Inc., to Ethelbert Stewart, April 22, 1922, box 10, entry 31, RG 257, NARA.

72. Clague, "Index of Productivity of Labor," 19; Anonymous, "Productivity of Labor in Eleven Industries," 49.

73. Clague, "Productivity and Wages in the United States," 290–292.

74. Ibid., 292.

75. Ibid., 292–293.

76. Ibid., 294–295.

77. Ibid., 295.

78. Ibid., 296.

79. Montgomery, *Workers' Control in America*, 123–124; Glickman, *A Living Wage*, 78–80.

80. Various blank schedules, box 4, RG 200, NARA.

81. Clague, *Reminiscences of Ewan Clague: Oral History, 1958*, 72–78. By 1933, Clague found that the data for the iron and steel study had been destroyed under the five-year rule.

Chapter 2

1. Köttgen, *Das wirtschaftliche Amerika*, 48.

2. This chapter is deeply inspired by Mary Nolan's groundbreaking work on Weimar German appropriations of American modernity. Nolan, *Visions of Modernity*. While Nolan analyzed the wide-ranging German debate about American culture and society, from labor relations and the educational system to family life and the role of women, this chapter focuses on the technological and economic aspects. For American efforts at promoting US ideas about wages, consumption, and the standard of living in Europe, see de Grazia, *Irresistible Empire*.

3. Allgemeiner Deutscher Gewerkschaftsbund, *Amerikareise deutscher Gewerkschaftsführer*, 254–255.

4. Jacoby, *Modern Manors*. While welfare capitalism may have been well suited for the American environment of large firms, weak unions, and small government, similarly paternalistic systems of labor relations emerged in other industrialized countries, including the German Empire (1871–1918), where many firms provided housing and created benefits programs—such as bonus payments and insurance programs—that sometimes transgressed the workers' private lives. Sachse, *Betriebliche Sozialpolitik als Familienpolitik*. For an example of the transatlantic circulation of welfare capitalist ideas, see Kohlrausch and Trischler, *Building Europe on Expertise*, 132–139.

5. Historians have reversed the perception that welfare capitalism found its demise in the 1930s and instead have argued that welfare capitalism survived in a modified, modernized version into the decades following World War II. Jacoby, *Modern Manors*; Cohen, *Making a New Deal*.

6. Like welfare capitalists, Taylorists typically avoided working with unions. But Weimar Germans related to the Taylorist corporatist promise of overcoming class through knowledge. Waring, *Taylorism Transformed*, 13–18; Maier, "Between Taylorism and Technocracy."

7. The term "Rationalisierung" is ascribed to the political economist Friedrich von Gottl-Ottlilienfeld, who also coined the term "Fordismus." Von Saldern and Hachtmann, "Das fordistische Jahrhundert." For a contemporary analysis, see Brady, "The Meaning of Rationalization." For technical appropriation, see Alexander, *The Mantra of Efficiency*, 101–125.

8. Weimar Germans used terms such as "work intensity" or "quantity of work" to assess the work output per unit of time—that is, productivity. Although historians have used the term "productivity" to describe the discussion of labor in Germany before the Second World War, the German term "Produktivität" was not utilized until the late 1940s.

9. In this interpretation, I follow Nolan, who also identified a plethora of German voices. Nolan, *Visions of Modernity*, 5. For a distinction between industrialist and labor positions, see Hausser, *Amerikanisierung der Arbeit?*

10. Jeremy, *Transatlantic Industrial Revolution*; Stapleton, *Transfer of Early Industrial Technologies*.

11. Rodgers, *Atlantic Crossings*. For graduate schools, see Hannaway, "The German Model of Chemical Education."

12. Wilkins, *The Emergence of Multinational Enterprise*.

13. Rosenberg, *Spreading the American Dream*, 38–86.

14. De Grazia, *Irresistible Empire*.

15. Jones, *Multinationals and Global Capitalism*; Wilkins, *The Maturing of Multinational Enterprise*.

16. For Ford's international expansion, see Wilkins and Hill, *American Business Abroad*.

17. Wilkins, *The Emergence of Multinational Enterprise*, 96.

18. Wilkins and Hill, *American Business Abroad*.

19. Ridgeway, *Merchants of Peace: Twenty Years*; Ridgeway, *Merchants of Peace: The History*.

20. Ridgeway, *Merchants of Peace: Twenty Years*; Rosenberg, *Spreading the American Dream*.

21. De Grazia, *Irresistible Empire*. For Filene's employment policies, see Jacobs, *Pocketbook Politics*, 50, 79–80.

22. Ridgeway, *Merchants of Peace: Twenty Years*, 384.

23. Maney, *The Maverick and His Machine*, 203–208; Watson and Petre, *Father, Son & Co*, 53–56.

24. While the history of IBM's domestic operations is burgeoning, some information about the company's international operations can be gleaned from Campbell-Kelly, *ICL*; Petzold, *Rechnende Maschinen*; Heide, *Punched-Card Systems*; Maney, *The Maverick and His Machine*.

25. Ford, *My Life and Work*. For the German reception, see Eifert, "Antisemit und Autokönig." For a popular but balanced biography of Ford, see Curcio, *Henry Ford*. For a critical biography based on extensive archival research, see Gelderman, *Henry Ford*.

26. Meyer, *The Five Dollar Day*.

27. For a historical analysis of the technological aspects of Ford Motor Company's production system, see Hounshell, *From the American System to Mass Production*.

28. Ford, *My Life and Work*, 116–120; Gelderman, *Henry Ford*, 23–24, 72–78, 172; Curcio, *Henry Ford*, 93–94, 103–104. Ford's critique of financial capitalism can also be seen in the context of his anti-Semitism.

29. Ford, *My Life and Work*, 87.

30. Ibid., 88. On the company's labor relations, see Meyer, *The Five Dollar Day*; Gelderman, *Henry Ford*, 330; Curcio, *Henry Ford*, 245–248.

31. Ford, *My Life and Work*, 100–101.

32. Ibid., 99, 106.

33. Ibid., 53, 100–101.

34. Ford was a vegan who tried to convince workers of a healthy lifestyle based on a soy diet. He didn't believe in charitable giving, and ran a boys' trade school and a hospital to prove that charity was unnecessary. He abhorred urbanization and agricultural concentration, and wanted workers to till small farms surrounding small plants. He was an ardent pacifist and participated in a futile peace mission during World War I. He also was anti-Semitic and allowed his company to become instrumental in the Nazi mobilization for World War II. Ford, *My Life and Work*, 129–131, 140–150, 168–169; Gelderman, *Henry Ford*, 239–240, 348–360; Curcio, *Henry Ford*, 155–156, 250–256.

35. Gelderman, *Henry Ford*, 321.

36. Ibid., 343–345; Curcio, *Henry Ford*, 245–250; Wilkins and Hill, *American Business Abroad*, 53; Ford, *My Life and Work*, 176.

37. While most writers reflected on their American experiences, some exclusively wrote about Ford, his production methods, and larger social and economic ideas. See, for example, von Gottl-Ottlilienfeld, *Fordismus*; Rieppel, *Ford-Betriebe und Ford-Methoden*; Faldix, *Henry Ford als Wirtschaftspolitiker*.

38. Riebensahm, *Der Zug nach U.S.A.*, 1.

39. Salomon, *Kultur im Werden*. While German publishing houses printed over two dozen book-length travel reports within the brief span of five years, many more reports circulated only internally, without finding an audience beyond the author's company or organization. The majority of Weimar travel reports were analytic, focusing on economic topics such as production methods, labor relations, and wage and price levels, rather than following the established narrative form of travel writing. See, for example, Holitscher, *Wiedersehen mit Amerika*.

40. Von Gottl-Ottlilienfeld used the term "Fordism" to define a system of efficient production methods coupled with the reinvestment of profits, high salaries, and low prices. Von Gottl-Ottlilienfeld, *Fordismus*, v–vi.

41. Riebensahm, *Der Zug nach U.S.A*, 5–7. One marker of the German bourgeoisie was education, particularly in the 1920s, when members of the lower bourgeoisie found it increasingly difficult to maintain an appropriate bourgeois lifestyle.

42. For an overview of the German rationalization movement, written from the perspective of a contemporary economist, see Brady, *The Rationalization Movement in German Industry*. See also Merkle, *Management and Ideology*, 172–207; Mai, "Politische Krise und Rationalisierungsdiskurs"; Shearer, "The Politics of Industrial Efficiency"; Homburg, *Rationalisierung und Industriearbeit*; von Freyberg, *Industrielle Rationalisierung*; Kleinschmidt, *Rationalisierung als Unternehmensstrategie*; Stahlmann, *Die Erste Revolution in der Autoindustrie*.

43. The first to raise this question was Brady, *The Rationalization Movement in German Industry*. J. Ronald Shearer compared actual and theoretical—that is, without rationalization methods—labor forces in mining in the 1920s, coming to the conclusion that technological mining rationalization contributed to rising unemployment rates by the late 1920s, in addition to the economic depression. Shearer, "The Politics of Industrial Efficiency," 380–397.

44. Von Gottl-Ottlilienfeld, *Fordismus*, 1–41.

45. This language was adopted by the Nazi movement. Von Gottl-Ottlilienfeld continued his career during the Third Reich, although he was considered an "old-school scholar" rather than an ardent ideologue.

46. The acronym "Weimar RKW" distinguishes the Reichskuratorium für Wirtschaftlichkeit from its post–World War II successor organization, the Rationalisierungskuratorium der Deutschen Wirtschaft (RKW). It should be noted that the "Weimar RKW" continued operations throughout the Nazi period. See chapter 5 for further discussion of RKW.

47. Pohl, "Die Geschichte der Rationalisierung," 87.

48. Shearer, "The Reichskuratorium für Wirtschaftlichkeit."

49. Maier, "Between Taylorism and Technocracy," 46–48.

50. Stollberg, *Die Rationalisierungsdebatte*, 72–73.

51. Maier, "Between Taylorism and Technocracy," 54.

52. Westermann, *Amerika wie ich es sah*, 9–10.

53. Bonn, *Geld und Geist*.

54. Riebensahm, *Der Zug nach U.S.A.*, 9–11, 17–19. Riebensahm's comments can be read as a snipe against Köttgen, who had sought a solution in a dense statistical report. Köttgen, *Das wirtschaftliche Amerika*, 7–29.

55. For Siemens's corporate history and its labor relations, see Feldenkirchen, *Siemens*, 185, 355–356; Kocka, *Unternehmensverwaltung und Angestelltenschaft*. For its rationalization process, see von Freyberg, *Industrielle Rationalisierung*.

56. Dienel, "Hier sauber und gründlich, dort husch-husch, fertig," In contrast, Joachim Radkau described the history of technology in Germany from the late nineteenth century onward as a succession of fascination with American production technologies and rejection of the same. Radkau, *Technik in Deutschland*.

57. In November 1918, as a result of the post–World War I revolution, the 8-hour day was introduced throughout German industry. In mining, reductions to 7.5 and 7 hours followed in March and April 1919, and pressure for the 6-hour workday persisted. By 1924, with rising unemployment, mines were able to return to longer hours, and unions called for new technologies to reduce unproductive work time. Shearer, "The Politics of Industrial Efficiency," 216, 221–222.

58. On US agriculture, agricultural machinery, and the scarcity of agricultural and industrial labor, see agricultural economist Aereboe, *Wirtschaft und Kultur in den Vereinigten Staaten*.

59. Köttgen, *Das wirtschaftliche Amerika*, 1–28.

60. By contrast, national economist Julius Hirsch proposed a European economic union to create a larger market that would support European economic development comparable to the United States. Hirsch, *Das amerikanische Wirtschaftswunder*.

61. Köttgen, *Das wirtschaftliche Amerika*, 30–32, 48.

62. Allgemeiner Deutscher Gewerkschaftsbund, *Amerikareise deutscher Gewerkschaftsführer*, 32. This was the only quote in the book that was fully attributed to its author.

63. Westermann, *Amerika wie ich es sah*, 6, 54.

64. Allgemeiner Deutscher Gewerkschaftsbund, *Amerikareise deutscher Gewerkschaftsführer*, 16.

65. This is in line with the overall positive response of German socialist unions to rationalization. By contrast, communist unions in Weimar Germany rejected rationalization. Stollberg, *Die Rationalisierungsdebatte*; von Freyberg, *Industrielle Rationalisierung*.

66. Allgemeiner Deutscher Gewerkschaftsbund, *Amerikareise deutscher Gewerkschaftsführer*, 17, 18–25, 27.

67. Ibid., 37, 38.

68. Ibid., 48 (my translation; emphasis in original).

69. See, in particular, Westermann, *Amerika wie ich es sah*; Allgemeiner Deutscher Gewerkschaftsbund, *Amerikareise deutscher Gewerkschaftsführer*, 37.

70. Allgemeiner Deutscher Gewerkschaftsbund, *Amerikareise deutscher Gewerkschaftsführer*, 156.

71. Riebensahm, *Der Zug nach U.S.A.*, 21–22.

72. Sombart, *Why Is There No Socialism in the United States?*

73. Allgemeiner Deutscher Gewerkschaftsbund, *Amerikareise deutscher Gewerkschaftsführer*, 132. Arthur Feiler, the business journalist and editor of the *Frankfurter Zeitung*, and engineer Paul Rieppel were among the few Weimar travelers who described welfare capitalist measures in their reports. Feiler, *America Seen through German Eyes*, 155; Rieppel, *Ford-Betriebe und Ford-Methoden*, 31–33. Some portrayed them as instrumental, suggesting that social measures such as shareholding and company unions aimed to increase corporate loyalty and created labor peace. Tänzler, *Aus dem Arbeitsleben Amerikas*, 166–167; Aereboe, *Wirtschaft und Kultur*, 13–16. But many writers simply dismissed benefits programs. See, for example, Köttgen, *Das wirtschaftliche Amerika*, 61; Riebensahm, *Der Zug nach U.S.A.*, 12.

74. Allgemeiner Deutscher Gewerkschaftsbund, *Amerikareise deutscher Gewerkschaftsführer*, 132, 238.

75. Ibid., 162. Von Gottl-Ottlilienfeld was one of the few Weimar commentators to address wages, insinuating that high wages appeared to contradict the attempt to cut costs in modern businesses. Von Gottl-Ottlilienfeld, *Fordismus*, 3.

76. Allgemeiner Deutscher Gewerkschaftsbund, *Amerikareise deutscher Gewerkschaftsführer*, 172–175. This insight into the incomparability of consumption patterns could have prevented an ill-fated study of the standard of living of Ford workers outside the United States that aimed at paying Ford workers abroad salaries that would allow them a standard of living comparable to the one of Ford workers in Detroit. Yet the common affordances of an automobile in Detroit, for example, would have been a marker of the upper class in continental Europe. De Grazia, *Irresistible Empire*, 79.

77. Allgemeiner Deutscher Gewerkschaftsbund, *Amerikareise deutscher Gewerkschaftsführer*, 142.

78. Ibid., 143–144.

79. Ibid., 145.

80. Hensel, *Aus Tagebüchern einer Reise*, 94, 139.

81. Feiler, *America Seen through German Eyes*, 167.

82. Westermann, *Amerika wie ich es sah*, 6, 44.

83. Salomon, *Character Is Destiny*, 144–145.

84. Allgemeiner Deutscher Gewerkschaftsbund, *Amerikareise deutscher Gewerkschaftsführer*, 135, 137, 205, 206 (English term "loyal employees" in German original).

85. Ibid., 207, 11, 121. See also Bates, *The Making of Black Detroit*.

86. Allgemeiner Deutscher Gewerkschaftsbund, *Amerikareise deutscher Gewerkschaftsführer*, 252–254, 255.

87. Throughout this book, I use the term "Nazi" rather than "National Socialist" because it has higher recognition among English-speaking audiences, although it may sound accusatory and even vulgar to German-speaking readers.

88. Hachtmann, "Die Begründer der amerikanischen Technik," 40, 44–45.

89. Gassert, *Amerika im Dritten Reich*, 15, 29–33; Hachtmann, "Die Begründer der amerikanischen Technik"; Welch, "Nazi Propaganda and the Volksgemeinschaft"; Baranowski, *Strength through Joy*, 5.

90. Baranowski, *Strength through Joy*, 75–92; Welch, "Nazi Propaganda and the Volksgemeinschaft," 223–225; Hau, *Performance Anxiety*.

91. König, *Volkswagen, Volksempfänger, Volksgemeinschaft*.

92. Spoerer, "Demontage eines Mythos?"; König, *Volkswagen, Volksempfänger, Volksgemeinschaft*, 172–174.

93. König, "Adolf Hitler vs. Henry Ford," 250–251. Since before the First World War, car companies had advertised their products as Volkswagen, and in the early 1930s offered small models for affordable prices, such as Opel's P-4 for 1,450 reichsmark, but the Nazi regime soon restricted the term "Volkswagen" to its own people's car project. For Porsche's US visits, see Mommsen and Grieger, *Das Volkswagenwerk und seine Arbeiter*, 167–168.

94. König, *Volkswagen, Volksempfänger, Volksgemeinschaft*, 173–181.

95. Ibid., 174.

96. Gassert, *Amerika im Dritten Reich*, 91.

97. König, *Volkswagen, Volksempfänger, Volksgemeinschaft*, 170: Mommsen and Grieger, *Das Volkswagenwerk und seine Arbeiter*.

98. For highway construction, see Zeller, *Driving Germany*.

99. Reich and Dowler, *Research Findings about Ford-Werke*; Wilkins and Hill, *American Business Abroad*, 270–285; König, "Adolf Hitler vs. Henry Ford," 252–254.

100. For IBM's involvement, see Black, *IBM and the Holocaust*.

101. Stahlmann, *Die Erste Revolution in der Autoindustrie*, 193–212; Homburg, *Rationalisierung und Industriearbeit*; Hachtmann, "Die Begründer der amerikanischen Technik," 44–45, 54, 58.

102. Gassert, *Amerika im Dritten Reich*, 30, 33; Schivelbusch, *Three New Deals*, 11–15.

Chapter 3

1. Everett H. Bellows, "Introductory Remarks," in Mutual Security Agency, "Work-Study Training Program," proceedings of a conference held on March 12, 1952, in Washington, DC, 2, box 1, entry 1058, RG 469, NARA. See also Everett H. Bellows, "Needed: A Second Industrial Revolution in Europe," March 27, 1952, 1, Bellows Papers, Truman Library.

2. An estimated 0.5 to 1.5 percent of Marshall aid was spent on technical assistance, which was a major component of the Productivity Program. The program also targeted problems of marketing, agricultural productivity, manpower utilization, public administration, tourism, transportation, and communications.

3. John W. Nickerson, "Report on Trip to Europe," in US Advisory Group on European Productivity, minutes, March 9, 1953, 2, box 1, entry 171, RG 469, NARA.

4. Bellows, *Oral History Interview*, 3–4, 8–14.

5. See Price, *The Marshall Plan and Its Meaning*; Abelshauser, *Wirtschaft in Westdeutschland*; Milward, *The Reconstruction of Western Europe*.

6. Hogan, *The Marshall Plan*; Lunestad, *"Empire" by Integration*. These postrevisionist historians have seen European integration as a way to create a larger European market that could serve as a trading partner to the United States in order to tie the aggressor Germany to the European community and strengthen Western Europe against Communist influence. Although John Krige emphasized American hegemony through economic, military, and scientific supremacy, like postrevisionist historians, he also acknowledged that Europeans "willingly cooperated," selectively appropriated and adapted US features, and thus coproduced American hegemony. Krige, *American Hegemony*, 4–5, 13.

7. Business historians and historians of technology have critically engaged with the process of transferring US economic ideas to Europe, challenging the notion of Americanization. They have emphasized that US business methods were not a unitary, coherent, and agreed-on model but instead formed an incoherent set of ideas, that the characteristics of the US model were disputed, and that US ideas changed over time. Also, they have drawn attention to the receiving side, arguing that Europeans accepted some US ideas and rejected others. Zeitlin and Herrigel, *Americanization and Its Limits*; Djelic, *Exporting the American Model*; Kipping and

Bjarnar, *The Americanisation of European Business*; McGlade, "The Illusion of Consensus"; Kleinschmidt, *Der produktive Blick*.

8. Lichtenstein and Harris, *Industrial Democracy in America*.

9. Cohen, *Making a New Deal*, 253, 257, 267, 286.

10. Ibid., 301, 315, 318.

11. Jacobs, *Pocketbook Politics*, 3–4.

12. Marshall, "The Marshall Plan Speech." The concrete outline of the Marshall Plan emerged slowly over the next ten months. Gimbel, *The Origins of the Marshall Plan*.

13. The Economic Cooperation Agency was subject to repeated reorganization, partly due to Cold War political pressures. It was established as an independent agency in April 1948, and subsumed by the State Department's Mutual Security Agency in 1950. The Mutual Security Agency's functions were transferred to the Foreign Operations Administration in 1953, and the State Department's International Cooperation Administration in 1955, the predecessor of today's US Agency for International Development. To avoid confusion, I refer to the Economic Cooperation Agency and the Mutual Security Agency as the "Marshall administration."

14. Hogan, *The Marshall Plan*; Maier, "Politics of Productivity."

15. Hoffman first encountered European production methods in Studebaker's London repair shops in the early 1920s. Hoffman, *Peace Can Be Won*, 100–101.

16. By now the Marshall administration was headed by Hoffman's successor and progressive business colleague, William C. Foster, at the State Department. For the increasing Cold War pressures on the Marshall administration, see McGlade, "The Illusion of Consensus"; McGlade, "From Business Reform Programme to Production Drive"; McGlade, "Americanization." Based on counterpart funds, productivity and technical exchange projects continued after the end of the Marshall Plan in 1952.

17. Stapleford, *The Cost of Living in America*, 184–252.

18. Lubin, "Social and Economic Adjustments."

19. BLS officers supported the Productivity Program in various ways, such as productivity surveys and negotiations in European countries, factory reports on US industries, and a technical digest service. Their support gave the Productivity Program the air of objectivity, in addition to substantive personnel and content assistance. "Memorandum of Agreement between Economic Cooperation Administration and the Department of Labor," March 9, 1949, box 5, entry 49, RG 469, NARA; "Outline of Projects to Be Undertaken under ECA–Department of Labor Memorandum of Agreement," March 9, 1949, box 5, entry 49, RG 469, NARA.

20. Wasser and Dolfman, "BLS and the Marshall Plan." For the BLS's productivity work during the 1930s, see Clague, *The Bureau of Labor Statistics*, 115–116.

21. Anglo-American Council on Productivity, "Committee 'C'—Productivity Measurements," December 22, 1948, Silberman Papers, Truman Library.

22. James M. Silberman, untitled manuscript, n.d., 3, Silberman Papers, Truman Library. Silberman returned to Europe in fall 1948 for more extensive productivity surveys of British and French industries; he participated in negotiations for the French Productivity Program; and he conducted a productivity survey in Austria in 1949.

23. "ECA Survey of French Productivity," n.d., 1, Silberman Papers, Truman Library.

24. James M. Silberman, untitled manuscript, n.d., 4, Silberman Papers, Truman Library.

25. Ibid., 4.

26. See, for example, these reports prepared for the Mutual Security Agency, Productivity and Technical Assistance Division, by the US Department of Labor, Bureau of Labor Statistics: *Men's Work Pants*; *Men's Goodyear Welt Dress Shoes*; *Radio and Television Manufacturing*; *Gray Iron Foundries*.

27. Wasser and Dolfman, "BLS and the Marshall Plan," 48, 50.

28. US Department of Labor, Bureau of Labor Statistics, *Men's Work Pants*. The same paragraph was repeated in each factory performance report, either on a first-page insert opposite the report title or in the introduction.

29. US Advisory Group on European Productivity, meeting minutes, March 10, 1952, 3, 9, box 1, entry 171, RG 469, NARA.

30. Richard H. Bissel Jr., "TOREP A-1584. Subject: ECA Production Assistance Drive," June 5, 1951, 2, box 21, entry 1202, RG 469, NARA.

31. Robert Oshins and Hall, "The Productivity and Technical Assistance Program for Europe," memorandum, December 17, 1952, 3, box 1, entry 171, RG 469, NARA.

32. Kasson, *Civilizing the Machine*; Marx, *The Machine in the Garden*; Smith, *Harpers Ferry Armory*; Hounshell, *From the American System to Mass Production*.

33. Hounshell, *From the American System to Mass Production*.

34. Strasser, *Satisfaction Guaranteed*; Laird, *Advertising Progress*; Marchand, *Advertising the American Dream*; Calder, *Financing the American Dream*. For transatlantic transfers of consumer culture and household technologies, see Oldenziel and Hård, *Consumers, Tinkerers, Rebels*; Oldenziel and Zachmann, *Cold War Kitchen*.

35. Fones-Wolf, *Selling Free Enterprise*; Harris, *The Right to Manage*. Within the Productivity Program, free enterprise entailed not only freedom from government and union interventions but also freedom from cartels and monopolies along with their price-setting practices, and freedom from import and export tariffs. These are two freedoms that US companies sometimes did not want to see applied to their own industries; for example, the NAM lobbied for higher import tariffs to protect smaller manufacturers with high labor costs against foreign competition.

36. Although government regulation slowly increased over the first half of the twentieth century, collective bargaining largely remained a private affair between the bargaining partners. The National Labor Relations Act of 1935 forbade employers from engaging in antiunion tactics and mandated collective bargaining procedures once the workers in a company had voted for union representation.

37. Government intervention was another issue of contention between conservatives, New Dealers, and business progressives, leading to a temporary compromise for the Marshall Plan. Hogan, *The Marshall Plan*; McGlade, "The Illusion of Consensus."

38. "Senate Aid Bill Asks Ban on Red Unions." For the two amendments, see "Moody Amendment to the Mutual Security Act," box 8, entry 1202, RG 469, NARA; "Benton Amendment to the Mutual Security Act," box 8, entry 1202, RG 469, NARA. See also Boel, *The European Productivity Agency*, 31–33; McGlade, "The Illusion of Consensus," 503, 513.

39. Hoffman, "Productivity," 42.

40. Richard H. Bissel Jr., "TOREP A-1584. Subject: ECA Production Assistance Drive," June 5, 1951, 2, box 21, entry 1202, RG 469, NARA.

41. Robert Oshins and Hall, "The Productivity and Technical Assistance Program for Europe," memorandum, December 17, 1952, 3, box 1, entry 171, RG 469, NARA.

42. Hoffman, "Productivity," 42.

43. "Procedures for Establishing Demonstration Projects," revised November 17, 1952, 1, box 34, entry 1058, RG 469, NARA. While US administrators and advisers initially considered US-operated or even US-owned training centers or plants as demonstration projects, they eventually turned existing European companies into demonstration plants.

44. Letter from Wallace Clark to John W. Nickerson, January 7, 1952, box 1, entry 171, RG 469, NARA.

45. Michael Harris, "Pilot Plants Help to Increase Productivity," for publication in a special issue of the *Rheinischer Merkur* newspaper, n.d., 3, box 20, entry 172, RG 469, NARA.

46. Economist Paul Fischer argued that "integrated market areas" versus individual firms should be targeted in demonstration projects so as to not only achieve lower prices and costs and higher wages but instead to turn a stagnant into a dynamically growing economy. Paul Fischer, "The 'Integrated Regional Approach' to Demonstration Projects," n.d., 3, box 48, entry 1058, RG 469, NARA. Labor advisers, by contrast, raised questions about the share-out agreements for demonstration projects. Memo from Frederick E. Scheven to Carl R. Mahder, January 6, 1953, box 18, entry 1202, RG 469, NARA. The Marshall administration never decided on the adoption of a particular share-out plan, although Allen W. Rucker, the developer of the Rucker Plan for profit sharing, was a member of the US Advisory Group on European Productivity.

47. Michael Harris, "Pilot Plants Help to Increase Productivity," for publication in a special issue of the *Rheinischer Merkur* newspaper, n.d., 3, box 20, entry 172, RG 469, NARA.

48. "Excerpts from the Minutes of the Productivity and Technical Assistance Regional Conference in Paris," October 1–4, 1952, 32, box 18, entry 1202, RG 469, NARA.

49. "Policy and Methods of Establishing Demonstration Projects," n.d., 2, box 34, entry 1058, RG 469, NARA. In German, the literal translation of the term "demonstration project" would have been *Demonstrationsprojekt*. German administrators, however, used the terms *Leitbetrieb* or *Förderbetrieb*, which can be translated as "leading firm" or "sponsored firm." For example, one of the German documents on demonstration projects was initially titled "Leitbetriebe," and the term "Förderbetriebe" was later added in handwriting. "Auswahl der Leitbetriebe," n.d., box 19, entry 1202, RG 469, NARA.

50. Michael Harris, "Pilot Plants Help to Increase Productivity," for publication in a special issue of the *Rheinischer Merkur* newspaper, n.d., 2, box 20, entry 172, RG 469, NARA.

51. "Excerpts from the Minutes of the Productivity and Technical Assistance Regional Conference in Paris," October 1–4, 1952, 26, 29, box 18, entry 1202, RG 469, NARA.

52. Richard H. Bissel Jr., "TOREP A-1584. Subject: ECA Production Assistance Drive," June 5, 1951, 2, box 21, entry 1202, RG 469, NARA.

53. "Development of Demonstration Plant Projects," November 26, 1952, box 34, entry 1058, RG 469, NARA.

54. Letter from Robert Bender to Carl R. Mahder, March 8, 1954, box 18, entry 1202, RG 469, NARA.

55. Letter from Carl R. Mahder to Spinnereien und Webereien im Wiesental AG, March 8, 1954, box 18, entry 1202, RG 469, NARA.

56. The archival records do not reveal which plants actually received the funding.

57. Memo from E. A. Wiesinger to Carl R. Maher, April 21, 1953, box 18, entry 1202, RG 469, NARA.

58. "Excerpts from the Minutes of the Productivity and Technical Assistance Regional Conference in Paris," October 1–4, 1952, 71–72, box 18, entry 1202, RG 469, NARA.

59. US Advisory Group on European Productivity, minutes, March 10, 1952, 1, box 1, entry 171, RG 469, NARA. Clague attended this AGEP meeting as a guest. For similar statements, see US Advisory Group on European Productivity, minutes, May 12, 1951, 6–7, box 1, entry 171, RG 469, NARA; Nickerson, "How Should Labor Participate in Gains through Technological Improvements"; "OSR Comments on TOREP A-1584," June 15, 1951, box 19, entry 1202, RG 469, NARA.

60. Harold B. Maynard to John W. Nickerson, January 16, 1952, box 1, entry 171, RG 469, NARA.

61. "The Productivity and Technical Assistance Programs for Europe," n.d., box 1, entry 171, RG 469, NARA.

62. US Advisory Group on European Productivity, minutes, February 11, 1952, box 1, entry 171, RG 469, NARA.

63. Ray M. Hudson to Everett H. Bellows, memo re: "Demonstrations," April 21, 1953, 1, box 34, entry 1058, RG 469, NARA. Hudson, who was Bellows's deputy, used phrases from the Parable of the Sower found in the New Testament: Matthew 13, Mark 4, and Luke 8. Nickerson considered Hudson's "cool wisdom and experience" as the perfect balance for Bellows's "eagerness and enthusiasm." US Advisory Group on European Productivity, minutes, March 9, 1953, 2, box 1, entry 171, RG 469, NARA.

64. "Policy and Methods of Establishing Demonstration Projects," n.d., 1, box 34, entry 1058, RG 469, NARA.

65. Robert Oshins and Hall, "The Productivity and Technical Assistance Program for Europe," memorandum, December 17, 1952, 3, box 1, entry 171, RG 469, NARA.

66. Everett H. Bellows, "Introductory Remarks," in Mutual Security Agency, "Work-Study Training Program," proceedings of a conference held on March 12, 1952 in Washington, DC, 5, box 1, entry 1058, RG 469, NARA; John W. Nickerson, "Introductory Remarks," in US Advisory Group on European Productivity, minutes, March 10, 1953, 1, box 1, entry 171, RG 469, NARA.

67. Everett H. Bellows, "Introductory Remarks," in Mutual Security Agency, "Work-Study Training Program," proceedings of a conference held on March 12, 1952 in Washington, DC, 4, box 1, entry 1058, RG 469, NARA; US Advisory Group on European Productivity, minutes, November 10, 1952, 3, box 1, entry 171, RG 469, NARA. This attitude among productivity officers and advisers coincides with historians' interpretation that the United States exerted influence in postwar Europe by "consensual hegemony" or "by invitation." Hogan, *The Marshall Plan*; Maier, "Politics of Productivity," 630; Lunestad, "Empire by Invitation?"

68. Charles Maier may have been the first to point to this religious metaphor, referring to the "messianic liberalism" and "buoyant belief" of American administrators. Maier, "Politics of Productivity," 612, 615. I argue that the missionary metaphor describes the productivity officers' approach as well as their self-understanding.

69. "WSTP Instructions to Colleges," n.d., 1, box 1, entry 1058, RG 469, NARA.

70. US Advisory Group on European Productivity, minutes, November 10, 1952, 3, box 1, entry 171, RG 469, NARA.

71. Everett H. Bellows, "The Productivity Program," memorandum, August 8, 1952, 1, box 1, entry 171, RG 469, NARA.

72. "International Council for Christian Leadership: Miscellaneous Sessions," n.d., box 1, entry 171-G, RG 469, NARA.

73. US Advisory Group on European Productivity, minutes, January 3, 1952, 7, box 1, entry 171, RG 469, NARA.

74. B. J. Johnson, "ICL Conference," report, February 9, 1953, 1, box 1, entry 171–G, RG 469, NARA.

75. J. O. P. Hummel, "Conference: International Christian Leadership," report, February 9, 1953, entry 171–G, RG 469, NARA.

76. Mutual Security Agency, "Work-Study Training Program," 10–11, box 1, entry 1058, RG 469, NARA.

77. Marshall, "The Marshall Plan Speech."

78. William H. Draper Jr. to Harold E. Stassen, April 2, 1953, box 2, entry 1058, RG 469, NARA.

79. "WSTP Instructions to Colleges," n.d., 3, box 1, entry 1058, RG 469, NARA.

80. Organization for European Economic Cooperation, "Symposium on Productivity through Technical Assistance," held at the Château de la Muette on January 30–31, 1952, February 7, 1952, 24, box 15, entry 172, RG 469, NARA.

81. "Productivity Plan Modified by ECA."

82. William Benton, "Address on E.C.A. by Senator William Benton (D. Conn.) Prepared for Delivery in the Senate," May 4, 1950, 1, box 1, entry 174–A, RG 469, NARA.

Chapter 4

1. Due to this conflation of wages, prices, and productivity, the brochure also appears to suggest that bicycles were relatively easy to obtain for Italian workers compared to US workers: Italian workers had to work less than three times as long for a bicycle, or 138 hours, as they did for a pair of overalls, at 48.5 hours. By contrast, a US worker needed to work more than fifteen times as long for a bicycle, or 35 hours, than the 2 hours for a pair of overalls. Such price relations, however, were results of wage and price distortions, because Italian workers used their wages for highly priced items of daily need such as food and clothing, leaving no disposable income for the purchase of consumer durables such as bicycles, and eliminating the market demand for such mass-produced items and reducing their prices.

2. Mutual Security Agency, "Waging the Peace . . . on Your Factory Floor," n.d., box 20, entry 172, RG 469, NARA.

3. Sometimes the notion of public diplomacy designates a government communicating directly to citizens in a foreign country. See, for example, Muir-Harmony, "Project Apollo." Here I use public diplomacy in the broader sense of enrolling US citizens and associations for the purposes of US foreign policy, building on a long history of US citizens engaging in international interactions. Rosenberg, *Spreading the American Dream*.

4. Maier, "Politics of Productivity," 613–615.

5. Hoffman, "Productivity," 41.

6. Contemporary Germans tended to use the terms "industrialists" and "workers," or "laborers," which implied class conflict; these terms were later replaced by the more neutral "employer" and "employee." By contrast, contemporary Americans rarely used the term "industrialists," and instead talked about managers and employees.

7. Bundesverband der Deutschen Industrie, *Erster Internationaler Industriellen-Kongress*.

8. All conference statements were prepared beforehand. For an assessment of the conference, see also Kipping, "Operation Impact."

9. The International Conference of Manufacturers initiated a series of similar conferences in Paris in 1954, in Hot Springs, Virginia, in 1956, and in London in 1960.

10. Sanford, *The American Business Community*, 75–77, 111–118.

11. Ibid., 263–277.

12. IBM Germany, "Menschen Maschinen Methoden," n.d., box 1, Firmenschriften IBM FS 003269, Deutsches Museum (my translation). See also IBM Germany, "Einführung in das IBM Lochkartenverfahren," 1961, 1963, box 071–079, Kleine Sammlung IBM 004, Deutsches Museum.

13. *New York Herald Tribune* (Paris), April 6, 1953, 9 (emphasis in original).

14. William Sanford Jr. lumped conservative and progressive members of the business community into one group, while Emily Rosenberg hinted at differentiations within the business community with regard to foreign policy already before World War II. Sanford, *The American Business Community*; Rosenberg, *Spreading the American Dream*, 137. For different voices even within NAM, see Delton, "The Triumph of Social Responsibility."

15. Sanford, *The American Business Community*, 263–277.

16. Committee on International Economic Relations, "Considerations Involved in Proposals for United States Aid to Other Countries to Facilitate Economic Recovery," October 29, 1947, box 76, NAM accession 1411, Hagley. The NAM did not seem to be concerned that the default of the private credits during the Great Depression created economic conditions, including high unemployment levels, that have been linked to the rise of the Nazi Party.

17. Herbert Schell, "Statement before House Foreign Relations Committee Presenting National Association of Manufacturers' Views on the Proposed European Economic Cooperation Act of 1948," n.d., 14, 7, box 76, NAM accession 1411, Hagley.

18. "NAM and International Relations," August 17, 1948, box 76, NAM accession 1411, Hagley; NAM Board, "Memorandum on the Scope of Activity in the International Field by the National Association of Manufacturers," September 17–18, 1948, 4, box 76, NAM accession 1411, Hagley.

19. International Relations Committee, "Statement on European Recovery Program," February 17, 1949, box 76, NAM accession 1411, Hagley; untitled report, February 2, 1949, box 76, NAM accession 1411, Hagley.

20. Jones, "Production Policies," 60–61.

21. Fones-Wolf, *Selling Free Enterprise*; Harris, *The Right to Manage*.

22. Free enterprise thus came to denote freedom from cartel agreements and monopoly practices in the international realm, not free trade practices. Reed, "The Challenge of This Conference," 213.

23. Randall, "Employer-Employee Relations," 137.

24. Mosher, "Response," 163.

25. Jacobsson, "Employee Relations Policies," 126.

26. Swedish Employers' Association, "Employee Relations Policies," 356, 358, 368–369.

27. National Association of Manufacturers, *Proceedings of the First International Conference of Manufacturers*, 139–161.

28. Randall, "Employer-Employee Relations," 136–137.

29. Jones, "Production Policies," 56.

30. Joyce, "Participation," 310.

31. Golden and Ruttenberg, *The Dynamics of Industrial Democracy*. For a historical assessment of industrial democracy, see Lichtenstein and Harris, *Industrial Democracy in America*.

32. Brooks, *Clint*, 1–37, 58, 87–107, 128–157, 213–223.

33. Romero, *The United States and the European Trade Union Movement*, 19–20. See also Angster, *Konsenskapitalismus und Sozialdemokratie*. Both the AFL and the CIO became involved in international politics and even in clandestine operations, despite their general aversion to government intervention.

34. Carew, *Labour under the Marshall Plan*, 73; Romero, *The United States and the European Trade Union Movement*, 109–111.

35. Clinton S. Golden, "Address before Roosevelt College," October 12, 1948, 1, box 4, entry 48, RG 469, NARA.

36. Carew, *Labour under the Marshall Plan*, 73–76, 80; Romero, *The United States and the European Trade Union Movement*, 109–113.

37. Clinton S. Golden, "Presentation to the ECA Public Advisory Board: The Problem of US Unemployment in Relation to Labor's Attitude about the Marshall Plan," May 25, 1949, 2, box 6, entry 48, RG 469, NARA.

38. Clinton S. Golden, "Second Statement to the ECA Public Advisory Board on the Subject: The Problem of US Unemployment in Relation to Labor's Attitude about the Marshall Plan," July 27, 1949, 2–3, box 6, entry 48, RG 469, NARA. See also Carew, *Labour under the Marshall Plan*, 81.

39. Clinton S. Golden, "Address before Roosevelt College," October 12, 1948, 1, box 4, entry 48, RG 469, NARA. Golden had recommended that labor officers be involved in top policy-making through central positions, unlike in the War Production Board, where labor officers had been relegated to their own unit and to labor problems. Clinton S. Golden, "Labor Participation in the Economic Cooperation Administration: Some Comments and Suggestions," April 6, 1948, box 4, entry 48, RG 469, NARA.

40. Clinton S. Golden, "Report before Public Advisory Board Meeting," August 25, 1948, 2–3, box 4, entry 48, RG 469, NARA.

41. Carew, *Labour under the Marshall Plan*, 80–89.

42. Letter from Clinton S. Golden to Averell Harriman, September 9, 1948, box 1, entry 48, RG 469, NARA.

43. Carew, *Labour under the Marshall Plan*, 111–118.

44. Correspondence, box 1, entry 49, RG 469, NARA; letter from Harvey W. Brown to Nelson H. Cruikshank, August 31, 1951, box 2, entry 49, RG 469, NARA; memo from Nelson Cruikshank to C. Tyler Wood, "Office Memorandum: Labor Division's Comments on the President's Draft Letter Outlining DMS Functions," February 28, 1952, box 4, entry 49, RG 469, NARA; cable from Bert Jewell and Clinton S. Golden to McCullough on European restrictive trade practices, March 15, 1950, box 2, entry 50, RG 469, NARA.

45. Romero, *The United States and the European Trade Union Movement*, 124–134. The American unions also inscribed anticolonial positions in the ICFTU constitution, while issues of conflict between European social democratic and American liberal capitalist positions were omitted from the ICFTU agenda; the ICFTU thus became an instrument of the Cold War.

46. Brooks, *Clint*, 204.

47. Golden, *Causes of Industrial Peace*, 8.

48. Golden and Ruttenberg, *The Dynamics of Industrial Democracy*, xxiv, xxvi, 3, 235–237.

49. Ibid., xi.

50. Golden later distanced himself from this principle. Brooks, *Clint*, 244.

51. Golden and Ruttenberg, *The Dynamics of Industrial Democracy*, 22. They conceded that by the early 1940s, even Ford had "learned."

52. Ibid., 164–169, 241–249.

53. Ibid., xxiv, 295–342. For further discussion of plant-level versus industry-wide collective bargaining, see chapters 6 and 8.

54. Ibid., 303–313.

55. Ibid., ix, 5. Academic researchers conducted a total of thirty case studies that the National Planning Association published individually between 1949 and 1953. Golden frequently promoted the studies in his speeches, and recommended a list of European recipients for the association's studies, which unfortunately is not

included in the archival record. Letter from Clinton S. Golden to John Coil, December 29, 1948, box 5, entry 48, RGA 469, NARA.

56. National Planning Association, "A Statement by the NPA Technical Advisory Committee (Draft)," June 21, 1948, 1–2, box 5, entry 48, RGA 469, NARA.

57. Scholars have critically engaged with Americanization arguments that implied that American culture was homogeneous and cultural transfers were one-way processes. While there are certain core characteristics to the American model, they have revealed that ambiguities and disparities characterize US industrial practices, and that local economic and institutional appropriations modified and changed the model. Zeitlin and Gary, *Americanization and Its Limits.* Historicizing this critique, Mary Nolan has asserted that transatlantic exchanges varied depending on the subject and period of research; a focus on cartel legislation, production methods, and the post–World War II decades revealed European assimilation to the American model, while a stress on codetermination, company structures, financing, training, and the late nineteenth and early twentieth centuries as well as the period since the 1970s lent itself to the discovery of persistent transatlantic differences. Nolan, "Varieties of Capitalism." For a further discussion of the reception of American culture and technology abroad, see the introduction to chapter 5.

58. Djelic, *Exporting the American Model,* 4–7. It remains an open question whether this assimilation process was an effect of the Productivity Program—as some scholars have argued—or other forces, such as the same competitive pressures on both sides of the Atlantic, or the same, albeit delayed but independent, development of mass consumer markets. These trends suggest a movement toward Chandlerian, multiunit businesses with managerial hierarchies. Chandler, *Visible Hand.*

59. Campbell-Kelly et al., *Computer.*

60. "Excerpts from the Minutes of the Productivity and Technical Assistance Regional Conference, Paris, October 1–4, 1952, 6–7, box 18, entry 1202, RG 469, NARA. An equitable sharing of productivity proceeds between owners, workers, and consumers was required by the 1952 Moody Agreement. See Boel, *The European Productivity Agency,* 33.

61. Letter from Paul H. Douglas to Averell Harriman, August 7, 1952, box 18, entry 1202, RG 469, NARA.

Chapter 5

1. These study groups usually comprised ten to twelve members, including businessmen, workers, union officials, officials from industry associations, public administrators, and sometimes politicans. Some were national groups, from a single country, and others were European groups, organized by the OEEC with participants

from different countries. Most study groups mixed business and labor representatives, although there were also some labor-only groups. Study group programs often included private dinner invitations to observe the American standard of living.

2. Mutual Security Agency, "The Work-Study Training for Productivity Program of the MSA," February 16, 1953, 4, box 29, entry 120, RG 469, NARA.

3. The WSTP program can be seen as a complement to Project Impact, the three-week company tour and conference trip for current European industrialists discussed in chapter 4.

4. "They See America."

5. "General Impressions of WSTP Students before and after, Etc.," n.d., box 5, entry 171–G, RG 469, NARA.

6. Bert Wessel, "Amerika-Tagebuch, 19.11–9.12.1951," n.d., 13, box 1, entry 1202, RG 469, NARA.

7. Edward L. Deuss, "Evaluation Meeting, TA 09-184, German Machine and Precision Tools Productivity Team, December 18, 1952, Park Sheraton Hotel, New York," January 2, 1953, 1, box 4, entry 171–G, RG 469, NARA.

8. Edward L. Deuss, "Evaluation Meeting, TA 09-101-2138, German Steel Heat Treating Study, October 31, 1952," November 10, 1952, box 1, entry 171–G, RG 469, NARA.

9. Excerpts from a letter to John W. Piercey, MSA/The Hague, to Trainee Branch, PTAD/Washington, DC, March 4, 1953, box 5, entry 171–G, RG 469, NARA.

10. In response to criticism of the Americanization paradigm, scholars have begun to study the European reception of American culture. See, for example, Poiger, *Jazz, Rock, and Rebels*; Kuisel, *Seducing the French*. A necessary corrective of former shortcomings, reception studies like these, however, present American culture as a given, offering itself to Europeans to be adopted or not. They neglect the American agency that often helped sway European perceptions.

   Agency on both sides often characterized transatlantic relations, and other studies have provided a more balanced account between American and European agency. For example, Victoria de Grazia analyzed how Europeans of different social classes reacted to American ideas about higher standards of living, and she also explored the motivations and intentions of Americans, from Rotary Club members to the department store magnate Filene. Similarly, Richard Pells paid equal attention to both sides when he described mutual misunderstandings and cultural stereotypes, and Brian McKenzie compared the goal of the Marshall Plan administration with the French reception of the program. John Krige, finally, has coined the term "consensual co-construction" to argue that European agreement was essential to restructuring basic science institutions in Europe after the American model was implemented during the postwar years. De Grazia, *Irresistible Empire*; Pells, *Not Like Us*; McKenzie, *Remaking France*; Krige, *American Hegemony*.

Historians of modern Germany devoted increased attention to the concept of Americanization in the 1990s and early 2000s, when the United States appeared as the winner of the Cold War conflict. Taking the German or European view, they emphasized European agency. Thus, Anselm Doering-Manteuffel proposed the notion of "Westernization" to describe the development of a common Western, trans-Atlantic set of political and economic values. Doering-Manteuffel, *Wie westlich sind die Deutschen?* See also Lüdtke, Marßolek and von Saldern, *Amerikanisierung*; Jarausch and Siegrist, *Amerikanisierung und Sowjetisierung*; Berghahn, "The Debate on 'Americanization'"; Logemann, "More Atlantic Crossings?"

Historians of technology have used the concepts of circulation and appropriation to investigate the homogenizing force of the global distribution of technology as well as the resistance and selective adaptation by local technology users that created heterogeneity. Misa and Schot, "Inventing Europe," 9. In this context, "circulation" denotes the process of moving technology from one locale to another, not the appropriation of technology or knowledge in a new locale, as in postcolonial studies and studies of knowledge transfers.

11. "Excerpts from the Minutes of the Productivity and Technical Assistance Regional Conference," held in Paris, October 1–4, 1952, 16, 18, 26, box 16, entry 172, RG 469, NARA. See also "RKW Vorausschau auf die Finanzlage im Haushalts-jahr 1954/55," September 14, 1953, file 1118/F8, Bestand 146, Bundesarchiv.

12. Funding for productivity projects did not cease with the end of Marshall Plan funding. Since productivity projects were paid for through counterpart funds and awarded on a loan basis, the Productivity Program created a pool of revolving funds. The RKW continued productivity projects through the 1960s, and the successor organization to the International Council for Youth Self-Help, the Carl Duisberg Gesellschaft, still offers vocational exchange programs. J. W. Funke, Internatio-naler Rat für Jugendselbsthilfe e.V., "Entwurf: Europäische Produktivitäts-Studien-Programm in USA," June 16, 1954, file 1118, Bestand 146, Bundesarchiv; "Die Carl Duisberg-Gesellschaft für Nachwuchsförderung e.V.," n.d., 2, box 5, entry 1202, RG 469, NARA.

13. "OSR Comments on TOREP A–1584," June 15, 1951, 3, box 19, entry 1202, RG 469, NARA. The productivity drive had been announced ten days earlier. Richard M. Bissell Jr., "ECA Production Assistance Drive, TOREP A–1584," box 19, entry 1202, RG 469, NARA.

14. Deutscher Gewerkschaftsbund, *Protokoll Gründungskongress*, 323.

15. LaVerne Baldwin, American consul general in Düsseldorf, "DGB Executive Com-mittee on Productivity Council," dispatch no. 555, enclosure no. 1, April 23, 1952, box 2, entry 786, RG 469, NARA.

16. LaVerne Baldwin, "DGB Executive Committee on Productivity Council," April 23, 1952, box 2, entry 786, RG 469, NARA.

17. The German economy comprised two distinct, regionally separate industrial systems: large-scale, vertically integrated enterprises with close ties to universal banks in the coal, steel, and chemical industries on the Ruhr and in the Rhineland, and small and medium-size companies with local networks of supporting institutions in the textiles, leather, wood, and metalworking industries as well as other specialty production industries in south and east Germany. Herrigel, *Industrial Constructions*.

18. Abelshauser, *Deutsche Wirtschaftsgeschichte*, 124, 333–335.

19. Bundesministerium für Wirtschaft, "Richtlinien für die Steigerung der Produktivität in Mittel- und Kleinbetrieben," July 15, 1953, file 5525, Bestand 115, Bundesarchiv; Bundesministerium für Wirtschaft, "Richtlinien zur Durchführung der Kreditaktion für Produktivitätssonderprojekte," July 15, 1953, file 5525, Bestand 115, Bundesarchiv.

20. "Notes on Meeting Held for the Purpose of Discussing Directives to Implement Productivity Program," June 1, 1953, 1, box 18, entry 1202, RG 469, NARA. See also Special Representative to Europe, Productivity and Technical Assistance Division, "SRE/PTAD Comments on Moody Amendment Programs," October 27, 1952, 13, box 1, entry 171, RG 469, NARA. During the Eisenhower administration, Productivity Program officers moved from share-out agreements to management and trade union education programs. See Boel, *The European Productivity Agency*, 150–151.

21. Pohl, "Geschichte der Rationalisierung," 93–97. The name of the post–World War II RKW translates to Rationalization Board of the German Economy.

22. Ibid., 90–93.

23. Everett H. Bellows, "Excerpts from the Minutes of the Productivity and Technical Assistance Regional Conference," conference held in Paris, October 1–4, 1952, 30, box 18, entry 1202, RG 469, NARA.

24. Letter from Dahlgrün to John W. Tuthill, draft of April 19, 1955, 2–3, file 5526, Bestand 115, Bundesarchiv; "6. Zwischenbericht über die Durchführung des Produktivitätsprogramms der Bundesregierung, Kreditaktion für Mittel- und Kleinbetriebe," September 3, 1954, file 4780, Bestand 115, Bundesarchiv.

25. Letter from John W. Tuthill to the minister of economic cooperation, October 11, 1954, file 5526, "Reprogrammierung 1956," Bestand 115, Bundesarchiv.

26. Letter from John W. Tuthill to Bundesminister für Wirtschaftliche Zusammenarbeit, May 26, 1955, file 5526, Bestand 115, Bundesarchiv.

27. For the legacy of the Productivity Program in German banking and government loan programs, see "Kreditanstalt für Wiederaufbau"; "6. Zwischenbericht über die

Durchführung des Produktivitätsprogramms der Bundesregierung, Kreditaktion für Mittel- und Kleinbetriebe," September 3, 1954, folder 4780, Bestand 115, Bundearchiv; Letter from Elson to Abt, March 2, 1956, file 5527, Bestand 115, Bundesarchiv.

28. Labor Productivity Branch, PTAD/SRE, "Organized Labor and Productivity in Western Europe," August 28, 1952, 1, box 38, entry 1058, RG 469, NARA. Notably, an earlier version of the report only mentioned labor's fear of technological unemployment; later, the surprising conclusion that labor was more open toward technological change than management was added. Labor Productivity Branch, "Draft—I," n.d., 1, box 2, entry 1058, RG 469, NARA.

29. H. C. Zulauf to C. R. Mahder, "Office Memorandum: Increased Labor's Share in GNP Versus 'and will assure Labor's share of GNP,'" February 12, 1953, box 18, entry 1202, RG 469, NARA (emphasis in original).

30. For the history of the labor movement in Germany, see Grebing, *Geschichte der deutschen Arbeiterbewegung*, 51–55. See also Kuhn, *Die deutsche Arbeiterbewegung*.

31. Internationaler Rat für Jugendselbsthilfe e.V. (hereafter Jugendselbsthilfe), "Auswertungskonferenz für die 6. Gruppe des WSTP-Programmes Königswinter 4.–6. April 1955. Diskussionsbeiträge des Plenums zum Bericht der Arbeitsgruppe III: Human Relations," 2–3, box 5, entry 1202, RG 469, NARA.

32. Jugendselbsthilfe, "Auswertungskonferenz," Bericht der Arbeitsgruppe V: Betriebs- und volkswirtschaftliche Fragen, 5, box 5, entry 1202, RG 469, NARA.

33. Schelsky, *Wandlungen der deutschen Familie*, 218–224. Helmut Schelsky's analysis of the German leveled middle-class society, however, differed from American society in that it was based on class configurations rather than individual mobility and identification, and did not promise the opportunity for individual uplift.

34. Rejecting Max Weber and Werner Sombart's definition of class as a social stratum, Dahrendorf built on Marx's theory, which he modified by assuming that classes were mobile; authority, property, and social status were not coupled with class; industrial and political conflict were not linked to class, and effective conflict regulation existed. Dahrendorf, *Class and Class Conflict*, 245, 257–262.

35. Logemann, *Trams or Tailfins?*, 98, 101–106.

36. Schelsky, *Wandlungen der deutschen Familie*, 232–237.

37. Parsons, *Essays in Sociological Theory*, 335, 339.

38. Scheuch developed a measure of "relative prestige" to determine social status. Scheuch and Rüschemeyer, "Scaling Social Status in Western Germany," 151–152. See also Scheuch, "Continuity and Change in German Social Structure."

39. Jugendselbsthilfe, "Auswertungskonferenz," Bericht der Arbeitsgruppe I: Die Grundausbildung und die berufsfördernden Massnahmen (Nachwuchsförderung) in Deutschland und den USA, 7–8, box 5, entry 1202, RG 469, NARA.

40. "Minutes of Meeting with Technical Delegation Representatives of the Foreign Embassies," October 6, 1952, 7, box 5, entry 171–G, RG 469, NARA. The Marshall Plan's labor division was charged with strengthening the European labor movement, and Americans usually described European labor unions as weak. Mutual Security Agency, "Work-Study Training Program," conference held on March 12, 1952, 94–95, box 5, entry 171–G, RG 469, NARA. Europeans, by contrast, considered their labor movements as strong. See, for example, Jacobsson, "Employee Relations Policies in Western Europe," 126–127.

41. Mutual Security Agency, "Work-Study Training Program," conference held on March 12, 1952, 96–97, box 5, entry 171–G, RG 469, NARA. See a similar statement by Everett Bellows: Mutual Security Agency, "Work-Study Training Program," conference held on March 12, 1952, 100–101, box 5, entry 171–G, RG 469, NARA.

42. Exchange students primarily stayed with smaller colleges in middling industrial centers in the Northeast and Midwest. "Minutes of Supervisors' Conference on Work-Study Training for Productivity Program," August 26–27, 1952, box 5, entry 171–G, RG 469, NARA; Mutual Security Agency, "Work Study Training Program," Conference Proceedings, March 12, 1952, 22–23, box 5, entry 171–G, RG 469, NARA. Colleges may have hoped for better contact to local labor organizations, a guaranteed source of tuition dollars through their participation in the WSTP program, or may have been motivated to aid in European economic reconstruction. Mutual Security Agency, "Work-Study Training Program," conference held on March 12, 1952, 21–23, 45, 59–60, box 5, entry 171–G, RG 469, NARA.

43. Productivity and Technical Assistance Division, SRE, "Excerpts from the Minutes of the Productivity and Technical Assistance Regional Conference," Paris, October 1–4, 1952, 51, box 16, entry 172, RG 469, NARA.

44. The RKW organized over two hundred productivity study groups to the United States between 1950 and 1959; about ninety trips resulted in published reports in the so-called "blue series" (Blaue Reihe), named after the color of the volumes' cover. Rationalisierungs-Kuratorium der Deutschen Wirtschaft, Internationaler Erfahrungsaustausch.

45. "General Impressions of WSTP Students before and after, Etc.," n. d., box 5, entry 171–G, RG 469, NARA; International Council for Youth Self-Help, "Reports by Participants in the Work Study Training for Productivity Program (WSTP)," 1952, box 5, entry 171–G, RG 469, NARA; Jugendselbsthilfe, "Auswertungskonferenz," box 5, entry 1202, RG 469, NARA. Students faced reintegration problems after their return from the United States, feeling frustrated by their inability to move ahead personally and resistance against their ideas.

46. International Council for Youth Self-Help, "Reports by Participants in the Work Study Training for Productivity Program (WSTP)," 1952, 30, box 5, entry 171–G, RG 469, NARA.

47. Logemann, *Trams or Tailfins?*, 76–78. Well-to-do American households in the 1920s were typically equipped with telephones, hot and cold running water, indoor plumbing, gas, and electricity, and owned one to two appliances while some families of industrial workers, small farmers, day laborers, and skilled craftsmen were trying to buy their own—albeit small—homes with amenities such as electricity, running water, a bathroom, gas, and a telephone, which most rural households still lacked. By the early 1940s, 53 percent of households had built-in bathing equipment, 52 percent of families owned a mechanical refrigerator, and 52 percent of families had access to washing machines in their homes; those numbers rose in the post–World War II decades. Cowan, *More Work for Mother*, 173, 183–190, 195–196.

48. Logemann, *Trams or Tailfins?*, 79–82.

49. Abelshauser, *Deutsche Wirtschaftsgeschichte*, 329.

50. Some students felt that good labor relations helped raise productivity in the United States and spilled out beyond corporate boundaries, leading to more equal, classless relations between workers and managers in their communities. "General Impressions of WSTP Students before and after, Etc.," n.d., box 5, entry 171–G, RG 469, NARA.

51. Jugendselbsthilfe, "Auswertungskonferenz," Bericht der Arbeitsgruppe III: Human Relations, 2, box 5, entry 1202, RG 469, NARA; Jugendselbsthilfe, "Auswertungskonferenz," Diskussionsbeiträge des Plenums zum Bericht der Arbeitsgruppe III, 1, box 5, entry 1202, RG 469, NARA.

52. Jugendselbsthilfe, "Auswertungskonferenz," Bericht der Arbeitsgruppe III: Human Relations, 1–2, 4, box 5, entry 1202, RG 469, NARA; Jugendselbsthilfe, "Auswertungskonferenz," Diskussionsbeiträge des Plenums zum Bericht der Arbeitsgruppe III, 2, box 5, entry 1202, RG 469, NARA.

53. Jugendselbsthilfe, "Auswertungskonferenz," Diskussionsbeiträge des Plenums zum Bericht der Arbeitsgruppe III, 2, box 5, entry 1202, RG 469, NARA.

54. The WSTP program had built-in tensions between two goals: individual professional advancement through technical and organizational skills, and forming future European labor leaders. "Minutes of Meeting with Technical Delegation Representatives of the Foreign Embassies," October 6, 1952, 2, box 5, entry 171–G, RG 469, NARA; Mutual Security Agency, "Guide for Use by Missions in Interpreting the Work-Study Training for Productivity Program of MSA," October 16, 1952, 1, box 5, entry 171–G, RG 469, NARA.

These tensions were reflected initially in problems in recruiting suitable exchange students: while the Productivity Program administration sought for bench workers and foremen—that is, workers in positions that in the United States, would be considered working-class, nonmanagerial positions—European students were chemists, engineers, office managers, and in some cases even the sons of company owners. Mutual Security Agency, "Work Study Training Program," conference proceedings, March 12, 1952, 47, box 5, entry 171–G, RG 469, NARA; "Minutes of Meeting with Technical Delegation Representatives of the Foreign Embassies," October 6, 1952, 4, box 5, entry 171–G, RG 469, NARA. European working-class students rarely received instruction in foreign languages, and the US colleges scrambled to set up basic English classes. Mutual Security Agency, "Work Study Training Program," conference proceedings, March 12, 1952, 22–24, box 5, entry 171–G, RG 469, NARA. Notably, none of the WSTP officers made a connection between working-class status and lack of access to education, including foreign language instruction, while the students commented on the educational opportunities in the United States.

55. Jugendselbsthilfe, "Auswertungskonferenz," Bericht der Arbeitsgruppe II: Labor Relations, 5, box 5, entry 1202, RG 469, NARA; Jugendselbsthilfe, "Auswertungskonferenz," Diskussionsbeiträge des Plenums zum Bericht der Arbeitsgruppe II, 1–2, box 5, entry 1202, RG 469, NARA.

56. Jugendselbsthilfe, "Auswertungskonferenz," Bericht der Arbeitsgruppe II: Labor Relations, 6, box 5, entry 1202, RG 469, NARA.

57. Ibid., 10.

58. Presumably, the blank replaces an expletive in a statement by J. W. Funke, a representative of the International Council for Youth Self-Help. Donald M. Typer and H. Carl Shugaar, "Summary and Report of Typer—Shugaar Visits to MSA Missions in Europe (February 20–April 17, 1953)," n.d., box 5, entry 171–G, RG 469, NARA.

59. Seeling, "Was mir in Amerika aufgefallen ist."

60. Bert Wessel, "Amerika-Tagebuch, 19.11–9.12.1951," n.d., 4, box 1, entry 1202, RG 469, NARA.

61. Wessel considered his travel diary a first general report on the economic situation and planned an additional technical report. Ibid., 29.

62. Ibid., 5. Wessel was also interested in office organization and technology, such as the use of a punch card system for accounting and payroll.

63. Edward L. Deuss, "Evaluation Meeting, TA 09-101-2138 German Steel Heat Treating Study, October 31, 1952," November 10, 1952, 1, box 1, entry 171–G, RG 469, NARA.

64. Ibid., 2.

65. Rationalisierungs-Kuratorium der Deutschen Wirtschaft, *Jahresbericht 1953/54 des Rationalisierungs-Kuratoriums der Deutschen Wirtschaft*, 11–12. A 1953 report on

"Productivity in the United States" marks the shift: Rationalisierungs-Kuratorium der Deutschen Wirtschaft, *Produktivität in USA*.

66. Rationalisierungs-Kuratorium der Deutschen Wirtschaft, *Produktivität und Fertigung*, 7.

67. US officers received these claims with cautious skepticism. Foreign Operations Administration, "Evaluation Meeting with the German Project Implementation Team, TA 09-227, April 8, 1954 at Washington, DC," April 20, 1954, 1, 3, box 3, entry 178–A, RG 469, NARA.

68. Rationalisierungs-Kuratorium der Deutschen Wirtschaft, *Human Relations in Industry*, 7. For a follow-up tour, see Rationalisierungs-Kuratorium der Deutschen Wirtschaft, *Gruppenarbeit und Produktivität*, 4, 9.

69. Edward L. Deuss, "Evaluation of the US Tour of the Mine Safety Study Group from Germany (TA 09-179), December 19, 1952, at New York, NY," January 23, 1953, box 4, entry 171–G, RG 469, NARA.

70. Edward L. Deuss, "Evaluation Meeting, TA 09-184, German Machine and Precision Tools Productivity Team, December 18, 1952, Park Sheraton Hotel, New York," January 2, 1953, 1, box 4, entry 171–G, RG 469, NARA; John L. Butler, "The Follow-up on the German Machine and Precision Tools Productivity Team," March 18, 1953, 1, box 4, entry 1202, RG 469, NARA.

71. Bier may have thought that younger workers were less productive because they were less disciplined or skilled, or because they were still in apprenticeships. "Report on the Evaluation Meeting with the German Trade Union Procedures Leather Workers Team—TA 09–209—at New York, NY, September 8, 1953," September 24, 1953, 3–4, box 3, entry 178–A, Entry 469, NARA.

72. Heinrich Krumm, "American Journey! November–December 1951," 3–5, box 11, entry 1202, RG 469, NARA.

73. For an analysis of the postwar German economy, see Abelshauser, *Deutsche Wirtschaftsgeschichte*. The Marshall Plan, however, created economic conditions in Europe that were beneficial to German economic recovery. See also Milward, *The Reconstruction of Western Europe*.

74. Heinrich Krumm, "American Journey! November–December 1951," 7, box 11, entry 1202, RG 469, NARA.

75. Ibid., 3, 7–8.

76. Seeling, "Was mir in Amerika aufgefallen ist." Seeling was a member of the conservative small business community in southern Germany that opposed codetermination. Müller, *Strukturwandel*, 170–171.

77. Seeling, "Was mir in Amerika aufgefallen ist."

78. Bert Wessel, "Amerika-Tagebuch, 19.11–9.12.1951," n.d., 24, box 1, entry 1202, RG 469, NARA. Similar models of workers' shareholding had been common counter-suggestions to codetermination. Müller, *Strukturwandel*. The similarities indicate that Wessel interpreted the Productivity Program's proposals for profit-sharing arrangements as corresponding to his opposition to codetermination in Germany.

79. "General Impressions of WSTP Students before and after, Etc.," n.d., box 5, entry 171–G, RG 469, NARA (emphasis in original).

80. "Analysis of the Impact of A-Projects: Preliminary Draft," n.d., 3, box 5, entry 171–G, RG 469, NARA; Lippitt and Watson, "Recommendations for Ways of Improving the 'Back Home' Impact and the Follow-up Evaluation of A-Projects," n.d., 1, box 5, entry 171–G, RG 469, NARA.

81. Everett H. Bellows, "Excerpts from the Minutes of the Productivity and Technical Assistance Regional Conference," conference held in Paris, October 1–4, 1952, 50–51, box 18, entry 1202, RG 469, NARA. For the tree-planting ceremony, see also "Minutes of Meeting with Technical Delegation Representatives of the Foreign Embassies," October 6, 1952, 9, box 5, entry 171–G, RG 469, NARA.

82. Jugendselbsthilfe "Auswertungskonferenz," Bericht der Arbeitsgruppe II: Labor Relations, 7, box 5, entry 1202, RG 469, NARA. Hoffman, the first head of the Marshall administration, had worked as president of Studebaker, and when the union proposed temporary wage cuts to reduce manufacturing costs, the company was hailed as an illustration of collaborative labor relations. Studebaker went out of business in the 1960s, and in hindsight may not appear as an economic model.

Chapter 6

1. Hughes, "German Solons Begin Debate on Union Bid."

2. Nicholls, *Freedom with Responsibility*, 223. See also Abelshauser, *Deutsche Wirtschaftsgeschichte*, 96.

3. Other Western economies also pursued planning approaches, like the Monnet Plan in France for economic recovery and modernization after World War II.

4. Hall and Soskice, *Varieties of Capitalism*.

5. Abelshauser, *Kulturkampf*. See also the comparable notions of competitive managerial capitalism and cooperative managerial capitalism. Chandler and Hikino, *Scale and Scope*. For a critique, see Hancké, Rhodes, and Thatcher, *Beyond Varieties of Capitalism*.

6. For changes of the social market economy from free market to the European social welfare state, see Van Hook, *Rebuilding Germany*. See also Abelshauser, *Kulturkampf*, 151.

7. Müller, *Mitbestimmung in der Nachkriegszeit.*

8. Adenauer's middle-class Catholicism proved a better unifying ground for a postwar Christian political party than the demands of Catholic trade unionists for nationalization, economic planning, and codetermination in the Ahlener Programm, mentioned shortly in the text. Mitchell, *The Origins of Christian Democracy,* 140–146.

9. The British Labor government pursued the nationalization of British industries in the postwar years and was probably sympathetic to the goals of codetermination, but a more conservative military implemented the British occupation policies in Germany.

10. Klessmann, "Betriebsräte und Gewerkschaften."

11. Nicholls, *Freedom with Responsibility;* Van Hook, *Rebuilding Germany.* See also Spicka, *Selling the Economic Miracle.*

12. For Erhard, the terms "social" and "free" were congruent. Heusgens, *Ludwig Erhards Lehre,* 171–173.

13. Nicholls, *Freedom with Responsibility,* 38, 62–63.

14. Ibid., 39–41, 58, 90.

15. For an irreverent but voluminous biography, see Hentschel, *Ludwig Erhard.* For a general audience, and a more adulatory work, see Mierzejewski, *Ludwig Erhard.*

16. Nicholls, *Freedom with Responsibility,* 71–73, 77, 103–109, 116–120.

17. Ibid., 44–48.

18. Ibid., 183–196.

19. Ibid., 238; Mitchell, *The Origins of Christian Democracy,* 146.

20. For example, the Catholic trade unionist Karl Arnold had formed a coalition government that included the CDU, the SPD, and the German Communist Party in North Rhine-Westphalia. Mitchell, *The Origins of Christian Democracy,* 146–148, 158–162. The *Düsseldorfer Leitsätze* quelled the influence of Catholic social teaching in the CDU. Nicholls, *Freedom with Responsibility,* 237–241.

21. Erhard created "Everybody's Program" (*Jedermannsprogramm*), which provided essential goods through mass production, low quality, and low prices to satisfy the immediate needs of the population. Nicholls, *Freedom with Responsibility,* 225–226.

22. Ibid., 210.

23. Müller, *Mitbestimmung in der Nachkriegszeit;* Müller, *Strukturwandel und Arbeitnehmerrechte;* Markovits, *The Politics of the West German Trade Unions.*

24. Abelshauser, *Kulturkampf*, 143–150. For the Weimar period, see also Müller, *Strukturwandel und Arbeitnehmerrechte*, 119–123.

25. Other European countries also had works councils, but they differed in their makeup from German codetermination. Rogers and Streeck, *Works Councils*.

26. Angster, *Konsenskapitalismus und Sozialdemokratie*, 57.

27. For the American interpretation of codetermination, see Herrigel, "American Occupation, Market Order, and Democracy," 379–380.

28. "Headed for a Ruhr Showdown."

29. Fowle, "Labor in Germany Aims at New Role."

30. "German Labor's Say in Industry Is Hit," 38.

31. Letter from Earl Bunting to Heinz L. Krekeler, August 23, 1950, box 141, series 1, NAM accession 1411, Hagley.

32. NAM, "For Release in Morning Papers of Monday, February 5, 1951," box 141, series 1, NAM accession 1411, Hagley.

33. Letter from Bunting to Heinz L. Krekeler, January 31, 1951, box 141, series 1, NAM accession 1411, Hagley.

34. Galantiere, "International Decorum."

35. Haynes and Michler, *Co-Determination*, 3–4.

36. Fones-Wolf, *Selling Free Enterprise*; Harris, *The Right to Manage*. For free enterprise, see also chapter 4.

37. National Association of Manufacturers, *NAM Position on Industrywide Bargaining*; Teplow, *Where Will Industry-Wide Bargaining Lead Us?* For the post–World War II controversy between business progressives and conservatives within NAM, see Delton, "The Triumph of Social Responsibility."

38. Harris, *The Right to Manage*, 131–135; Fones-Wolf, *Selling Free Enterprise*.

39. National Association of Manufacturers, *Labor Monopoly and Industry-Wide Bargaining*, 8–11; Bunting, *Labor Monopoly and the Public Interest*, 2–4.

40. National Association of Manufacturers, *NAM Position on Industrywide Bargaining*. The NAM's position allowed for informal collaboration between employers, such as the exchange of information on wages, hours, and working conditions.

41. National Association of Manufacturers, *Labor Monopoly and Industry-Wide Bargaining*, 16.

42. Teplow, *Where Will Industry-Wide Bargaining Lead Us?*, 4–6.

43. Backman, *Multi-Employer Bargaining*, 6–11, 50–51.

44. Meyer Bernstein, "Memorandum re Eldridge Haynes Letters," June 19, 1951, 5, 8, box 2, entry 49, RG 469, NARA. Other critical reporting in the German labor press followed.

45. Ibid.

46. Congress of Industrial Organizations, "CIO Pres. Murray Urges US Mediation of German Labor Management Dispute on 'Co-Determination' Issue," press release, January 25, 1951, box 141, series I, NAM accession 1411, Hagley.

47. Paul Fisher, "Memo: Assistance to German Labor Organizations in the Administration of Economic Co-Determination," May 3, 1951, 3, box 2, entry 49, RG 469, NARA. Haynes complained about a left-leaning tendency in the German press, although there is no evidence of another public CIO statement, for example. Haynes and Michler, *Co-Determination*, 8.

48. Brooks, *Clint*, 325–326. Based on the timing of the labor officers' response to Golden's suggestions, Golden must have traveled in the early phase of the debate. Letter from Harvey W. Brown to Clinton S. Golden, March 23, 1951, box 2, entry 49, RG 469, NARA.

49. Brooks, *Clint*, 325–326.

50. Fisher, "Labor Codetermination in Germany," 472.

51. Congress of Industrial Organizations, "CIO Pres. Murray Urges US Mediation of German Labor Management Dispute on 'Co-Determination' Issue," press release, January 25, 1951, box 141, series 1, NAM accession 1411, Hagley.

52. Internationaler Rat für Jugendselbsthilfe e.V., "Auswertungskonferenz," Bericht der Arbeitsgruppe II: Labor Relations, 5–6, box 5, entry 1202, RG 469, NARA.

53. Brown served as the director of the US office of labor affairs in Germany. Brown, "'Mitbestimmung' (Codetermination): Labor's New Responsibility in Germany," 237 (emphasis in original). Brown claims that McCloy's statement of neutrality helped bring German negotiating parties back to the table under Adenauer's purview in January 1951.

54. Klessmann, "Betriebsräte und Gewerkschaften."

55. Meyer Bernstein, "Memorandum re Eldridge Haynes Letters," June 19, 1951, 2–4, box 2, entry 49, RG 469, NARA. To accelerate reporting, Bernstein's initial report on the legislation was based on this read-aloud version and mistakenly omitted a six-word phrase—the reason why Haynes later accused him of wrongful information.

56. Abelshauser, "The First Post-Liberal Nation." See also Abelshauser, *Kulturkampf*, 97, 102–106; Thelen, *Union of Parts*.

57. Abelshauser, *Deutsche Wirtschaftsgeschichte*, 28–32. The 1957 cartel legislation introduced a similar corporatist element to the West German economic order, although US Productivity Program officers and Erhard supported stronger antitrust legislation. Teupe, "Verhandelte Grenzüberschreitungen." Other historians have disagreed with Abelshauser's position on corporatism. See Nicholls, *Freedom with Responsibility*, 9; Van Hook, *Rebuilding Germany*, 194–195, 230.

58. Meyer Bernstein, "Diary, Monday, April 2, 1951," box 2, entry 49, RG 469, NARA; letter from Meyer Bernstein to Clinton S. Golden, April 3, 1951, box 2, entry 49, RG 469, NARA; letter from Harvey W. Brown to Clinton S. Golden, March 23, 1951, box 2, entry 49, RG 469, NARA; Paul Fisher, "Memo: Assistance to German Labor Organizations in the Administration of Economic Co-Determination," May 3, 1951, box 2, entry 49, RG 469, NARA.

59. Paul Fisher, "Memo: Assistance to German Labor Organizations in the Administration of Economic Co-Determination," May 3, 1951, 2, box 2, entry 49, RG 469, NARA.

60. Ibid., 3.

61. Letter from Harvey W. Brown to Clinton S. Golden, March 23, 1951, box 2, entry 49, RG 469, NARA. While the US labor advisers in Germany felt that a request for support needed to come from the German side, they considered ascertaining if American assistance was welcome. Paul Fisher, "Memo: Assistance to German Labor Organizations in the Administration of Economic Co-Determination," May 3, 1951, 3, box 2, entry 49, RG 469, NARA.

62. "Germans' System of Labor Studied," 36.

63. Maier, "Politics of Productivity," 628.

64. Haynes and Michler, *Co-Determination*, 12. This suggestion went against the NAM's demands that US companies be allowed to operate abroad under the usually more permissive local cartel legislation in foreign countries. National Association of Manufacturers, Research Department, *NAM Looks at Cartels*, 3. See also Wells, *Antitrust and the Formation of the Postwar World*.

Chapter 7

1. "Memorandum of Telephone Call between Mr. Thomas J. Watson and Mr. H. K. Chauncey, Paris, April 26, 1950," 1, folder 7, box 852, Watson Sr. Papers, IBM Archive.

2. Council of Europe, Consultative Assembly, Committee on Science and Technology, "Report on the Computer Industry in Europe: Hardware Manufacturing (Rapporteur: Mr. Lloyd)," document 2893, 1971, 9, 21, folder 5, box 11, CBI 62, CBI.

Market share value here is measured as either the value of yearly deliveries or the value of the stock of installations.

3. Jacoby, *Modern Manors*; Cohen, *Making a New Deal*.

4. Stebenne, "IBM's 'New Deal.'"

5. Medina, "Big Blue in the Bottomless Pit," 32.

6. The rich literature on IBM, often written from historic, journalistic, or economic angles, has focused on Watson and his personality. Yet the company's welfare capitalist features have usually been ignored or presented as unique to IBM, rather than integrating them in the larger historiography of US welfare capitalist companies such as Eastman Kodak, Sears Roebuck, and the National Cash Register Company, where Watson experienced welfare capitalist labor relations as a young executive. Also, until about a decade ago, publications on IBM suffered from a lack of access to corporate archival sources, with the exception of Kevin Maney's biography of Watson Sr. Maney, *The Maverick and His Machine*. On IBM's management, competitive position, and technology development, see, for example, DeLamarter, *Big Blue*; Usselman, "IBM and Its Imitators"; Pugh, *Building IBM*. For an early history, see Cortada, *Before the Computer*. For IBM's international operations, see Foy, *The Sun Never Sets on IBM*; Medina, "Big Blue in the Bottomless Pit"; Schlombs, "Engineering International Expansion"; Paju and Haigh, "IBM Rebuilds Europe."

7. IBM Archives, "IBM Rally Song, Ever Onward Transcript"; Hutchison, "Tripping through IBM's Astonishingly Insane 1937 Corporate Songbook."

8. Maney, *The Maverick and His Machine*, 3–18.

9. Ibid., 21–25, 56–57.

10. Global operations again increased slowly in the decades following World War II, to take off only in the last quarter of the twentieth century. Jones, *Transnational Corporations*; Wilkins, *The Maturing of Multinational Enterprise*.

11. During his 1937 visit to Germany, Watson accepted the Cross of Merit, Nazi Germany's second-highest honor for foreigners. For this decision, see Maney, *The Maverick and His Machine*, 203–208; Watson and Petre, *Father, Son & Co.*, 53–56.

12. Jacoby, *Modern Manors*.

13. Stebenne, "IBM's 'New Deal.'"

14. Maney, *The Maverick and His Machine*, 144, 163–168.

15. Ibid., 165.

16. Ibid., 39–47, 89.

17. Campbell-Kelly, *ICL*; Petzold, *Rechnende Maschinen*.

18. Maney, *The Maverick and His Machine*; Watson and Petre, *Father, Son & Co.*

19. I have been unable to locate any documentation of the reasons for IBM's reorganization. Archival materials indicate that IBM first reorganized the South American business in the mid- to late 1940s, and that tax and labor regulations as well as the goal to simplify operations played a major role in the reorganization. Compilation of documents on IBM in South America, folder 5, box 108, Record Group 6, IBM Archive. IBM again considered the same issues when expanding into African markets in the early 1950s. John N. Irwin II, memorandum to Thomas J. Watson Sr., July 8, 1952, folder 3, box 855, Watson Sr. Papers, IBM Archive.

20. The Marshall Plan was one of the primary expressions of this policy. Campbell-Kelly, *ICL*, 140.

21. Since the mid-1920s, IBM's foreign business comprised about a third of its overall business, both in terms of employees and sales. While foreign sales and employment grew significantly in absolute numbers, they remained in line with IBM's domestic operations. George L. Ridgeway, "Growth of IBM World Trade," memorandum to Arthur K. Watson, May 31, 1950, folder 8, box 852, Watson Sr. Papers, IBM Archive.

22. In his speech, Watson Sr. may have overemphasized developments in European countries and the growth of IBM's foreign business. Undated manuscript of Watson Sr.'s speech, folder 2, box 852, Watson Sr. Papers, IBM Archive.

23. Documentation of the 1949 telephone hookup, folders 2–3, box 852, Watson Sr. Papers, IBM Archive; *IBM World Trade Corporation News* 1, no. 1 (October 1949): 1–5, HzGD.

24. Maney, *The Maverick and His Machine*, 21, 26.

25. She stayed in the marriage because Watson Sr. became distressed when she asked for a divorce. Maney, *The Maverick and His Machine*, 136; Watson and Petre, *Father, Son & Co.*, 21.

26. Maney, *The Maverick and His Machine*, 21–31, 263–265.

27. "Frau Thos. J. Watson ist Mitglied des Aufsichtsrats," *IBM World Trade Corporation News* 1, no. 1 (October 1949): 2, HzGD. The lighter keyboard allowed bedridden veterans, for example, to hold it on their laps and type to produce text on a full typewriter to which the keyboard was connected with a cord; it was developed as a prototype model. IBM Archives, "Remote Control Keyboard."

28. Schlombs, "The 'IBM Family.'"

29. Watson and Petre, *Father, Son & Co.*, 149–152.

30. "Gedanken zum Besuch von Thos. J. Watson," *IBM Deutschland* (August 1953): 3, HzGD.

31. May, *Homeward Bound*; Weiss, *To Have and to Hold*; Kessler-Harris, *In Pursuit of Equity*; Gordon, *New Deals*.

32. "Die Rede von Thomas J. Watson," *IBM Deutschland* (August 1953): 4, HzGD.

33. "Mitmenschliche Beziehungen," *IBM World Trade News* (July 1955): 2, HzGD.

34. In West Germany, the rate of women between fifteen and fifty-nine years in salaried employment steadily increased during the 1950s, from 47.4 percent in 1950, to 51.8 percent in 1955, and 54.1 percent in 1960. Zu Castell, "Die demographischen Konsequenzen des Ersten und Zweiten Weltkrieges."

35. With legislation on the so-called housewife marriage (*Hausfrauenehe*) in 1957, a woman no longer needed her husband's approval to work, but remained tied to her primary duties as housewife and mother. Moeller, *Protecting Motherhood*.

36. The strong emphasis on the nuclear family in West German legislation must be seen in the larger Cold War context. Heineman, *What Difference Does a Husband Make?*, 146–147. See also Frevert, *Women in German History*, 265; Moeller, *Protecting Motherhood*, 71.

37. "Unsere Feier auf dem Killesberg," *IBM Deutschland* (June 1954): 4, HzGD.

38. While no demographic information on IBM's workforce in Germany is available, numerous images in the employee magazines document that significant numbers of women worked in clerical as well as manufacturing positions.

39. "Gedanken zum Besuch von Thos. J. Watson," *IBM Deutschland* (August 1953): 3, HzGD.

40. Ibid., 3.

41. "Mehr Verbesserungs-Vorschläge im April 1954 als in der ersten Jahreshälfte 1953 zusammen," *IBM Deutschland* (August 1954): 1, HzGD. By contrast, IBM's open-door policy did not seem to have found wide acceptance in Germany, even by the 1960s. Correspondence between Gunnar Patt and Thomas J. Watson Jr., folder 6, box 264, Watson Jr. Papers, IBM Archive.

42. Harrison Chauncey to Thomas Watson Sr., February 13, 1951, folder 1, box 854, Watson Sr. Papers, IBM Archive.

43. On other occasions, Stender identified as an "old REFA man"—that is, a member of the technology-centered German rationalization movement discussed in chapters 2 and 5.

44. "Minutes of Meeting with Members of Works Council—Boeblingen—Aug. 14, 1953," 1–3, 7, folder 3, box 856, Watson Sr. Papers, IBM Archive.

45. "Unsere Feier auf dem Killesberg," *IBM Deutschland* (June 1954): 5, HzGD.

46. "Das Haus und das Werk," *IBM Deutschland* (December 1954): 2, HzGD.

47. W. B., "Proletarier oder Persönlichkeiten?," *IBM Deutschland* (August 1954): 2, HzGD. Signed with initials only, the editorial was likely authored by W. Berger. In a later statement, the works council referred to the Catholic philosopher Josef Pieper. "Entproletarisieren! Ein Diskussionsbeitrag des Betriebsrates Sindelfingen/Böblingen," *IBM Deutschland* (February 1955): 3, HzGD.

48. Grebing, *Geschichte der deutschen Arbeiterbewegung*.

49. "Betriebsversammlungen in Berlin und Sindelfingen," *IBM Deutschland* (July 1953): 1, HzGD.

50. "Unsere Betriebsversammlungen am 18. Dezember 1953," *IBM Deutschland* (January–March 1953): 12, HzGD; "Wir sprechen miteinander über unsere betrieblichen Tagesfragen," *IBM Deutschland* (September 1954): 2, HzGD.

51. Thomas Watson Sr. to Hector McNeil, August 13, 1955, folder 5, box 858, Watson Sr. Papers, IBM Archive.

52. Ibid.

53. Unfortunately, permission to reproduce the described image was not granted.

54. Sociologist Erwin K. Scheuch formalized "prestige" as a statistical measure of social groups. Scheuch and Rüschemeyer, "Scaling Social Status in Western Germany." For further discussion, see also chapter 6.

55. "Wie teuer ist das Prestige?," *IBM Deuschland* (July 1958): 16, HzGD.

56. Watson Jr. placed this policy change in the tradition of his father's policies; Watson Sr. had eliminated piecework pay for manufacturing employees in favor of hourly wages in 1934, during a decade when the company also introduced other welfare capitalist measures. IBM Archives, "IBM's 100 Icons of Progress." The new policy had an effect similar to the guaranteed annual wage, the bargaining objective of the United Autoworker union in 1955, discussed in more detail in chapter 8.

57. "Der 1. August 1958," *IBM Deutschland* (August 1958), 3, HzGD.

58. "Mehr Geltung"; "Berlin gehört unser ganzes Vertrauen," *IBM Deutschland* (January 1959): 3, HzGD.

59. "Ab 1. August 1958 haben alle IBM Mitarbeiter die gleichen sozialen Rechte und Pflichten," *IBM Deutschland* (August 1958): 3, HzGD; "Die Arbeit des Betriebsrats im Dienste der Betriebsangehörigen: Der Rechenschaftsbericht für das 2. Quartal 1958," *IBM Deutschland* (August 1958): 6, HzGD.

60. "Die neuen sozialen Leistungen der IBM Deutschland," *IBM Deutschland* (August 1958): 7, HzGD.

61. Unfortunately, permission to reproduce the described image was not granted. For the social history of middle-class white-collar professionals in Germany, see Kocka, *Die Angestellten in der deutschen Geschichte*. The guarded response of white-collar employees stood in stark contrast to the jubilant smiles of blue-collar workers at, for instance, the July 1955 employee assembly, when management announced a new pension plan—an achievement that Stender and the works council had sought for several years. "Aussprache von Herrn Hörrmann und Bekanntgabe des Altersversorgungs-Planes," *IBM Deutschland* (July 1955): 4, HzGD.

62. "Der 1. August 1958," *IBM Deutschland* (August 1958): 3, HzGD.

63. "Ab 1. August 1958 haben alle IBM Mitarbeiter die gleichen sozialen Rechte und Pflichten: Der Betriebsrat berichtet über seine Arbeit," *IBM Deutschland* (August 1958): 3, HzGD.

64. "Berlin gehört unser ganzes Vertrauen," *IBM Deutschland* (January 1959): 3, HzGD.

65. "Der 1. August 1958," *IBM Deutschland* (August 1958): 3, HzGD.

66. German original: "Schutz und Schirm in den Wechselfällen des Lebens." "Der 1. August 1958," *IBM Deutschland*, August (1958): 3, HzGD. The line alluded to the request for relief and care through Mary: "Maria breit den Mantel aus, mach Schirm und Schild für uns daraus, . . . bis alle Stürm vorüber gehn." Literal translation: "Mary spread your coat, make it our umbrella and our shield, . . . until all storms have passed."

67. "Berlin gehört unser ganzes Vertrauen," *IBM Deutschland* (January 1959): 3, HzGD; "Betriebsversammlung in Sindelfingen," *IBM Deutschland* (April 1959): 7, HzGD; and "Betriebsversammlung in Sindelfingen," *IBM Deutschland* (January 1960): 5, HzGD.

68. "Zum Bericht der Geschäftleitung: Mahnung zum Masshalten," *IBM Deutschland* (August 1958): 7–8, HzGD.

69. Schlombs, "Engineering International Expansion."

70. In 1952, in order to build an electronic laboratory, IBM hired Karl E. Ganzhorn, a young physics PhD who would later direct all IBM's European laboratories. In Germany, IBM primarily competed for talent against Siemens, a long-established company with a strong research reputation. Ganzhorn, *The IBM Laboratories Boeblingen*, 13–14, 24; Ganzhorn, oral history interview. Being recognized as a local company also may have helped IBM acquire federal research funding in West Germany. Petzold, *Rechnende Maschinen*.

Chapter 8

1. Ted F. Silvey, "Memorandum about Four Different Color Ball Point Pencils, Merchandized with Pocket Clip Hair Comb, in Colored Plastic Case, in Relation to Automation," June 20, 1956, 2, folder 11, box 48, Accession 625, Reuther Library; Silvey, "A Ball-Point Pencil—Is a Computer?" The text of the article was identical to Silvey's 1956 memorandum, and only the new title connected the ballpoint pens and computers. Lacking an editorial note or other explanation, the exact relation between the two technologies remains an open question.

2. Silvey, born in 1904 into a New Hampshire working-class family, grew up in Zanesville, Ohio. When his family needed extra support after his father died from an industrial accident, he took up odd jobs and completed school through evening classes.

3. Diebold, *Automation*, 161–163.

4. Hunt Brown, "Office Automation," loose-leaf collection, part 4, section A: Social consequences of automation, 1, 3, folder 26, box 69, CBI 55, CBI; Hunt Brown, "Office Automation," loose-leaf collection, part 4, section C: Humanities of automation, 1, folder 26, box 69, CBI 55, CBI.

5. Noble, *The Forces of Production*; Bix, *Inventing Ourselves Out of Jobs?*

6. Schlombs, "Univac."

7. Stollberg, *Die Rationalisierungsdebatte*, 96.

8. Brady, *The Rationalization Movement in German Industry*, xi.

9. Shearer, "The Politics of Industrial Efficiency," 387–394.

10. The use of the term "electronic brain" (*Elektronenhirn*) was widespread in the German local and national press. See press review, Kuratorensammlung, Deutsches Museum.

11. Wiener, *Cybernetics*. For the German reception, see Aumann, *Mode und Methode*.

12. "Beängstigend menschlich," 37.

13. "Elektronengehirne: Die Magie der Roboter," 52.

14. "Elektronengehirn für intellektuelle Arbeit," 33.

15. "Elektronengehirne: Die Magie der Roboter," 44.

16. "Automation: Die Revolution der Roboter," 29.

17. "Berg-Friede."

18. Balke, "Automation as an Aid to Scientific Plant Management," copy in folder 40, box 14, CBI 3, CBI.

19. In a presentation to the German society for business administration (*Deutsche Gesellschaft für Betriebswirtschaft*), the Siemens engineer Vollbert argued that due to the different ratio between machine and personnel costs, in the United States, a large-scale electronic computer only needed to release 80 to 120 clerks in order to pay back rental costs, while in Germany, the same computer would have to release 200 to 300 employees in order to be cost efficient. "Büro-Automatisierung im Anfangsstadium."

20. Balke, "Automation as an Aid to Scientific Plant Management."

21. Gresmann, "Revolution oder nicht Revolution?" See also Salin, "Drei Schritte, die die Welt verändern."

22. Boel, *The European Productivity Agency*.

23. Airgram USRO Paris to ICA Washington, "Subject: EPA Requests Data on Automation Courses in US," July 22, 1955, box 40, entry 1058, RG 469, NARA; Airgram USRO Paris to ICA Washington, "Subject: EPA Requests Data on Automation Courses in US," September 23, 1955, box 40, entry 1058, RG 469, NARA; Airgram USRO Paris to ICA Washington, "Subject: US Position vis-à-vis Work in Automation," November 26, 1955, box 40, entry 1058, RG 469, NARA.

24. Carew, *Labour under the Marshall Plan*, 166.

25. OEEC, EPA, "Automatic Processes in Industry," May 3, 1955, 28, 29, 33, box 40, entry 1058, RG 469, NARA.

26. Ibid.; Carew, *Labour under the Marshall Plan*, 166.

27. Romero, *The United States and the European Trade Union Movement*, 131–137. See also the discussion in chapter 4.

28. Diebold, *Automation*.

29. Letter from John Harlan Jr. to Donald B. MacPhail, July 5, 1955, box 40, entry 1058, RG 469, NARA. A few months later, Diebold provided the lead testimony for the US congressional hearings on automation.

30. Airgram USRO Paris to ICA Washington, "Subject: US Position vis-à-vis Work in Automation," November 26, 1955, box 40, entry 1058, RG 469, NARA; Airgram USRO Paris to ICA Washington, "Subject: EPA Requests Data on Automation Courses in US," July 22, 1955, box 40, entry 1058, RG 469, NARA.

31. Marx, *The Machine in the Garden*; Bix, *Inventing Ourselves Out of Jobs?*.

32. For a bibliography of the US automation debate, see Ted F. Silvey, "Automation Listings in the New York Times Index from First Appearance of This Subject through

October 15, 1955," November 30, 1955, folder 16, box 48, Accession 625, Reuther Library.

33. Barclay, "Metals Industry Speeds Conversion"; "'Automatism' Eases Manpower Shortage"; Freeman, "New Way Devised to Make Abrasives."

34. National Manpower Council, *A Policy for Scientific and Professional Manpower.* The National Manpower Council, headed by James D. Zellerbach, helped create the notion of the "shortage of brainpower" and resulted in changes to military deferment policy. See also National Manpower Council, *A Report on the National Manpower Council*, 24–27.

35. "Sidelights on the Financial and Business Developments of the Day"; Kaempffert, "A New 'Tinkertoy' Speeds Production."

36. "Emerson Extends Automation's Use"; "New RCA Device Aids Automation."

37. "Machine, Job Rise, Linked"; Rusk, "Industry Grows in Humanity."

38. Rusk, "Industry Grows in Humanity," 62. Industry experts also exposed the economic disadvantages of automatic control, such as long downtimes for tooling changes and competition from cheap global labor. "Idle Machine Time Bars Robot Plants"; "Financial and Business Sidelights of the Day."

39. "Financial and Business Sidelights."

40. "Presto!"

41. Garfinkel, "Technological Unemployment"; Kaempffert, "Walter Reuther's Fears of 'Automation'"; Freeman, "Automation Aims at New Freedom."

42. "Machine, Job Rise Linked"; Rusk, "Industry Grows in Humanity."

43. "Reuther Assails Robot-Job Trend.

44. "Fairless Upholds 'Automation' Aims."

45. Ibid.; Porter, "Menzies in US for Talks on Asia"; "Reuther Insists Weeks Apologize."

46. Congress of Industrial Organizations, Committee on Economic Policy, *Challenge of Automation.*

47. *Automation—Engineering for Tomorrow: General Sessions Addresses from the Centennial Symposium of the School of Engineering, Michigan State University, May 13, 1955,* folder 40, box 26, Accession 625, Reuther Library; Yale University, *Man and Automation*; Silvey and Ryan, "Labor and Management."

48. Weinberg, *A Case Study of a Company Manufacturing Electronic Equipment*; Van Auken, *The Introduction of an Electronic Computer in a Large Insurance Company*; Rothberg, *A Case Study of a Large Mechanized Bakery*; Weinberg and Rothberg, *A Case Study*

of a Modernized Petroleum Refinery; Weinberg and Jakubauskas, *A Case Study of an Automatic Airline Reservation System.*

49. US Congress, Joint Committee on the Economic Report, *Automation and Technological Change.*

50. Congress of Industrial Organizations, Automation and the New Technology: Resolution Adopted by UAW-CIO, Fifteenth Constitutional Convention, March 27–April 1, 1955, Cleveland, 2, folder 36, box 26, Accession 625, Reuther Library.

51. Crane, "New Margin Rule Urged by Banker."

52. Letter from Norbert Wiener to Walter Reuther, August 13, 1949, Accession 1168, Reuther Library.

53. Letter from Ted F. Silvey to Walter Reuther, October 7, 1953, folder 3, box 40, Accession 625, Reuther Library. Silvey had served as field organizer, edited a union newspaper, served as the first elected secretary treasurer of the CIO, held several elected union offices, and by 1953, had joined the staff of the national CIO headquarters in Washington, DC.

54. Lichtenstein, *The Most Dangerous Man,* 5, 25–46, 55.

55. Ibid., 280–281. Sliding scales tied to the cost of living had their origins in Trotskyist proposals for the transition of the Soviet economy in the 1930s.

56. Ibid, 280–281. Murray, then still CIO president, envisioned a continuation of the wartime tripartite collective bargaining. Ibid., 226.

57. Ibid., 287.

58. UAW-CIO Education Department, *Automation: A Report to the UAW-CIO Economic and Collective Bargaining Conference.*

59. Ibid., 6.

60. Maier, "Politics of Productivity."

61. "Railroad—What Technological Change Can Mean"; International Association of Machinists, *Meeting the Problems of Automation through Collective Bargaining; Steelworkers and Automation*, folder 26, box 1, Accession 625, Reuther Library; Shugart, *Your Place in the Atom (Concerning Musicians).*

62. Congress of Industrial Organizations, "Automation and the New Technology: Resolution Adopted by UAW-CIO," Fifteenth Constitutional Convention, March 27–April 1, 1955, Cleveland, 4, folder 36, box 26, Accession 625, Reuther Library.

63. Noble, *The Forces of Production.* Noble's analysis suggests that by not influencing how automation technology was implemented, unions allowed management to increase its control over shop floor operations through automation technology; they

also abandoned efforts at having a say over automation technology. Rifkin, *The End of Work*, 84–88.

64. Weinberg, *A Case Study of a Company Manufacturing Electronic Equipment*; Van Auken, *The Introduction of an Electronic Computer in a Large Insurance Company*; Rothberg, *A Case Study of a Large Mechanized Bakery*; Weinberg and Rothberg, *A Case Study of a Modernized Petroleum Refinery*; Weinberg and Jakubauskas, *A Case Study of an Automatic Airline Reservation System*; Baldwin and Schultz, "Automation: A New Dimension to Old Problems."

65. US Congress, Joint Committee on the Economic Report, *Automation and Technological Change*, 2, 636.

66. Deutscher Gewerkschaftsbund, *Protokoll Gründungskongress*, 323.

67. Deutscher Gewerkschaftsbund, *Protokoll 2*, 244–245; Deutscher Gewerkschaftsbund, *Protokoll 4*, 371.

68. For the keynote address, see Deutscher Gewerkschaftsbund, *Automation*. For working group results, see Deutscher Gewerkschaftsbund, *Arbeitnehmer und Automation*.

69. Deutscher Gewerkschaftsbund, *Automation*, 16, 19. Other speakers offered similar assurances. See Deutscher Gewerkschaftsbund, *Arbeitnehmer und Automation*, 5, 21, 32.

70. Deutscher Gewerkschaftsbund, *Automation*, 29.

71. Herbert Vogel, *Bericht über eine Studienreise "Automation in USA,"* October 1957, Archiv der Sozialen Demokratie, FES. Hans Matthöfer, then an IG Metall officer, served as team leader of the group; he also received one of Silvey's ballpoint pen sets. Ted F. Silvey, handwritten list of recipients of ballpoint pen sets, n.d., folder 11, box 48, Accession 625, Reuther Library.

72. Herbert Vogel, *Bericht über eine Studienreise "Automation in USA,"* October 1957, 22, Archiv der Sozialen Demokratie, FES.

73. Ibid., 6.

74. Ibid., 13–16.

75. Ibid., 16.

76. Deutscher Gewerkschaftsbund, *Arbeitnehmer und Automation*, 30–31. See also Deutscher Gewerkschaftsbund, "Ergebnisprotokoll über die Sitzung des Unterausschusses 'Betriebs- und volkswirtschaftliche Auswirkungen der Automation' am 7. November 1957 in Düsseldorf," November 13, 1957, file 5/DGAN000670, Archiv der Sozialen Demokratie, FES.

77. Deutscher Gewerkschaftsbund, *Arbeitnehmer und Automation*, 32.

78. Deutscher Gewerkschaftsbund, *Automation*, 35.

79. Markovits, *The Politics of the West German Trade Unions*.

80. "Neue Aufgaben des Betriebsrates bei technischen Umstellungen der Produktion," reprint from *Der Gewerkschafter*, nos. 8–9, 1961, file 5/HBVH910715B, Archiv der Sozialen Demokratie, FES; "Automation und technischer Fortschritt in Deutschland und in den USA: Internationale Arbeitstagung der Industriegewerkschaft Metall für die Bundesrepublik Deutschland, 3. bis 5. Juli 1963, Frankfurt am Main," file 5/HBVH910715B, Archiv der Sozialen Demokratie, FES.

81. The West German trade unionists also debated pursuing macroeconomic—not only company-wide—measures in response to automation. Lohmann, *Die Angestellten im Zeitalter der Automation*, 14.

82. "DGB-Tagung 'Automation—Gewinn oder Gefahr?'" DGB Nachrichtendienst, January 27, 1958, 2, file 5/DGAN000700, Archiv der Sozialen Demokratie, FES. According to the conference proceedings, however, Rosenberg used the term "defenseless slaves"—not "barbarism"—in his address, a more direct critique of the fraught history of US race relations that the Weimar union group had already pointed out, and that communists continued to level against the United States during the Cold War. Deutscher Gewerkschaftsbund, *Automation*, 46.

83. For the transatlantic histories of these two machines, see Schlombs, "Univac"; Schlombs, "A Cost-Saving Machine."

84. Grier, *When Computers Were Human*.

85. Von Neumann, "First Draft of a Report on EDVAC"; Berkeley, *Giant Brains*.

86. "Automation bedeutet Fortschritt," *IBM World Trade News* (December 1955): 5, HzGD.

87. Woitschach, "Können Maschinen denken?," 228.

88. Unfortunately, permission to reproduce the described image was not granted. Woitschach, "Kommt das automatische Büro?," 330. Lively discussions of automation occurred in IBM's employee magazines in the mid- to late 1950s.

89. Woitschach, "Können Maschinen denken?," 239–240; Woitschach, "Kommt das automatische Büro?," 322–326, 329.

90. Eggenkämper, Modert, and Pretzlik, *Bits and Bytes*. For Allianz's rationalization during the 1920s, see Arps, *Wechselvolle Zeiten*, 227.

91. Heinz-Leo Müller-Lutz, "Gutachten zur Frage der Automation," ca. 1956, 9, 14, box 1, Entwicklung der EDV, Allianz.

92. Ibid., 33–34.

93. Schlombs, "A Cost-Saving Machine," 48. See also Schlombs, "Built on the Hands of Women."

94. Lukoff, *From Dits to Bits*, 139.

95. Schlombs, "A Cost-Saving Machine."

96. Bundesagentur für Arbeit, *Arbeitsmarkt in Zahlen*.

97. Kennedy, press conference. For the unemployment statistics, see US Bureau of Labor Statistics, "Employment Status of the Civilian Noninstitutional Population."

98. Council of Europe, Consultative Assembly, Committee on Science and Technology, "Report on the Computer Industry in Europe: Hardware Manufacturing (Rapporteur: Mr. Lloyd)," document 2893, 1971, 9, 21, folder 5, box 11, CBI 62, CBI.

Conclusion

1. Luce, "The American Century"; Brinkley, *The Publisher*.

2. Pinch and Bijker, "Social Construction of Facts and Artifacts," 17–50.

3. Feenberg, *Alternative Modernity*.

4. Maier, "Politics of Productivity."

5. Hall and Soskice, *Varieties of Capitalism*.

6. In his call to integrate technology into the history of US foreign relations, for example, LaFeber suggested that the information technologies of the third Industrial Revolution began to play a role in foreign relations from the 1970s onward. LaFeber, "Technology and U.S. Foreign Relations."

7. Sociologist Manuel Castells, for example, dated the beginnings of the "network society" to the 1970s. At this time, Castells argued, the global dissemination of networked information technologies caused the demise of nation-states, and created global corporate organizations and social movements. Castells, *The Rise of the Network Society*.

8. Lenger, *Werner Sombart*, 145–148; Appel, *Werner Sombart*.

9. Sombart, *Why Is There No Socialism in the United States?*

10. Kocka, *Capitalism Is Not Democratic*, 15.

11. Sewell, "The Capitalist Epoch."

12. Ekbladh, *The Great American Mission*. For a rare glimpse at electronic computers in global development, see Tinn, "Modeling Computers and Computer Models."

13. World Inequality Lab, *World Inequality Report 2018*, 4–5, 8.

14. Sewell, "The Capitalist Epoch," 4–5.

15. Ensmenger, "The Environmental History of Computing."

16. Berg and Seeber, *Slow Professor*.

17. Powers, *Hamlet's Blackberry*; Pang, *Distraction Addiction*.

Bibliography

ARCHIVAL COLLECTIONS

Allianz Firmenarchiv, Munich, Germany

Entwicklung der EDV

Bundesarchiv Koblenz, Germany

Bestand 115, Ministerielle Bundesvermögensverwaltung—Verwaltung des ERP-Sondervermögens

Bestand 146, Bundesministerium für den Marshallplan

Charles Babbage Institute (CBI), Minneapolis, Minnesota

CBI 3, Carl Hammer Papers

CBI 55, Market and Product Reports Collection

CBI 62, International Computing Collection, 1939–1990

Deutsches Museum, Munich, Germany

Firmenschriften IBM FS 003269

Kleine Sammlung IBM 004

Kuratorensammlung (curator's collection)

Friedrich-Ebert-Stiftung (FES), Bonn, Germany

Archiv der Sozialen Demokratie

IBM Corporate Archive, Somers, New York

Record Group 6, Legal Records

Record Group 11, Employees: Thomas J. Watson Jr. Papers

Thomas J. Watson Sr. Papers

Hagley Museum and Library, Wilmington, Delaware

Imprints Collection

NAM accession 1411

Harry S. Truman Library and Museum, Independence, Missouri

James M. Silberman Personal Papers

Everett H. Bellows Personal Papers

Haus zur Geschichte der IBM Datenverarbeitung (HzDG), Sindelfingen, Germany

IBM Deutschland (employee magazine)

IBM World Trade Corporation News (international employee magazine, renamed *IBM World Trade News*)

National Archives and Records Administration (NARA) II, College Park, Maryland

Record Group (RG) 200, National Archives Gift Collection, Papers of Ewan Clague, Personal File 1919–1979

Record Group (RG) 257, Records of the Bureau of Labor Statistics
    Entry 31, Office Files of Ethelbart Stewart, 1904–1931

Record Group (RG) 469, Records of US Foreign Assistance Agencies, 1948–1961
    Entry 48, General Correspondence of Clinton S. Golden, Labor Adviser, 1948–1951
    Entry 49, Office of Labor Advisers, Subject Files of Nelson Cruikshank and Robert Oliver, Labor Advisers, 1948–1952
    Entry 50, Subject Files of Ted F. Silvey, Executive Assistant, 1948–1950
    Entry 171, Assistant Administrator for Production, Productivity and Technical Assistance Division, Records Relating to US Advisory Group on European Productivity, 1952–1953
    Entry 171–G, Assistant Administrator for Production, Productivity and Technical Assistance Division, Subject Files of Joseph A. Smith, Assistant Director for US Operations, 1949–1953
    Entry 172, Productivity and Technical Assistance Division, Office of the Director, Subject Files, 1950–1951

Entry 174–A, Office of the Assistant Administrator of Production, Productivity and Technical Assistance Division, Project Review Branch, Subject Files, 1949–1953

Entry 178–A, Assistant Administration for Production, Productivity and Technical Assistance Division, Technical Assistance Overseas Branch, Country and Regional File, 1949–1954

Entry 178–C, Assistant Administration for Production, Productivity and Technical Assistance Division, Military Projects Branch, Country Files, 1949–1954

Entry 786, Office of Deputy Director of Technical Services, Office of Labor Affairs, Labor Programs Division, Records Relating to Foreign Technical Assistance Projects, 1948–1954

Entry 1058, SRE Special Representative in Europe, Deputy for Economic Affairs, Productivity and Technical Assistance Division, Subject Files, 1950–1956

Entry 1202, Mission to Germany, Productivity and Technical Assistance Division, Subject Files, 1952–1954

Entry 1203, Mission to Germany, Audio-Visual Aids Section, Records Relating to the Film Program, 1950–1956

Walter P. Reuther Library, Detroit, Michigan

Accession 625, Ted F. Silvey Papers

Accession 1168, Norbert Wiener Papers

Core Primary Sources

See https://productivitymachines.com

PUBLISHED SOURCES

Abelshauser, Werner. *Deutsche Wirtschaftsgeschichte: Von 1945 bis zur Gegenwart. Zweite Auflage.* Munich: C. H. Beck, 2011. First published 2004.

Abelshauser, Werner. "The First Post-Liberal Nation: Stages in the Development of Modern Corporatism in Germany." *European History Quarterly* 14 (1984): 285–318.

Abelshauser, Werner. *Kulturkampf: Der deutsche Weg in die Neue Wirtschaft und die amerikanische Herausforderung.* Berlin: Kulturverlag Kadmos, 2003.

Abelshauser, Werner. *Wirtschaft in Westdeutschland 1945–1948: Rekonstruktion und Wachstumsbedingungen in der amerikanischen und britischen Zone.* Stuttgart: Deutsche Verlags-Anstalt, 1975.

Adas, Michael. *Dominance by Design: Technological Imperatives and America's Civilizing Mission.* Cambridge, MA: Harvard University Press, 2006.

Abraham, Itty. "The Ambivalence of Nuclear Histories." *OSIRIS* 21 (2006): 49–65.

Adas, Michael. *Machines as the Measure of Man: Science, Technology, and Ideologies of Western Dominance*. Ithaca, NY: Cornell University Press, 1989.

Aereboe, F. *Wirtschaft und Kultur in den Vereinigten Staaten von Nordamerika*. Berlin: Verlagsbuchhandlung Paul Parey, 1930.

Alexander, Jennifer K. *The Mantra of Efficiency: From Waterwheel to Social Control*. Baltimore: Johns Hopkins University Press, 2008.

Allgemeiner Deutscher Gewerkschaftsbund. *Amerikareise deutscher Gewerkschaftsführer*. Berlin: Verlagsgesellschaft des Allgemeinen Deutschen Gewerkschaftsbundes, 1926.

Angster, Julia. *Konsenskapitalismus und Sozialdemokratie: Die Westernisierung von SPD und DGB*. Munich: Oldenbourg Verlag, 2003.

Anonymous. "Productivity of Labor in Eleven Industries." *Monthly Labor Review* 24 (January 1927): 35–49.

Anonymous. "Productivity of Labor in Slaughtering and Meat Packing and in Petroleum Refining." *Monthly Labor Review* 23 (November 1926): 30–40.

Anonymous. "Productivity of Labor in the Cement, Leather, Flour, and Sugar-Refining Industries, 1914 to 1925." *Monthly Labor Review* 23 (October 1926): 10–34.

Anonymous. "Productivity of Labor in the Rubber Tire and the Iron and Steel (Revised) Industries." *Monthly Labor Review* 23 (December 1926): 28–36.

Appel, Michael. *Werner Sombart: Historiker und Theoretiker des modernen Kapitalismus*. Marburg: Metropolis-Verlag, 1992.

Archer, Robin. *Why Is There No Labor Party in the United States?* Princeton, NJ: Princeton University Press, 2007.

Arps, Ludwig. *Wechselvolle Zeiten: 75 Jahre Allianz Versicherung 1890–1965*. Munich: Allianz Versicherungs-AG, 1965.

Aspray, William. *Technological Competitiveness: Contemporary and Historical Perspectives on the Electrical, Electronics, and Computer Industries*. New York: Institute of Electrical and Electronics Engineers, 1993.

Aumann, Philipp. *Mode und Methode: Die Kybernetik in der Bundesrepublik Deutschland*. Göttingen: Wallstein, 2009.

"Automation: Die Revolution der Roboter." *Der Spiegel*, July 27, 1955, 20–30.

"'Automatism' Eases Manpower Shortage." *New York Times*, September 16, 1951, 146.

Backman, Jules. *Multi-Employer Bargaining*. New York: New York University, Institute of Labor Relations and Social Security, 1951.

Baldwin, George B., and George P. Schultz. "Automation: A New Dimension to Old Problems." *Monthly Labor Review* 78, no. 2 (1955): 165–169.

Balke, Siegfried. "Automation as an Aid to Scientific Plant Management." In *Inauguration of the Remington Rand Computing Center Europe: Speeches*, Frankfurt/Main, 1956.

Baranowski, Shelley. *Strength through Joy: Consumerism and Mass Tourism in the Third Reich*. Cambridge: Cambridge University Press, 2004.

Barclay, Hartley W. "Metals Industry Speeds Conversion." *New York Times*, July 30, 1950, 111.

Bates, Beth T. *The Making of Black Detroit in the Age of Henry Ford*. Chapel Hill: University of North Carolina Press, 2012.

"Beängstigend menschlich." *Der Spiegel*, July 13, 1950, 37–39.

Beckert, Sven. "History of American Capitalism." In *American History Now*, edited by Eric Foner and Lisa McGirr, 314–335. Philadelphia: Temple University Press, 2011.

Beilharz, Peter. *Socialism and Modernity*. Minneapolis: University of Minnesota Press, 2009.

Bellows, Everett H. *Oral History Interview*. Foreign Affairs Oral History Project, 1989. http://www.adst.org/OH%20TOCs/Bellows,%20Everett.toc.pdf.

Berg, Maggie, and Barbara K. Seeber. *Slow Professor: Challenging the Culture of Speed in the Academy*. Toronto: University of Toronto Press, 2016.

"Berg-Friede." *ZEIT*, October 11, 1956, 8.

Berghahn, Volker R. "The Debate on 'Americanization' among Economic and Cultural Historians." *Cold War History* 10, no. 1 (2010): 107–130.

Berkeley, Edmund C. *Giant Brains: Or, Machines That Think*. New York: Wiley, 1949.

Bix, Amy S. *Inventing Ourselves Out of Jobs? America's Debate over Technological Unemployment, 1929–1981*. Baltimore: Johns Hopkins University Press, 2000.

Black, Edwin. *IBM and the Holocaust: The Strategic Alliance between Nazi Germany and America's Most Powerful Corporation*. New York: Crown Publishers, 2001.

Bloemen, Erik. "The Movement for Scientific Management in Europe between the Wars." In *Scientific Management: Frederick Taylor's Gift to the World?*, edited by J.-C. Spender and Hugo J. Kijne, 111–131. Boston: Kluwer, 1991.

Boel, Bent. *The European Productivity Agency and Transatlantic Relations, 1953–1961*. Copenhagen: Museum Tusculanum/University of Copenhagen, 2003.

Bonn, Moritz. *Geld und Geist: Vom Wesen und Werden der Amerikanischen Welt*. Berlin: S. Fischer, 1927.

Brady, Robert A. "The Meaning of Rationalization: An Analysis of the Literature." *Quarterly Journal of Economics* 46, no. 3 (May 1932): 526–540.

Brady, Robert A. *The Rationalization Movement in German Industry: A Study in the Evolution of Economic Planning*. Berkeley: University of California Press, 1933.

Braverman, Harry. *Labor and Monopoly Capital: The Degradation of Work in the Twentieth Century*. New York: Monthly Review Press, 1974.

Brinkley, Alan. *The Publisher: Henry Luce and His American Century*. New York: Knopf, 2010.

Brooks, Thomas R. *Clint: A Biography of a Labor Intellectual*. New York: Atheneum, 1978.

Brown, Harvey W. "'Mitbestimmung' (Codetermination): Labor's New Responsibility in Germany." *Machinist Monthly*, (August 1951): 236–237.

Bundesagentur für Arbeit. *Arbeitsmarkt in Zahlen: Arbeitslosigkeit im Zeitverlauf*. Nuremberg: Bundesagentur für Arbeit, 2017. http://www.arbeitsagentur.de.

Bundesverband der Deutschen Industrie. *Erster Internationaler Industriellen-Kongress, New York, 2.–4. Dezember 1951: Dokumente und Berichte*. Köln: Bundesverband der Deutschen Industrie, 1952.

Bunting, Earl. *Labor Monopoly and the Public Interest: Taken from Statement before the Subcommittee on Study of Monopoly Power of the Committee of the Judiciary, House of Representatives, November 15, 1949*. New York: National Association of Manufacturers, 1949.

"Büro-Automatisierung im Anfangsstadium: Die Lohnabrechnung ein geeignetes Versuchsfeld." *Frankfurter Allgemeine Zeitung*, December 17, 1956, 11.

Calder, Lendol. *Financing the American Dream: A Cultural History of Consumer Credit*. Princeton, NJ: Princeton University Press, 1999.

Campbell-Kelly, Martin. *ICL: A Business and Technical History*. Oxford: Clarendon Press, 1989.

Campbell-Kelly, Martin, William Aspray, Nathan Ensmenger, and Jeffrey R. Yost. *Computer: A History of the Information Machine*. 3rd ed. Boulder, CO: Westview Press, 2013.

Campbell-Kelly, Martin, and Daniel D. Garcia-Swartz. *From Mainframes to Smartphones: A History of the International Computer Industry*. Cambridge, MA: Harvard University Press, 2015.

Carew, Anthony. *Labour under the Marshall Plan: The Politics of Productivity and the Marketing of Management Science.* Detroit, MI: Wayne State University Press, 1987.

Castells, Manuel. *The Rise of the Network Society.* 2nd ed. Oxford: Blackwell, 2000.

Ceruzzi, Paul E. *A History of Modern Computing.* Cambridge, MA: MIT Press, 2000.

Chandler, Alfred D. *The Visible Hand: The Managerial Revolution in American Business.* Cambridge, MA: Belknap Press, 1977.

Chandler, Alfred D., and Takashi Hikino. *Scale and Scope: The Dynamics of Industrial Capitalism.* Cambridge, MA: Belknap Press, 1990.

Clague, Ewan. *The Bureau of Labor Statistics.* New York: Praeger, 1968.

Clague, Ewan. "Index of Productivity of Labor in the Steel, Automobile, Shoe and Paper Industries." *Monthly Labor Review* 23, no. 1 (July 1926): 1–19.

Clague, Ewan. "Productivity and Wages in the United States." *American Federationist* (March 1927): 285–296.

Clague, Ewan. *Reminiscences of Ewan Clague: Oral History, 1958.* New York: Columbia Center for Oral History, 1958.

Clague, Ewan. *Reminiscences of Ewan Clague: Oral History, 1966.* New York: Columbia Center for Oral History, 1966.

Coates, David. *Varieties of Capitalism, Varieties of Approaches.* Basingstoke, UK: Palgrave Macmillan, 2007.

Cohen, Lizabeth. *Making a New Deal: Industrial Workers in Chicago, 1919–1939.* New York: Cambridge University Press, 1990.

Congress of Industrial Organizations, Committee on Economic Policy. *Challenge of Automation.* Washington, DC: Public Affairs Press, 1955.

Coopey, Richard, ed. *Information Technology Policy: An International History.* Oxford: Oxford University Press, 2004.

Cortada, James W. *All the Facts: A History of Information in the United States since 1870.* New York: Oxford University Press, 2016.

Cortada, James W. *Before the Computer: IBM, NCR, Burroughs, and Remington Rand and the Industry They Created, 1865–1956.* Princeton, NJ: Princeton University Press, 1993.

Cortada, James W. *The Digital Flood: The Diffusion of Information Technology across the U.S., Europe, and Asia.* New York: Oxford University Press, 2012.

Cowan, Ruth Schwartz. *More Work for Mother: The Ironies of Household Technology from the Open Hearth to the Microwave.* New York, Basic Books: 1983.

Crane, Burton. "New Margin Rule Urged by Banker." *New York Times*, March 18, 1955, 1, 20.

Curcio, Vincent. *Henry Ford*. Oxford: Oxford University Press, 2013.

Dahrendorf, Ralf. *Class and Class Conflict in Industrial Society*. Stanford, CA: Stanford University Press, 1959. First published 1957.

de Grazia, Victoria. *Irresistible Empire: America's Advance through Twentieth-Century Europe*. Cambridge, MA: Belknap Press, 2005.

DeLamarter, Richard T. *Big Blue: IBM's Use and Abuse of Power*. New York: Dodd, Mead, 1986.

Delton, Jennifer. "The Triumph of Social Responsibility in the National Association of Manufacturers in the 1950s." In *Capital Gains: Business and Politics in Twentieth-Century America*, edited by Richard R. John and Kim Phillips-Fein, 181–196. Philadelphia: University of Pennsylvania Press, 2017.

Desrosières, Alain. *The Politics of Large Numbers: A History of Statistical Reasoning*. Cambridge, MA: Harvard University Press, 1998.

Deutscher Gewerkschaftsbund. *Arbeitnehmer und Automation: Ergebnisse einer Arbeitstagung des Deutschen Gewerkschaftsbundes am 23. und 24. Januar 1958 in Essen*. Düsseldorf: Deutscher Gewerkschaftsbund, 1958.

Deutscher Gewerkschaftsbund. *Automation—Gewinn oder Gefahr? Arbeitstagung des Deutschen Gewerkschaftsbundes am 23. und 24. Januar 1958 in Essen*. Düsseldorf: Deutscher Gewerkschaftsbund, 1958.

Deutscher Gewerkschaftsbund. *Protokoll 4. Ordentlicher Bundeskongress Hamburg 1. bis 6. Oktober 1956*. Cologne: Druckhaus Deutz, 1956.

Deutscher Gewerkschaftsbund. *Protokoll Gründungskongress des Deutschen Gewerkschaftsbundes, München, 12.–14. Oktober 1949*. Cologne: Bund-Verlag, 1949[?].

Deutscher Gewerkschaftsbund. *Protokoll 2. Ordentlicher Bundeskongress Berlin 13. bis 17. Oktober 1952*. Düsseldorf: Deutscher Gewerkschaftsbund, 1952.

Devinat, Paul. *Scientific Management in Europe*. Geneva: International Labor Office, 1927.

"DGB-Tagung 'Automation—Gewinn oder Gefahr?'" *DGB Nachrichtendienst*, January 27, 1958, 2.

Diebold, John. *Automation: The Advent of the Automatic Factory*. New York: Van Nostrand Company, 1952.

Dienel, Hans-Luidger. "'Hier sauber und gründlich, dort husch-husch, fertig': Deutsche Vorbehalte gegen amerikanische Produktionsmethoden, 1870–1930." *Blätter für Technikgeschichte* 55 (1993): 11–29.

Djelic, Marie-Laure. *Exporting the American Model: The Postwar Transformation of European Business.* New York: Oxford University Press, 1998.

Doering-Manteuffel, Anselm. *Wie westlich sind die Deutschen? Amerikanisierung und Westernisierung im 20. Jahrhundert.* Göttingen: Vandenhoeck and Ruprecht, 1999.

Eggenkämper, Barbara, Gerd Modert, and Stefan Pretzlik. *Bits and Bytes for Business: 50 Jahre EDV bei der Allianz.* Munich: Allianz Deutschland AG, 2006.

Eifert, Christiane. "Antisemit und Autokönig: Henry Fords Autobiographie und ihre deutsche Rezeption in den 1920er Jahren." *Zeithistorische Forschungen* 6, no. 2 (2009): 209–229.

Ekbladh, David. *The Great American Mission: Modernization and the Construction of an American World Order.* Princeton, NJ: Princeton University Press, 2010.

"Elektronengehirn für intellektuelle Arbeit." *Der Spiegel,* June 18, 1952, 32–33.

"Elektronengehirne: Die Magie der Roboter." *Der Spiegel,* October 3, 1956, 42–53.

"Emerson Extends Automation's Use: President at Annual Meeting Describes Savings by New Radio Assembly Methods." *New York Times,* February 3, 1955, 35.

Ensmenger, Nathan. "The Environmental History of Computing." *Technology and Culture* 59, no. 4 supplement (2018): S7–S33.

Epstein, Katherine. *Torpedo: Inventing the Military-Industrial Complex in the United States and Great Britain.* Cambridge, MA: Harvard University Press, 2014.

"Ewan Clague, Ex-Labor Data Official, Dies." *Los Angeles Times,* April 16, 1987. http://articles.latimes.com/1987-04-16/news/mn-502_1_ewan-clague.

"Ewan Clague, Former Chief of Labor Statistics, Dies," *Washington Post,* April 15, 1987, B4.

"Fairless Upholds 'Automation' Aims: US Steel Chairman Asserts Fears of Joblessness Are 'Vicious Propaganda.'" *New York Times,* February 12, 1955, 27.

Faldix, G. *Henry Ford als Wirtschaftspolitiker.* Munich: Pfeiffer, 1925.

Feenberg, Andrew. *Alternative Modernity: The Technological Turn in Philosophy and Social Theory.* Berkeley: University of California Press, 1995.

Feenberg, Andrew. *Critical Theory of Technology.* New York: Oxford University Press, 1991.

Feiler, Arthur. *America Seen through German Eyes*. New York: Arno Press, 1974. First published 1928.

Feldenkirchen, Wilfried. *Siemens, 1918–1945*. Columbus: Ohio State University Press, 1999.

"Financial and Business Sidelights." *New York Times*, April 7, 1954, 49.

"Financial and Business Sidelights of the Day: An Achilles Heel." *New York Times*, March 4, 1954, 37.

Fisher, Paul. "Labor Codetermination in Germany." *Social Research* 18, no. 4 (1951): 449–485.

Flamm, Kenneth. *Targeting the Computer: Government Support and International Competition*. Washington, DC: Brookings Institution, 1987.

Foner, Eric. "Why Is There No Socialism in the United States?" *History Workshop* 17 (1984): 57–80.

Fones-Wolf, Elizabeth A. *Selling Free Enterprise: The Business Assault on Labor and Liberalism, 1945–60*. Urbana: University of Illinois Press, 1994.

Ford, Henry. *My Life and Work: An Autobiography of Henry Ford*. N.p.: Snowball Publishing, 2012.

Fowle, Farnsworth. "Labor in Germany Aims at New Role." *New York Times*, February 5, 1951, 5.

Foy, Nancy. *The Sun Never Sets on IBM*. New York: William Morrow, 1975.

Freeman, William M. "Automation Aims at New Freedom: Devices That Run Factories Promise to Release Men for Richer Living." *New York Times*, January 3, 1955, 5.

Freeman, William M. "New Way Devised to Make Abrasives: Carborundum's Process Use of Beta Rays with Fully Automatic Controls." *New York Times*, June 8, 1952, 171.

Frevert, Ute. *Women in German History: From Bourgeois Emancipation to Sexual Liberation*. Oxford: Berg, 1990.

Galantiere, Lewis. "International Decorum." Letter to the editor. *New York Times*, February 11, 1951, 140.

Ganzhorn, Karl E. *The IBM Laboratories Boeblingen: Foundation and Build-up. A Personal Review*. Sindelfingen: Röhm, 2000.

Ganzhorn, Karl E. Oral history interview with the author. April 27, 2007.

Garfinkel, Herbert. "Technological Unemployment." Letter to the editor. *New York Times*, December 11, 1954, 12.

Gassert, Philipp. *Amerika im Dritten Reich: Ideologie, Propaganda und Volksmeinung 1933–1945.* Stuttgart: Franz Steiner Verlag, 1997.

Gelderman, Carol. *Henry Ford: The Wayward Capitalist.* New York: Dial Press, 1981.

Geppert, Dominik, ed. *The Postwar Challenge: Cultural, Social, and Political Change in Western Europe, 1945–58.* Oxford: Oxford University Press, 2003.

"German Labor's Say in Industry Is Hit." *New York Times,* February 5, 1951, 38, 40.

"Germans' System of Labor Studied: Representation on Company Boards Found Less Radical Than Had Been Expected." *New York Times,* October 15, 1956, 36.

Gerovitch, Slava. *From Newspeak to Cyberspeak: A History of Soviet Cybernetics.* Cambridge, MA: MIT Press, 2002.

Gimbel, John. *The Origins of the Marshall Plan.* Stanford, CA: Stanford University Press, 1976.

Glickman, Lawrence B. *A Living Wage: American Workers and the Making of Consumer Society.* Ithaca, NY: Cornell University Press, 1997.

Golden, Clinton S., ed. *Causes of Industrial Peace under Collective Bargaining.* New York: Harper, 1955.

Golden, Clinton S., and Harold J. Ruttenberg. *The Dynamics of Industrial Democracy.* New York: Harper, 1942.

Gordon, Colin. *New Deals: Business, Labor, and Politics in America, 1920–1935.* New York: Cambridge University Press, 1994.

Grebing, Helga. *Geschichte der deutschen Arbeiterbewegung: Von der Revolution 1848 bis ins 21. Jahrhundert.* Berlin: Vorwärts, 2007.

Gresmann, Hans. "Revolution oder nicht Revolution? Zur Automatisierungs-Tagung der List-Gesellschaft in Frankfurt." *ZEIT,* October 17, 1957, 4

Grier, David A. *When Computers Were Human.* Princeton, NJ: Princeton University Press, 2005.

Guillén, Mauro F. *Models of Management: Work, Authority, and Organization in a Comparative Perspective.* Chicago: University of Chicago Press, 1994.

Hachtmann, Rüdiger. "'Die Begründer der amerikanischen Technik sind fast lauter schwäbisch-allemannische Menschen': Nazi-Deutschland, der Blick auf die USA und die 'Amerikanisierung' der industriellen Produktionsstrukturen im 'Dritten Reich.'" In *Amerikanisierung: Traum und Albtraum im Deutschland des 20. Jahrhunderts,* edited by Alf Lüdtke, Inge Marßolek, and Adelheid von Saldern, 37–66. Stuttgart: Franz Steiner Verlag, 1996.

Hall, Peter A., and David Soskice. *Varieties of Capitalism: The Institutional Foundations of Comparative Advantage.* Oxford: Oxford University Press, 2001.

Hancké, Bob, Martin Rhodes, and Mark Thatcher, *Beyond Varieties of Capitalism: Conflict, Contradictions, and Complementarities in the European Economy.* Oxford: Oxford University Press, 2010.

Hannaway, Owen. "The German Model of Chemical Education in America: Ira Remsen at Johns Hopkins." *Ambix* 23 (1976): 145–164.

Harris, Howell J. *The Right to Manage: Industrial Relations Policies of American Business in the 1940s.* Madison: University of Wisconsin Press, 1982.

Hau, Michael. *Performance Anxiety: Sport and Work in Germany from the Empire to Nazism.* Toronto: University of Toronto Press, 2017.

Hausser, Christian. *Amerikanisierung der Arbeit? Deutsche Wirtschaftsführer und Gewerkschaften im Streit um Ford und Taylor (1919–1932).* Stuttgart: Ibidem, 2008.

Haynes, Eldridge, and Gordon Michler. *Co-Determination: A Report Prepared for the National Association of Manufacturers.* New York: Industrial Relations Division, National Association of Manufacturers, 1956.

"Headed for a Ruhr Showdown." *Business Week*, January 20, 1951.

Headrick, Daniel R. *The Tools of Empire: Technology and European Imperialism in the Nineteenth Century.* New York: Oxford University Press, 1981.

Hecht, Gabrielle, ed. *Entangled Geographies: Empire and Technopolitics in the Global Cold War.* Cambridge, MA: MIT Press, 2011.

Hecht, Gabrielle. *The Radiance of France: Nuclear Power and National Identity after World War II.* Cambridge, MA: MIT Press, 1998.

Heide, Lars. *Punched-Card Systems and the Early Information Explosion, 1880–1945.* Baltimore: Johns Hopkins University Press, 2009.

Heineman, Elizabeth D. *What Difference Does a Husband Make? Women and Marital Status in Nazi and Postwar Germany.* Berkeley: University of California Press, 1999.

Hensel, Rudolf. *Aus Tagebüchern einer Reise.* Hellerau bei Dresden: Jakob Hegner, 1929.

Hentschel, Volker. *Ludwig Erhard: Ein Politikerleben.* Munich: Olzog, 1996.

Herrigel, Gary. "American Occupation, Market Order, and Democracy: Reconfiguring the Steel Industry in Japan and Germany after the Second World War." In *Americanization and Its Limits: Reworking US Technology and Management in Post-War Europe and Japan,* edited by Jonathan Zeitlin and Gary Herrigel, 340–399. New York: Oxford University Press, 2000.

Herrigel, Gary. *Industrial Constructions: The Sources of German Industrial Power.* Cambridge: Cambridge University Press, 1996.

Heusgens, Christoph. *Ludwig Erhards Lehre von der Sozialen Marktwirtschaft: Ursprünge, Kerngehalt und Wandlungen.* Bern: Verlag Paul Haupt, 1981.

Hicks, Marie. *Programmed Inequality: How Britain Discarded Women Technologists and Lost Its Edge in Computing.* Cambridge, MA: MIT Press, 2017.

Hirsch, Julius. Das amerikanische Wirtschaftswunder. Berlin: Fischer, 1928.

Hoffman, Paul G. *Peace Can Be Won.* Garden City, NY: Doubleday, 1951.

Hoffman, Paul G. "Productivity—Buttress to Freedom." In *Proceedings of the First International Conference of Manufacturers, December 3–5, 1951,* edited by National Association of Manufacturers, 39–48. New York: National Association of Manufacturers, 1952.

Hogan, Michael J. *The Marshall Plan: America, Britain, and the Reconstruction of Western Europe, 1947–1952.* Cambridge: Cambridge University Press, 1987.

Holitscher, Arthur. *Wiedersehen mit Amerika.* Berlin: S. Fischer, 1930.

Homburg, Heidrun. *Rationalisierung und Industriearbeit: Arbeitsmarkt—Management—Arbeiterschaft im Siemens-Konzern Berlin, 1900–1939.* Berlin: Haude and Spener, 1991.

Hughes, Edward. "German Solons Begin Debate on Union Bid for 'Codetermination.'" *Wall Street Journal,* April 4, 1951.

Hounshell, David A. *From the American System to Mass Production, 1800–1932: The Development of Manufacturing Technology in the United States.* Baltimore: Johns Hopkins University Press, 1984.

Hutchison, Lee. "Tripping through IBM's Astonishingly Insane 1937 Corporate Songbook." https://arstechnica.com/business/2014/08/tripping-through-ibms-astonishingly-insane-1937-corporate-songbook/.

International Association of Machinists. *Meeting the Problems of Automation through Collective Bargaining: A Compilation of Contract Clauses and Case Descriptions.* IAM Research Department, 1960.

IBM Archives. "IBM Rally Song, Ever Onward Transcript." https://www-03.ibm.com/ibm/history/multimedia/everonward_trans.html.

IBM Archives. "IBM's 100 Icons of Progress: The First Salaried Workforce." https://www-03.ibm.com/ibm/history/ibm100/us/en/icons/workforce/.

IBM Archives. "Remote Control Keyboard." https://www-03.ibm.com/ibm/history/exhibits/specialprod1/specialprod1_6.html.

"Idle Machine Time Bars Robot Plants: Automatic Factory Operations Only Half of Day Won't Pay, Tool Engineers Are Told." *New York Times*, March 20, 1953, 35.

Igo, Sarah E. *The Averaged American: Surveys, Citizens, and the Making of a Mass Public.* Cambridge, MA: Harvard University Press, 2007.

Jacobs, Meg. *Pocketbook Politics: Economic Citizenship in Twentieth-Century America.* Princeton, NJ: Princeton University Press, 2005.

Jacobsson, Carl A. "Employee Relations Policies in Western Europe." In *Proceedings of the First International Conference of Manufacturers, December 3–5, 1951*, edited by National Association of Manufacturers, 125–131. New York: National Association of Manufacturers, 1952.

Jacoby, Sanford M. *Modern Manors: Welfare Capitalism since the New Deal.* Princeton, NJ: Princeton University Press, 1997.

Jarausch, Konrad, and Hannes Siegrist, eds. *Amerikanisierung und Sowjetisierung in Deutschland 1945–1970.* Frankfurt: Campus Verlag, 1997.

Jeremy, David J. *Transatlantic Industrial Revolution: The Diffusion of Textile Technologies between Britain and America, 1790–1830s.* Cambridge, MA: MIT Press, 1982.

Jones, Geoffrey. *Multinationals and Global Capitalism: From the Nineteenth to the Twenty-First Century.* Oxford: Oxford University Press, 2005.

Jones, Geoffrey. *Transnational Corporations: A Historical Perspective.* London: Routledge, 1993.

Jones, Thomas R. "Production Policies Which Contribute Directly to Increased Output and a Reduction in Time and Effort." In *Proceedings of the First International Conference of Manufacturers, December 3–5, 1951*, edited by National Association of Manufacturers, 53–62. New York: National Association of Manufacturers, 1952.

Joyce, William H. "Participation." In *Proceedings of the First International Conference of Manufacturers, December 3–5, 1951*, edited by National Association of Manufacturers, 306–314. New York: National Association of Manufacturers, 1952.

Kaempffert, Waldemar. "A New 'Tinkertoy' Speeds Production: Navy and Bureau of Standards Reveal System for Automatic Factory in Electronics." *New York Times*, September 20, 1953, 46.

Kaempffert, Waldemar. "Walter Reuther's Fears of 'Automation' Stir New Interest in Factory of the Future." *New York Times*, December 12, 1954, 184.

Kasson, John F. *Civilizing the Machine: Technology and Republican Values in America, 1776–1900.* New York: Grossman, 1978.

Kennedy, John F. Press conference. February 14, 1962. https://www.jfklibrary.org/
Research/Research-Aids/Ready-Reference/Press-Conferences/News-Conference-24
.aspx.

Kessler-Harris, Alice. *In Pursuit of Equity: Women, Men, and the Quest for Economic Citizenship in 20th Century America*. Oxford: Oxford University Press, 2001.

Kipping, Matthias. "'Operation Impact': Converting European Employers to the American Creed." In *The Americanisation of European Business: The Marshall Plan and the Transfer of US Management Models*, edited by Matthias Kipping and Ove Bjarnar, 55–73. London: Routledge, 1998.

Kipping, Matthias, and Ove Bjarnar, eds. *The Americanisation of European Business: The Marshall Plan and the Transfer of US Management Models*. London: Routledge, 1998.

Kleinschmidt, Christian. *Der produktive Blick: Wahrnehmung amerikanischer und japanischer Management- und Produktionsmethoden durch deutsche Unternehmer 1950–1985*. Berlin: Akademie-Verlag, 2002.

Kleinschmidt, Christian. *Rationalisierung als Unternehmensstrategie: Die Eisen- und Stahlindustrie des Ruhrgebiets zwischen Jahrhundertwende und Weltwirtschaftskrise*. Essen: Klartext, 1993.

Klessmann, Christoph. "Betriebsräte und Gewerkschaften in Deutschland 1945–1952." In *Politische Weichenstellungen im Nachkriegsdeutschland, 1945–1953. Geschichte und Gesellschaft. Zeitschrift für Historische Sozialwissenschaft. Sonderheft 5*, edited by Heinrich A. Winkler, 44–73. Göttingen: Vandenhoeck and Ruprecht, 1979.

Kocka, Jürgen. *Die Angestellten in der deutschen Geschichte 1850–1980*. Göttingen: Vandenhoeck and Ruprecht, 1981.

Kocka, Jürgen. *Capitalism Is Not Democratic and Democracy Is Not Capitalistic*. Florence: Firenze University Press, 2015.

Kocka, Jürgen. *Unternehmensverwaltung und Angestelltenschaft am Beispiel Siemens, 1847–1914: Zum Verhältnis von Kapitalismus und Bürokratie in der deutschen Industrialisierung*. Stuttgart: Ernst Klett Verlag, 1969.

König, Wolfgang. "Adolf Hitler vs. Henry Ford: The Volkswagen, the Role of America as a Model, and the Failure of a Nazi Consumer Society." *German Studies Review* 27, no. 2 (2004): 249–268.

König, Wolfgang. *Volkswagen, Volksempfänger, Volksgemeinschaft: "Volksprodukte" im Dritten Reich: Vom Scheitern einer nationalsozialistischen Konsumgesellschaft*. Paderborn: Schöningh, 2004.

Köttgen, Carl. *Das wirtschaftliche Amerika*. Berlin: VDI-Verlag, 1925.

Kohlrausch, Martin, and Helmuth Trischler. *Building Europe on Expertise: Innovators, Organizers, Networkers*. Basingstoke, UK: Palgrave Macmillan, 2014.

"Kreditanstalt für Wiederaufbau." *Der Volkswirt*, April 14, 1955, 34.

Krige, John. *American Hegemony and the Postwar Reconstruction of Science in Europe*. Cambridge, MA: MIT Press, 2006.

Krige, John, and Kai-Henrik Barth, eds. *Global Power Knowledge: Historical Perspectives on Science, Technology, and International Affairs*. OSIRIS 21 (2006).

Kuhn, Axel. *Die deutsche Arbeiterbewegung*. Stuttgart: Reclam, 2004.

Kuisel, Richard F. *Seducing the French: The Dilemma of Americanization*. Berkeley: University of California Press, 1993.

LaFeber, Walter. "Technology and U.S. Foreign Relations." *Diplomatic History* 24, no. 1 (2000): 1–19.

Laird, Pamela W. *Advertising Progress: American Business and the Rise of Consumer Marketing*. Baltimore: Johns Hopkins University Press, 1998.

Lenger, Friedrich. *Werner Sombart, 1863–1941: Eine Biographie*. Munich: C. H. Beck, 1994.

Lichtenstein, Nelson. *The Most Dangerous Man in Detroit: Walter Reuther and the Fate of American Labor*. New York: Basic Books, 1995.

Lichtenstein, Nelson, and Howell J. Harris, eds. *Industrial Democracy in America: The Ambiguous Promise*. New York: Cambridge University Press, 1993.

Lipset, Seymour Martin, and Gary Marks. *It Didn't Happen Here: Why Socialism Failed in the United States*. New York: W. W. Norton, 2000.

Logemann, Jan L. "More Atlantic Crossings? European Voices and the Postwar Atlantic Community." *Bulletin of the German Historical Institute* 54, supplement no. 10 (2014): 7–18.

Logemann, Jan L. *Trams or Tailfins? Public and Private Prosperity in Postwar West Germany and the United States*. Chicago: University of Chicago Press, 2012.

Logemann, Jan L., and Mary Nolan, eds. *More Atlantic Crossings? European Voices in the Postwar Atlantic Community*. Washington, DC: German Historical Institute, 2014.

Lohmann, Martin. *Die Angestellten im Zeitalter der Automation: Vortrag gehalten auf dem 2. Landesbezirks-Angestelltentag des DGB am 9. Juni 1956 in Villingen/Schwarzwald*. Stuttgart: Deutscher Gewerkschaftsbund, Landesbezirk Baden-Württemberg, 1956.

Lubin, Isador. "Social and Economic Adjustments in a Democratic World." *Journal of the American Statistical Association* 42, no. 237 (March 1947): 11–19.

Luce, Henry R. "The American Century." *Life*, February 17, 1941, 61–65.

Lüdtke, Alf, Inge Marßolek and Adelheit von Saldern, eds. *Amerikanisierung: Traum und Alptraum im Deutschland des 20. Jahrhunderts*. Stuttgart: Franz Steiner Verlag, 1996.

Lukoff, Herman. *From Dits to Bits: A Personal History of the Electronic Computer*. Portland, OR: Robotics Press, 1979.

Lunestad, Geir. *"Empire" by Integration: The United States and European Integration, 1945–1997*. New York: Oxford University Press, 1998.

Lunestad, Geir. "Empire by Invitation? The United States and Western Europe, 1945–1952." *Journal of Peace Research* 23, no. 3 (1986): 263–277.

"Machine, Job Rise, Linked: Study Finds That Labor Saving Devices Aid Employment." *New York Times*, September 10, 1950, 129.

Mai, Gunther. "Politische Krise und Rationalisierungsdiskurs in den zwanziger Jahren." *Technikgeschichte* 62, no. 4 (1995): 317–332.

Maier, Charles S. "Between Taylorism and Technocracy: European Ideologies and the Vision of Industrial Productivity in the 1920s." *Journal of Contemporary History* 5, no. 2 (1970): 27–61.

Maier, Charles S. "Politics of Productivity: Foundations of American International Economic Policy after World War II." *International Organisation* 31, no. 4 (1977): 607–633.

Maney, Kevin. *The Maverick and His Machine: Thomas Watson, Sr. and the Making of IBM*. Hoboken, NJ: Wiley, 2003.

Marchand, Roland. *Advertising the American Dream: Making Way for Modernity, 1920–1940*. Berkeley: University of California Press, 1985.

Markovits, Andrei S. *The Politics of the West German Trade Unions: Strategies of Class and Interest Representation in Growth and Crisis*. Cambridge: Cambridge University Press, 1986.

Marshall, George C. "The Marshall Plan Speech." Delivered at Harvard University on June 5, 1947. http://marshallfoundation.org/marshall/the-marshall-plan/marshall-plan-speech/.

Marx, Leo. *The Machine in the Garden: Technology and the Pastoral Ideal in America*. New York: Oxford University Press, 1964.

May, Elaine Tyler. *Homeward Bound: American Families in the Cold War Era*. New York: Basic Books, 2008. First published 1988.

McGlade, Jacqueline. "Americanization: Ideology or Process? The Case of the United States Technical Assistance and Productivity Programme." In *Americanization and Its Limits: Reworking US Technology and Management in Post-War Europe and Japan*, edited by Jonathan Zeitlin and Gary Herrigel, 53–75. New York: Oxford University Press, 2000.

McGlade, Jacqueline. "From Business Reform Programme to Production Drive: The Transformation of US Technical Assistance to Western Europe." In *The Americanisation of European Business: The Marshall Plan and the Transfer of US Management Models*, edited by Matthias Kipping and Ove Bjarnar, 18–34. London: Routledge, 1998.

McGlade, Jacqueline. "The Illusion of Consensus: American Business, Cold War Aid, and the Industrial Recovery of Western Europe, 1948–1958." PhD diss., George Washington University, 1995.

McKenzie, Brian A. *Remaking France: Americanization, Public Diplomacy, and the Marshall Plan*. New York: Berghahn Books, 2005.

Medina, Eden. "Big Blue in the Bottomless Pit: The Early Years of IBM Chile." *IEEE Annals of the History of Computing* 30, no. 4 (2008): 26–41.

Medina, Eden. *Cybernetic Revolutionaries: Technology and Politics in Allende's Chile*. Cambridge, MA: MIT Press, 2011.

"Mehr Geltung." *Spiegel*, October 1, 1958, 32–34.

Merkle, Judith A. *Management and Ideology: The Legacy of the International Scientific Management Movement*. Berkeley: University of California Press, 1980.

Meyer, Stephen, III. *The Five Dollar Day: Labor Management and Social Control in the Ford Motor Company, 1908–1921*. Albany: SUNY Press, 1981.

Mierzejewski, Alfred. *Ludwig Erhard: A Biography*. Chapel Hill: University of North Carolina Press, 2004.

Milward, Alan S. *The Reconstruction of Western Europe, 1945–51*. Berkeley: University of California Press, 1984.

Misa, Thomas J., and Johan Schot. "Inventing Europe: Technology and the Hidden Integration of Europe." *History and Technology* 21, no. 1 (2005): 1–19.

Mitchell, Maria D. *The Origins of Christian Democracy: Politics and Confession in Modern Germany*. Ann Arbor: University of Michigan Press, 2012.

Moeller, Robert G. *Protecting Motherhood: Women and the Family in the Politics of Post-war West Germany*. Berkeley: University of California Press, 1993.

Mommsen, Hans, and Manfred Grieger. *Das Volkswagenwerk und seine Arbeiter im Dritten Reich*. Düsseldorf: ECON, 1996.

Montgomery, David. *Workers' Control in America: Studies in the History of Work, Technology, and Labor Struggles.* Cambridge: Cambridge University Press, 1979.

Mosher, Ira. "Response." In *Proceedings of the First International Conference of Manufacturers, December 3–5, 1951,* edited by National Association of Manufacturers, 163–165. New York: National Association of Manufacturers, 1952.

Mowery, David C., ed. *The International Computer Software Industry: A Comparative Study of Industry Evolution and Structure.* New York: Oxford University Press, 1996.

Muir-Harmony, Teasel. "Project Apollo, Cold War Diplomacy and the American Framing of Global Interdependence." PhD diss., Massachusetts Institute of Technology, 2014.

Mullaney, Thomas. *The Chinese Typewriter: A History.* Cambridge, MA: MIT Press, 2017.

Müller, Gloria. *Mitbestimmung in der Nachkriegszeit: Britische Besatzungsmacht—Unternehmer—Gewerkschaften.* Düsseldorf: Schwann, 1987.

Müller, Gloria. *Strukturwandel und Arbeitnehmerrechte: Die wirtschaftliche Mitbestimmung in der Eisen- und Stahlindustrie 1945–1975.* Essen: Klartext-Verlag, 1991.

National Association of Manufacturers. *Labor Monopoly and Industry-Wide Bargaining: Text of Statement by the National Association of Manufacturers Filed Monday August 8, 1949 with the Senate Banking and Currency Committee.* New York: National Association of Manufacturers, 1949.

National Association of Manufacturers. *Proceedings of the First International Conference of Manufacturers.* New York: National Association of Manufacturers, 1952.

National Association of Manufacturers, Industrial Relations Committee. *NAM Position on Industrywide Bargaining Adopted by the NAM Board of Directors, May 26, 1949.* New York: National Association of Manufacturers, 1949.

National Association of Manufacturers, Research Department. *NAM Looks at Cartels: Positions Formulated by the Committee on International Economic Regulations and Approved by the Board of Directors Together with an Analysis of the Economic Aspects of Cartels, Prepared by the Research Department, November 1946.* New York: National Association of Manufacturers, 1946.

National Manpower Council. *A Policy for Scientific and Professional Manpower.* New York: Columbia University Press, 1953.

National Manpower Council. *A Report on the National Manpower Council.* New York: Graduate School of Business, Columbia University, 1954.

"New RCA Device Aids Automation: Printed Circuit Board Punch Said to Cut Time and Costs in Radio and TV Making." *New York Times,* February 4, 1955, 28.

Nicholls, Anthony J. *Freedom with Responsibility: The Social Market Economy in Germany, 1818–1963*. Oxford: Clarendon Press, 1994.

Nickerson, John W. "How Should Labor Participate in Gains through Technological Improvements." *Advanced Management* (June 1952): 2–7.

Noble, David F. *The Forces of Production: A Social History of Automation*. New York: Knopf, 1984.

Nolan, Mary. *The Transatlantic Century: Europe and America, 1890–2010*. Cambridge: Cambridge University Press, 2012.

Nolan, Mary. "'Varieties of Capitalism' und Versionen der Amerikanisierung." In *Gibt es einen deutschen Kapitalismus? Tradition und globale Perspektiven der sozialen Marktwirtschaft*, edited by Volker Berghahn and Sigurt Vitols, 96–112. Frankfurt: Campus Verlag, 2006.

Nolan, Mary. *Visions of Modernity: American Business and the Modernization of Germany*. New York: Oxford University Press, 1994.

Oldenziel, Ruth, and Mikael Hård. *Consumers, Tinkerers, Rebels: The People Who Shaped Europe*. Basingstoke, UK: Palgrave Macmillan, 2013.

Oldenziel, Ruth, and Karin Zachmann, eds. *Cold War Kitchen: Americanization, Technology, and European Users*. Cambridge, MA: MIT Press, 2009.

Paju, Petri, and Thomas Haigh. "IBM Rebuilds Europe: The Curious Case of the Transnational Typewriter." *Enterprise and Society* 17, no. 2 (2016): 265–300.

Pang, Alex Soojung-Kim. *The Distraction Addiction: Getting the Information You Need and the Communication You Want, without Enraging Your Family, Annoying Your Colleagues, and Destroying Your Soul*. New York: Little, Brown and Co., 2014.

Parsons, Talcott. *Essays in Sociological Theory Pure and Applied*. Glencoe, IL: Free Press, 1949.

Pells, Richard. *Not Like Us: How Europeans Have Loved, Hated, and Transformed American Culture since World War II*. New York: Basic Books, 1997.

Peters, Benjamin. *How Not to Network a Nation: The Uneasy History of the Soviet Internet*. Cambridge, MA: MIT Press, 2016.

Petzold, Hartmut. *Rechnende Maschinen: Eine historische Untersuchung ihre Herstellung und Anwendung vom Kaiserreich bis zur Bundesrepublik*. Düsseldorf: VDI-Verlag, 1985.

Pinch, Trevor J., and Wiebe E. Bijker. "The Social Construction of Facts and Artifacts: Or How the Sociology of Science and the Sociology of Technology Might Benefit Each Other." In *The Social Construction of Technological Systems: New Directions in the Sociology and History of Technology*, edited by Wiebe E. Bijker, Thomas P. Hughes, and Trevor J. Pinch, 17–50. Cambridge, MA: MIT Press, 1997.

Pohl, Manfred. "Geschichte der Rationalisierung: Das RKW 1921 bis 1996." In *Rationalisierung sichert Zukunft: 75 Jahre RKW*, edited by Rationalisierungskuratorium der Deutschen Wirtschaft, 85–115. Frankfurt: Rationalisierungskuratorium der Deutschen Wirtschaft, 1996. https://www.rkw-kompetenzzentrum.de/das-rkw/das-rkw/die-geschichte-des-rkw.

Poiger, Uta G. *Jazz, Rock, and Rebels: Cold War Politics and American Culture in a Divided Germany*. Berkeley: University of California Press, 2000.

Porter, Russell. "Menzies in US for Talks on Asia: Warns on Reds' Divisive Tactics." *New York Times*, March 6, 1955, 1, 4.

Porter, Theodore. *Trust in Numbers: The Pursuit of Objectivity in Science and Public Life*. Princeton, NJ: Princeton University Press, 1995.

Powers, William. *Hamlet's Blackberry: A Practical Philosophy for Building a Good Life in the Digital Age*. Melbourne: Scribe, 2013.

"Presto! And a Block of Metal Becomes a V-8 Engine." *New York Times*, January 4, 1954, 58.

Price, Harry B. *The Marshall Plan and Its Meaning*. Ithaca, NY: Cornell University Press, 1955.

"Productivity Plan Modified by ECA." *New York Times*, September 5, 1951, 14.

Pugh, Emerson W. *Building IBM: Shaping an Industry and Its Technology*. Cambridge, MA: MIT Press, 1995.

Radkau, Joachim. *Technik in Deutschland: Vom 18. Jahrhundert bis heute*. Frankfurt: Campus Verlag, 2008.

"Railroad—What Technological Change Can Mean." *Washington Window*, October 8, 1955.

Randall, Clarence B. "Employer-Employee Relations in the United States." In *Proceedings of the First International Conference of Manufacturers, December 3–5, 1951*, edited by National Association of Manufacturers, 132–138. New York: National Association of Manufacturers, 1952.

Rationalisierungs-Kuratorium der Deutschen Wirtschaft. *Gruppenarbeit und Produktivität: Bericht über eine Studienreise in USA*. Heft 72. Munich: Carl Hanser Verlag, 1958.

Rationalisierungs-Kuratorium der Deutschen Wirtschaft. *Human Relations in Industry—Die menschlichen Beziehungen in der Industrie: Beobachtungen einer deutschen Studiengruppe in USA*. Heft 41. Munich: Carl Hanser Verlag, 1956.

Rationalisierungs-Kuratorium der Deutschen Wirtschaft. *Internationaler Erfahrungsaustausch: Zusammenstellung der Studienreisen und deren Teilnehmer im Bereich der gewerblichen Wirtschaft*. Berlin: Beuth-Vertrieb, 1957.

Rationalisierungs-Kuratorium der Deutschen Wirtschaft. *Jahresbericht 1953/54 des Rationalisierungs-Kuratoriums der Deutschen Wirtschaft.* Frankfurt: Rationalisierungs-kuratorium der Deutschen Wirtschaft, 1954.

Rationalisierungs-Kuratorium der Deutschen Wirtschaft. *Produktivität in USA: Einige Eindrücke einer deutschen Studiengruppe von einer Reise durch USA. Heft 20.* Munich: Carl Hanser Verlag, 1953.

Rationalisierungs-Kuratorium der Deutschen Wirtschaft. *Produktivität und Fertigung: Eindrücke einer deutschen Studiengruppe der Werkzeugindustrie von einer Reise durch USA. Heft 31.* Munich: Carl Hanser Verlag, 1955.

Reed, Philip D. "The Challenge of This Conference." In *Proceedings of the First International Conference of Manufacturers, December 3–5, 1951,* edited by National Association of Manufacturers, 211–218. New York: National Association of Manufacturers, 1952.

Reich, Simon, and Lawrence Dowler. *Research Findings about Ford-Werke under the Nazi Regime.* Dearborn, MI: Ford Motor Company, 2001.

"Reuther Assails Robot-Job Trend: CIO Chief Says President's Economic Council Ignores Impact of 'Automation.'" *New York Times,* February 11, 1955, 46.

"Reuther Insists Weeks Apologize: Assails Secretary for Charge of Automation Scare." *New York Times,* March 13, 1955, 77.

Ridgeway, George L. *Merchants of Peace: The History of the International Chamber of Commerce.* Boston: Little, Brown, 1959.

Ridgeway, George L. *Merchants of Peace: Twenty Years of Business Diplomacy through the International Chamber of Commerce, 1919–1938.* New York: Columbia University Press, 1938.

Riebensahm, Paul. *Der Zug nach U.S.A.: Gedanken nach einer Amerika-Reise 1924.* Berlin: Springer, 1925.

Rieppel, Paul. *Ford-Betriebe und Ford-Methoden.* Munich: Oldenbourg, 1925.

Rifkin, Jeremy. *The End of Work: The Decline of the Global Labor Force and the Dawn of the Post-Market Era.* New York: Penguin, 1995.

Rockman, Seth. "What Makes the History of Capitalism Newsworth?" *Journal of the Early Republic* 34, no. 4 (2014): 439–466.

Rodgers, Daniel T. *Atlantic Crossings: Social Politics in a Progressive Age.* Cambridge, MA: Belknap Press, 1998.

Rogers, Joel, and Wolfgang Streeck, eds. *Works Councils: Consultation, Representation, and Cooperation in Industrial Relations.* Chicago: University of Chicago Press, 1995.

Romero, Federico. *The United States and the European Trade Union Movement, 1944–1951*. Chapel Hill: University of North Carolina Press, 1992. First published 1989.

Rosenberg, Emily S. *Spreading the American Dream: American Economic and Cultural Expansion, 1890–1945*. New York: Hill and Wang, 1982.

Rothberg, Herman J. *A Case Study of a Large Mechanized Bakery*. Washington, DC: US Bureau of Labor Statistics, 1956.

Rusk, Howard A. "Industry Grows in Humanity and Discovers That It Pays." *New York Times*, January 5, 1953, 49, 62.

Sachse, Carola. *Betriebliche Sozialpolitik als Familienpolitik in der Weimarer Republik und im Nationalisozialismus: Mit einer Fallstudie über die Firma Siemens, Berlin*. Hamburg: Forschungsberichte des Hamburger Instituts für Sozialforschung, 1987.

Salin, Edgar. "Drei Schritte, die die Welt verändern." *ZEIT*, February 9, 16, 23, 1956.

Salomon, Alice. *Character Is Destiny: The Autobiography of Alice Salomon*. Ann Arbor: University of Michigan Press, 2004. First published 1924.

Salomon, Alice. *Kultur im Werden: Amerikanische Reiseeindrücke*. Berlin: Ullstein, 1924.

Sanford, William F., Jr. *The American Business Community and the European Recovery Program, 1947–1952*. New York: Garland Publishing, 1987.

Schelsky, Helmut. *Wandlungen der deutschen Familie in der Gegenwart*. Dortmund: Andrey Verlag, 1953.

Scheuch, Erwin K. "Continuity and Change in German Social Structure: Germany: An Enigma?" *Historical Social Research* 13, no. 2 (1988): 31–121.

Scheuch, Erwin K., and Dietrich Rüschemeyer. "Scaling Social Status in Western Germany." *British Journal of Sociology* 11, no. 2 (1960): 151–168.

Schivelbusch, Wolfgang. *Three New Deals: Reflections on Roosevelt's America, Mussolini's Italy, and Hitler's Germany, 1933–1939*. New York: Picador, 2007.

Schlombs, Corinna. "Built on the Hands of Women: Gender and Office Automation in the West German Banking and Insurance Industries." *Technology and Culture*, forthcoming.

Schlombs, Corinna. "A Cost-Saving Machine: Computing at the German Allianz Insurance Company." *Information and Culture* 52, no. 1 (2017): 31–63.

Schlombs, Corinna. "Engineering International Expansion: IBM and Remington Rand in European Computer Markets." *IEEE Annals of the History of Computing* 30, no. 4 (2008): 42–58.

Schlombs, Corinna. "The 'IBM Family': American Welfare Capitalism, Labor, and Gender in Postwar Germany." *IEEE Annals of the History of Computing* 39, no. 4 (2017): 12–26.

Schlombs, Corinna. "Toward International Computing History." *IEEE Annals for the History of Computing* 28, no. 1 (2006): 107–108.

Schlombs, Corinna. "Univac—Ein internationales Meisterwerk in der Computerge-schichte." *Kultur und Technik* 32, no. 1 (2008): 24–29.

Schot, Johan, and Phil Scranton, eds. *Making Europe: Technology and Transformations, 1850–2000.* 6 vols. Basingstoke, UK: Palgrave Macmillan, 2013–2018.

Seeling, Otto. "Was mir in Amerika aufgefallen ist." *Frankfurter Allgemeine Zeitung,* June 24, 1952.

"Senate Aid Bill Asks Ban on Red Unions." *New York Times,* September 2, 1951, 25.

Sewell, William H., Jr. "The Capitalist Epoch." *Social Science History* 38, no. 1–2 (2014): 1–11.

Shearer, J. Ronald. "The Politics of Industrial Efficiency in the Weimar Republic: Technological Innovation, Economic Efficiency, and Their Social Consequences in the Ruhr Coal Mining Industry, 1918–1929." PhD diss., University of Pennsylvania, 1989.

Shearer, J. Ronald "The Reichskuratorium für Wirtschaftlichkeit: Fordism and Orga-nized Capitalism in Germany, 1918–1945." *Business History Review* 71 (Winter 1997): 569–602.

Shugart, Kelly. *Your Place in the Atom (Concerning Musicians).* Los Angeles: Los Ange-les Musicians Union, Local 47, American Federation of Musicians, n.d.

"Sidelights on the Financial and Business Developments of the Day: Look, No Hands!" *New York Times,* May 13, 1952, 33.

Silvey, Ted F. "A Ball-Point Pencil—Is a Computer?" *Computers and Automation* 7, no. 2 (1958): 19–20.

Silvey, Ted F., and James X. Ryan. "Labor and Management: How They Look at Automation." *Tool Engineer* (1956): 73–79.

Smith, Merritt Roe. *Harpers Ferry Armory and the New Technology: The Challenge of Change.* Ithaca, NY: Cornell University Press, 1977.

Sombart, Werner. *Why Is There No Socialism in the United States?* London: Macmillan, 1976. First published 1906.

Spender, J.-C., and Hugo J. Kijne, eds. *Scientific Management: Frederick Taylor's Gift to the World?* Boston: Kluwer, 1991.

Spicka, Mark E. *Selling the Economic Miracle: Economic Reconstruction and Politics in West Germany, 1949–1957*. New York: Berghahn Books, 2007.

Spoerer, Mark. "Demontage eines Mythos? Zu der Kontroverse über das national-sozialistische 'Wirtschaftswunder.'" *Geschichte und Gesellschaft* 31, no. 3 (2005): 415–438.

Stahlmann, Michael. *Die Erste Revolution in der Autoindustrie: Management und Arbeitspolitik von 1900–1940*. Frankfurt: Campus Verlag, 1993.

Stapleford, Thomas A. *The Cost of Living in America: A Political History of Economic Statistics, 1880–2000*. Cambridge: Cambridge University Press, 2009.

Stapleton, Darwin H. *The Transfer of Early Industrial Technologies to America*. Philadelphia: American Philosophical Society, 1987.

Stebenne, David L. "IBM's 'New Deal': Employment Policies of the International Business Machines Corporation, 1933–1956." *Journal of the Historical Society* 5, no. 1 (2005): 47–77.

Stewart, Ethelbert. "Standardization of Output by Agreement." *Monthly Labor Review* 13, no. 2 (August 1921): 263–264.

Stollberg, Gunnar. *Die Rationalisierungsdebatte, 1908–1933: Freie Gewerkschaften zwischen Mitwirkung und Gegenwehr*. Frankfurt: Campus Verlag, 1981.

Strasser, Susan. *Satisfaction Guaranteed: The Making of the American Mass Market*. Washington, DC: Smithsonian Institution Press, 1995.

Swedish Employers' Association. "Employee Relations Policies: European Report No. 3." In *Proceedings of the First International Conference of Manufacturers, December 3–5, 1951*, edited by National Association of Manufacturers, 353–369. New York: National Association of Manufacturers, 1952.

Tänzler, Fritz. *Aus dem Arbeitsleben Amerikas: Arbeitsverhältnisse, Arbeitsmethoden und Sozialpolitik in den Vereinigten Staaten von Amerika*. Berlin: Reimar Hobbing, 1927.

Teplow, Leo. *Where Will Industry-Wide Bargaining Lead Us? Address by Leo Teplow, Associate Director, Industrial Relations Division, NAM, before Amherst College, Amherst, Massachusetts, April 28, 1950*. New York: National Association of Manufacturers, 1950.

Teupe, Sebastian. "Verhandelte Grenzüberschreitungen. Die Entwicklungstrends der deutschen Kartellrechtspraxis am Beispiel der Fernsehgeräteindustrie, 1950–1990." In *Tatort Unternehmen: Zur Geschichte der Wirtschaftskriminalität im 20. und 21. Jahrhundert*, edited by Hartmut Berghoff, Cornelia Rauh, and Thomas Welskopp, 101–127. Berlin: De Gruyter Oldenbourg, 2016.

Thelen, Kathleen A. *Union of Parts: Labor Politics in Postwar Germany.* Ithaca, NY: Cornell University Press, 1991.

"They See America." *Studebaker Spotlight* 17, no. 5 (June 1953): 3–6.

Tinn, Honghong. "Modeling Computers and Computer Models: Manufacturing Economic-Planning Projects in Cold War Taiwan, 1959–1968." *Technology and Culture* 59, no. 4 supplement (2018): S66–S99.

UAW-CIO Education Department. *Automation: A Report to the UAW-CIO Economic and Collective Bargaining Conference, November 12–13, 1954.* Detroit, Michigan: CIO, 1955.

US Bureau of Labor Statistics. "Employment Status of the Civilian Noninstitutional Population, 1940 to Date." Washington, DC, 2017. http://www.bls.gov/cps/cpsaat1.pdf.

US Congress, Joint Committee on the Economic Report. *Automation and Technological Change: Hearings before the Subcommittee on Economic Stabilization of the Joint Committee on the Economic Report.* Washington, DC: US Government Printing Office, 1955.

US Department of Labor, Bureau of Labor Statistics. *Gray Iron Foundries: Case Study Data on Productivity and Factory Performance.* Washington, DC, August 1951.

US Department of Labor, Bureau of Labor Statistics. *Men's Goodyear Welt Dress Shoes: Case Study Data on Productivity and Factory Performance.* Washington, DC, 1951.

US Department of Labor, Bureau of Labor Statistics. *Men's Work Pants: Case Study Data on Productivity and Factory Performance.* Washington, DC, 1951.

US Department of Labor, Bureau of Labor Statistics. *Radio and Television Manufacturing: Case Study Data on Productivity and Factory Performance.* Washington, DC, 1952.

Usselman, Steven W. "IBM and Its Imitators: Organizational Capabilities and the Emergence of the International Computer Industry." *Business and Economic History* 22, no. 2 (1993): 1–35.

Van Auken, Kenneth G. *The Introduction of an Electronic Computer in a Large Insurance Company.* Washington, DC: US Bureau of Labor Statistics, 1955.

Van Hook, James C. *Rebuilding Germany: The Creation of the Social Market Economy, 1945–1957.* Cambridge: Cambridge University Press, 2004.

Van Vleck, Jenifer. *Empire of the Air: Aviation and the American Ascendancy.* Cambridge, MA: Harvard University Press, 2013.

von Freyberg, Thomas. *Industrielle Rationalisierung in der Weimarer Republik: Untersucht an Beispielen aus dem Maschinenbau und der Elektroindustrie.* Frankfurt: Campus Verlag, 1989.

von Gottl-Ottlilienfeld, Friedrich. *Fordismus: Über Industrie und Technische Vernunft. Dritte, um den Abdruck verwandter Arbeiten erweiterte Auflage.* Jena: Gustav Fischer, 1926.

von Neumann, John. "First Draft of a Report on EDVAC." 1945. http://www.wiley. com/legacy/wileychi/wang_archi/supp/appendix_a.pdf.

von Saldern, Adelheid, and Rüdiger Hachtmann. "Das fordistische Jahrhundert: Eine Einleitung." *Zeithistorische Forschungen* 6, no. 2 (2009): 174–185.

Waring, Stephen P. *Taylorism Transformed: Scientific Management Theory since 1945.* Chapel Hill: University of North Carolina Press, 1991.

Wasser, Solidelle F., and Michael. L. Dolfman. "BLS and the Marshall Plan: The Forgotten Story." *Monthly Labor Review* 128, no. 6 (2005): 44–52.

Watson, Thomas J., Jr., and Peter Petre. *Father, Son & Co: My Life at IBM and Beyond.* New York: Bantam, 1990.

Weinberg, Edgar. *A Case Study of a Company Manufacturing Electronic Equipment.* Washington, DC: US Bureau of Labor Statistics, 1955.

Weinberg, Edgar, and Edward Jakubauskas. *A Case Study of an Automatic Airline Reservation System.* Washington, DC: US Bureau of Labor Statistics, 1958.

Weinberg, Edgar, and Herman Rothberg. *A Case Study of a Modernized Petroleum Refinery.* Washington, DC: US Bureau of Labor Statistics, 1957.

Weiss, Jessica. *To Have and to Hold: Marriage, the Baby Boom, and Social Change.* Chicago: University of Chicago Press, 2000.

Welch, David. "Nazi Propaganda and the Volksgemeinschaft: Constructing a People's Community." *Journal of Contemporary History* 39, no. 2 (2004): 213–238.

Wells, Wyatt. *Antitrust and the Formation of the Postwar World.* New York: Columbia University Press, 2002.

Westad, Odd Arne. "Bernath Lecture: The New International History of the Cold War: Three (Possible) Paradigms." *Diplomatic History* 24, no. 4 (2000): 551–565.

Westermann, Franz. *Amerika wie ich es sah: Reiseskizzen eines Ingenieurs, 2. Verbesserte Auflage.* Halberstadt: H. Meyer's Buchdruckerei, 1926.

Wiener, Norbert. *Cybernetics, or, Control and Communication in the Animal and the Machine.* New York: John Wiley, 1948.

White, Donald W. *The American Century: The Rise and Decline of the United States as a World Power.* New Haven, CT: Yale University Press, 1996.

Wilkins, Mira. *The Emergence of Multinational Enterprise: American Business Abroad from the Colonial Era to 1914.* Cambridge, MA: Harvard University Press, 1981.

Wilkins, Mira. *The Maturing of Multinational Enterprise: American Business Abroad from 1914 to 1970*. Cambridge, MA: Harvard University Press, 1974.

Wilkins, Mira, and Frank E. Hill. *American Business Abroad: Ford on Six Continents*. Detroit: Wayne State University Press, 1964.

Winkler, Jonathan R. "Technology and the Environment in the Global Economy." In *America in the World: The Historiography of American Foreign Relations since 1941*, edited by Frank Costigliola and Michael J. Hogan, 284–306. 2nd ed. Cambridge: Cambridge University Press, 2014.

Winner, Langdon. "Do Artifacts Have Politics?" *Daedalus* 109, no. 1 (1980): 121–136.

Woitschach, Max. "Kommt das automatische Büro?" *IBM Nachrichten* 126 (May 1956): 320–330.

Woitschach, Max. "Können Maschinen denken? Elektronisches Rechnen und Entscheiden." *IBM Nachrichten* 123 (July 1955): 216–241.

World Inequality Lab. *World Inequality Report 2018: Executive Summary*. https://wir2018.wid.world/files/download/wir2018-summary-english.pdf.

Wright, Carroll D. *Hand and Machine Labor: 13th Annual Report of the Commissioner of Labor, 1898*. Washington, DC: Government Printing Office, 1899.

Yale University. *Man and Automation: Report of the Proceedings of a Conference Sponsored by the Society of Applied Anthropology at Yale University, December 27, 28, 1955*. New Haven, CT: Yale University, 1956.

Zeitlin, Jonathan, and Gary Herrigel, eds. *Americanization and Its Limits: Reworking US Technology and Management in Post-War Europe and Japan*. New York: Oxford University Press, 2000.

Zeller, Thomas. *Driving Germany: The Landscape of the German Autobahn, 1930–1970*. New York: Berghahn Books, 2007.

zu Castell, Adelheid. "Die demographischen Konsequenzen des Ersten und Zweiten Weltkrieges für das Deutsche Reich, die Deutsche Demokratische Republik und die Bundesrepublik Deutschland." In *Zweiter Weltkrieg und sozialer Wandel: Achsenmächte und besetzte Länder*, edited by Waclaw Dlugoborski, 117–137. Göttingen: Vandenhoeck, 1981.

Index

Note: Page numbers followed by *f* or *t* refer to figures and tables, respectively.

Abel, I. W., 234
Abelshauser, Werner, 172
Adenauer, Konrad, 176–177, 179
ADGB. *See* General German Union Association
Advisory Group on European Productivity (AGEP), 88–89, 95, 100, 134. *See also* Marshall Plan Productivity Program
AEG-Telefunken, 9
AFL. *See* American Federation of Labor (AFL)
AGEP. *See* Advisory Group on European Productivity (AGEP)
Allgemeiner Deutscher Gewerkschaftsbund (ADGB). *See* General German Union Association
Allianz insurance company, 68, 243–245
Allied Control Council, 173
American Century, 2, 249, 253
American exceptionalism, 101, 253–254
American Federationist, 15–16, 35
American Federation of Labor (AFL), 35, 81, 146
  automation and, 227, 237
  codetermination and, 187
  international union organization and, 123

labor advisors for Productivity Program, 109, 122–126
Americanization, 74, 80, 285n57, 286–287n10
American Statistical Society, 85
Anglo-American Council on Productivity, 83, 86
Automation, computing technology, 105–106
Automation—Benefit or Danger? conference, Germany, 239–240
  BLS case studies, 233
  codetermination and, 191
  constructionist approach, 4–5, 8–9
  defined, 220
  DGB and, 221, 238–240, 242
  economic trends, unemployment and, 246–247
  Germany, unions' response, 238–243
  history of computing, 5–6
  IBM market share, 247–248
  labor in US, 229–233
  labor relations and, 226–229
  Reuther on, 232–239
  Silvey's ballpoint pen campaign, 219–220
  Tinkertoy project, 231
  transatlantic dialogue, 226–229, 245–246

Automation, computing technology
  (cont.)
  transfer modes, 9–10
  unions' reforms and, 22
  US and German public debate,
    221–222
  US unions response, guaranteed
    annual wage, 233–239
  West German debate, 222–226
  worker productivity and, 1–2

Balke, Siegfried, 225–226
Beckmann, Olaf, 155–156
Bellows, Everett H., 77–79, 105
Benton, William, 93, 105
Benton Amendment, 93. See also Mar-
  shall Plan Productivity Program
Berg, Fritz, 225
Berger, W., 209–211
Berkeley, Edmund, 244
Bernstein, Meyer, 127–128, 188
Bier, Willi, 162
Bissell, Richard, 113, 114, 227
BLS. See US Bureau of Labor Statistics
  (BLS)
Bonn, Moritz, 59
Borsdorf, Johannes H., 204, 207, 212,
  213–217
Bosch, 231
Brenner, Otto, 242–243
Bundesverband der deutschen Industrie
  (BDI), 225
Bunting, Earl, 180, 188, 231
Business Advisory Council, 113
Business Week, 180

Capitalism, 11–12. See also Varieties of
  capitalism
  American system, 89–92
  automation and, 234, 239
  Berger on, 210–211
  Cold War and capitalist "West," 10–
    11, 172, 193, 249

rationale, economic growth, 1,
    254–255
Carey, James, 124, 234
Carl Duisberg-Gesellschaft, 168
Cartels, 250
  codetermination and, 171
  demonstration projects and, 77, 95, 98
  free enterprise and, 91, 119, 121
  occupation forces in Germany and,
    8, 173
  Productivity Program and, 84, 93, 107
  social market economy and, 172, 174,
    176–177
  US and European entrepreneurs on,
    Project Impact and, 117–118
Catholic social ethic, 147, 173, 179,
    206, 211
CDU. See Christian Democratic Union
  (CDU)
Census of Manufactures, 24, 30
Central Working Group (Zentralarbeits-
  gemeinschaft), 178
Chauncey, Harrison, 195, 209
Christian Democratic Union (CDU),
    173, 176, 209, 225
Christian labor unions, German, 145–
    148, 197, 206, 209, 211
Christian Leadership Movement,
    102–103
CIO. See Congress of Industrial Organi-
  zations (CIO)
Citroen, 228
Clague, Ewan, 13–14, 15–16, 27–40,
    77, 85
Clark, Myron, 102
Clark, Wallace, 95–96
Class relations, 4, 10, 110, 251
  American views on, in Germany, 145,
    148
  comparative living standards and,
    66–67
  German students in United States and,
    136–137

German unions and, 146–148
German views on, in United States,
55–57, 63, 65–69, 75, 153–155,
163–165
IBM and, 197, 212–214, 216–218
industrialists' views on, Project Impact
and, 117, 120, 151
in Nazi Germany, 70–71, 73
prestige as marker of, status symbols,
148–150, 213
Productivity Program and, 110, 127,
137, 252
sociological analysis of, in Germany,
150–151
in United States, 11, 69, 253
Codetermination, 3, 141, 165, 172, 192,
209, 217
American business, unions, and government on, 169–170, 181–188
automation and, 221, 238, 242
collaborative labor relations versus, 4,
10, 169–172
corporatism and, 188–191
debate, Germany, 179–180, 188
definition, Weimar origins, 178–179
occupation forces in Germany and,
173
Collaborative labor relations, 92–93,
110, 127–128, 130–133, 137–138,
151–153, 158, 193
Collective bargaining, 40, 118, 121–122,
132–133
General Motors and United Auto
Workers, 40, 182, 235–236
industrial peace through, 127–132
NAM on, 182–183
plant-level versus industry-wide or
national, 141, 182, 183
Committee for Economic Development,
83, 113
Commons, John R., 27, 37–38
Communism, 77, 95–96, 100, 123, 147–
148, 152, 195

Computers and Automation, 219
Computing technology. See Automation, computing technology
Congress of Industrial Organizations
(CIO), 81, 146
automation and, 219–220, 227, 233–
234, 236–237, 246–247
codetermination and, 186–187
international union organization and,
123, 126
labor advisors for Productivity Program, 109, 121–123, 125–126, 152
Marshall Plan and, 123–125
Conway, John, 227
Coordinated market economy. See Varieties of capitalism
Corporate governance. See Codetermination; Collaborative labor relations;
Collective bargaining; Industrial
democracy; Welfare capitalism
Corporate Governance Law (Betriebsverfassungsgesetz), 189
Cortada, James, 6
Cost-of-living index, 14–15, 17–18, 28,
35–36, 40–41, 85
Crown Zellerbach Corporation, 131
Cruikshank, Nelson, 102–103, 153, 167

Dahrendorf, Ralf, 150
Davis, James, 69
Dawes Plan (1924), 49, 56
Demonstration projects, 77–78, 94–101,
105–106, 141–142, 144–145. See also
Marshall Plan Productivity Program
Der Spiegel, 224, 229
Desk Set, 238
DGB (Deutscher Gewerkschaftsbund),
147–148, 157, 169, 195
automation and, 140, 221, 238–240,
242
codetermination and, 178, 185, 189
Productivity Program and, 140–141
Diebold, John, 220–221, 228

Digital Flood (Cortada), 6
DINTA. *See* German Institute for Technical Vocational Training (DINTA)
Dorndorf Schuhfabrik, 98–99, 149*f*, 162
Draper, William H., Jr, 104
Dynamics of Industrial Democracy, The (Golden and Ruttenberg), 121, 128–130

Ebert, Friedrich, 69
Economic growth. *See* Capitalism, rationale, economic growth
Economic planning, 58, 71, 126–127, 130–132, 170–171, 174, 193, 251–252
Efficiency, 8–9. *See also* Productivity
EPA. *See* European Productivity Agency (EPA)
Erhard, Ludwig, 170, 175–178, 225, 251–252
Eucken, Walter, 174–175
European Productivity Agency (EPA), 226–227
European Recovery Program. *See* Marshall Plan

Factory Performance Reports (BLS), 87–89
Fairless, Benjamin F., 233
Feenberg, Andrew, 4–5, 250
Feiler, Arthur, 46, 68
Filene, Edward, 44, 49–50
Filene Cooperative Association, 49–50
First International Conference of Manufacturers, 111–112, 113, 118–120, 133
Ford, Henry, 44, 51–54, 250
Ford Motor Company
  American economic system and, 51–54
  automation and, 220, 227–228, 236–237
  German perceptions of, 55–56, 60–61, 65, 67–68, 159, 164

international operations, 48–49
labor relations and salaries, 53, 67–68
Nazi regime and, 70, 73–74
production processes, 52–54
Productivity Program and, 133
welfare capitalism and, 44–45, 54
Foster, William C., 114
Free enterprise, 91–92, 117–121, 170, 182, 192–193

Galantiere, Lewis, 180–181
General German Union Association, 62, 162
  organization of German unions, pre-1945, 146–147
  positive view of rationalization technology, 140
  report on US visit, 43–44, 62–70, 75, 254
General Motors, 40, 182, 235
German Institute for Technical Vocational Training (DINTA), 57, 74, 142
German Labor Front (Deutsche Arbeitsfront, DAF), 71–72
German transatlantic visitors. *See also* General German Union Association; Köttgen, Carl; Study groups on productivity; Work-Study Training for Productivity (WSTP)
  from German Empire, 253
  through Marshall Plan Productivity Program, 98, 112–115, 135–136, 166–167, 240–241
  Nazi, 70
  Weimar, 55–56, 59–66, 68, 76
Giant Brains (Berkeley), 244
Golden, Clinton S., 110, 121–125, 127–132, 187, 191, 237

Hamann, C. F., 103
Hand and Machine Labor (BLS), 18–21
Harlan, G., Jr., 228
Harriman, Averell, 113–114

Harriman Report, 83, 124
Haynes, Eldridge, 169–170, 181–182,
   184–185, 190, 192–193
Hensel, Rudolf, 68
Hinsch, Walter, 160–161
Hirsch-Dunckersche Vereine, 146–148
Hitler, Adolf, 50, 70, 72–73. *See also*
   Rationalization
Hoffman, Paul G., 82–83, 93–94, 111,
   113–114, 136
Homann Stove Works, 99–100
Hörrmann, Oskar, 204, 207, 213–217
Hughes, Thomas, 6
Human Use of Human Beings (Wiener),
   229

IBM. *See* International Business
   Machines Corporation (IBM)
IBM Deutschland, 207, 214
IBM Nachrichten, 244
IBM World Trade Corporation, 51, 195,
   201, 203
ICC. *See* International Chamber of
   Commerce (ICC)
IG Metall, 99, 242
Imperial Committee for the Determina-
   tion of Working Hours (REFA), 58,
   74, 142, 163, 301n43
Industrial democracy, 58, 80–81, 121,
   128–130
Innocent Upsetter, The, 128
International Business Machines Corpo-
   ration (IBM), 50, 90, 106, 111–112,
   114–115, 133
   automation and, 243–244
   class elimination, Germany, 212–217
   German Works Council, 208–212
   IBM 650, 228, 247
   international operations, 48, 200–203
   progressive corporate culture, US,
      197–200
   Watson's IBM family, Germany,
      202–208

welfare capitalist labor relations, 196
   women's careers at, 199
International Chamber of Commerce
   (ICC), 49–50, 198–199
International Christian Leadership Con-
   ference, 102, 103
International Congress of Free Trade
   Unions (ICFTU), 126–127
International Council for Youth Self-
   Help (Internationaler Rat für Jugend-
   selbsthilfe), 153

Jacobsson, Carl, 119
Jewell, Bert, 122, 125
Jones, Thomas R., 116–117, 120
Joyce, William H., 104, 105, 120

Karp (DGB delegate), 157
Kennedy, John F., 247
Köttgen, Carl, 43, 46, 59–63, 142
Krumm, Heinrich, 162–164

Labor productivity index of eleven US
   industries, 29–32, 33t, 34
Labor relations. *See* Collaborative labor
   relations; Industrial democracy
LaFeber, Walter, 6, 252
League of Nations, 49
Liberal market economy. *See* Varieties of
   capitalism
Life, 2
Living wage, 22–23, 39
Lubin, Isador, 85
Luce, Henry, 2, 249

Mahder, Carl R., 96, 100, 144
Maier, Charles, 110, 192, 251
Market economies, liberal versus
   coordinated. *See* Varieties of
   capitalism
Marshall, George, 82, 104, 113
Marshall Plan, 2–3, 7–8, 110–111
   administration of, 82–83, 100–101

Marshall Plan (cont.)
    US business community and, 113–114,
        116
    US labor unions and, 123–127
Marshall Plan Productivity Program,
    3–4, 7, 77–80, 82–84. *See also* Advisory Group on European Productivity (AGEP); Demonstration projects;
    Project Impact; Study groups on productivity; Work-Study Training for
    Productivity (WSTP)
    American productivity values, 89–94
    Benton Amendment, 93
    consumerist labor policies and, 80–82
    exchange programs, 135 (*see also* German transatlantic visitors)
    exchange students, German Trade
        Unions, 156–158
    German labor and class relations and,
        145–156
    German perceptions, 135–139
    in Germany, 139–145
    methods, demonstration projects,
        94–101
    productivity definition, measurement,
        cooperation with US Bureau of Labor
        Statistics, 84–89
    Project Impact, 110–117
    public-private partnerships and, 109–
        110, 132
    RKW administration of, 139, 142–145
    US labor unions and, 121–127
    US productivity officers as Marshall
        missionaries, 101–105
McCloy, John, 186–188
McLean, Arthur, 103
Meany, George, 124
Metropolitan Life Insurance, 227–228
Michler, Gordon, 169–170, 181, 184–
    185, 192–193
Mitbestimmung, 179. *See also*
    Codetermination
Monthly Labor Review, 28, 34

Morgan, J. P., 49
Mosher, Ira, 118
Müller-Armack, Alfred, 174–175
Müller-Lutz, Heinz-Leo, 244–245
Murray, Philip, 124, 127, 186–187
My Life and Work (Ford), 51–54

NAM. *See* National Association of Manufacturers (NAM)
Naphtali, Fritz, 58, 179
National Association of Manufacturers
    (NAM), 92
    automation and, 231
    codetermination and, 169–170, 180–
        181, 185, 188–189
    Marshall Plan and, 115–116
    Productivity Program, 109
    Project Impact, 111
National Cash Register (NCR), 198–199
National Foreign Trade Council (NFTC),
    169–170, 181–185
National Labor Relations Act (Wagner
    Act), 81, 91–92, 117
National Management Council, 109,
    114, 190
National Planning Association, 131,
    171. *See also* Economic planning
Nazi government, 8
    class relations, Volksgemeinschaft,
        70–71, 74
    consumer program, Volksprodukte, 71
    Ford Motor Company and, 73–74
    German business community and,
        171, 173
    German labor unions and, 147, 171,
        173, 238, 242
    rationalization, technological unemployment, 57, 70, 74–76, 144, 223
    social market economists under,
        175–178
    Volkswagen, 71–73
    Watson, Thomas, and, 50
NCR. *See* National Cash Register (NCR)

New Deal, 81, 85, 117, 151, 170–171
New York Times, 179, 180, 229
NFTC. See National Foreign Trade Council (NFTC)
Nickerson, John W., 88, 100, 102, 134
Nixdorf, Heinz, 9

Opel, 73–74
Operation Bootstrap, 126
Organization for Economic Cooperation and Development (OECD), 168
Organization for European Economic Cooperation (OEEC), 8, 168, 219

Personality (Personalität), 147, 205–206, 211, 216. See also Catholic social ethic
Player Piano (Vonnegut), 229
Point Four Program, 255
Porter, Theodore, 16
Prestige, 213–214. See also Class relations
Productivity
BLS calculations and history, 13–17
critical views, US, 158–166
earnings, United States 1910–1940, 39f
history, literature, and research, 4–9
interpretive flexibility of, 3, 41–42, 249–255
measurements and methodological history, 15–16
social impact, 11
standard of living, economic growth and, 255–257
United States versus European countries, 107–111
wages and, 35–40
Productivity index, calculation, 30–35
Productivity Program. See Marshall Plan Productivity Program
Project Impact, 110–117. See also Marshall Plan Productivity Program

Race relations and class relations, 64–69
Rand, Marcel N., 111, 159
Randall, Clarence B., 118, 119–120
Rathenau, Walther, 58
Rationalisierungskuratorium der Deutschen Wirtschaft (RKW), 139, 142–145
Rationalization, 101. See also Köttgen, Carl
definition and history, 45, 55–59
German unions and, 64, 140–141, 192, 239
under Nazism, 70–74
post–World War II RKW and, 142–143
at Siemens, 60
unemployment and, 222–223
REFA. See Imperial Committee for the Determination of Working Hours (REFA)
Reichskuratorium für Wirtschaftlichkeit (Weimar RKW), 57–58, 59, 142, 144
Remington Rand, 106, 112, 114–115, 133, 158–159, 222, 225–226, 243
Rerum Novarum, 147, 211
Reuther, Walter, 124, 232–239
Riebensahm, Paul, 56, 59
RKW. See Rationalisierungskuratorium der Deutschen Wirtschaft (RKW); Reichskuratorium für Wirtschaftlichkeit (Weimar RKW)
Robinson, Moncure, 47
Rodnick, David, 103
Roosevelt, Franklin D., 81
Röpke, Wilhelm, 174–176
Rotary International, 48
Ruffin, William H., 183
Rüstow, Alexander, 174–176
Ruttenberg, Harold J., 121, 128–130

Salomon, Alice, 55, 68
Schacht, Hjalmar, 175
Schell, Herbert, 116
Schelsky, Helmut, 150

Scheuch, Erwin K., 150–151
Schilling, A., 59
Schmid, Carlo, 226
Schuman Plan, 179
Scientific management, 8–9, 22, 39, 45,
    57
Seeling, Otto, 158, 164–165
Seip, A. J., 166
Shugaar, Carl, 152
Siemens, 9, 43, 59–60
Silberman, James M., 85–87
Silvey, Ted F., 219–220, 234
Social Democratic Party (SPD), 69, 146–
    148, 177
Socialist trade union associations, Ger-
    many, 43–44, 146–148. *See also* DGB
    (Deutscher Gewerkschaftsbund);
    General German Union Association
Social market economy, Germany, 7,
    170–178, 190, 192, 251–252
Sombart, Werner, 11, 65, 253
Spinnerei und Weberei Wiesental, 98
Standard of living, 3, 6, 10, 71, 75, 110,
    154–155, 182, 221, 239, 244, 246,
    250–251
Sternberg, Fritz, 239–240
Stewart, Ethelbert, 27, 36
Studebaker, 82, 136–137, 275n15,
    294n82
Study groups on productivity, 285–
    286n1, 290n44. *See also* German
    transatlantic visitors; Marshall Plan
    Productivity Program
    arrangements in United States, "Wag-
    ing the Peace" brochure, 107–109,
    114–115
    automation and, 227, 240
    effectiveness of, 138, 153
    reports, examples, 159–161, 166–167

Taft-Hartley legislation, 92, 117, 151,
    235
Taylor, Frederick W., 8–9, 45, 57

Taylorism, 8–9. *See also* Scientific
    management
Third way, 170, 174
Transatlantic exchanges, history, 11–12,
    47–61. *See also* German transatlantic
    visitors
Truman, Harry S., 82, 255
Tuthill, John W., 145
Twentieth Century Fund, 50

Unemployment, technological, 10
    computing and, 221, 222–223, 228,
    244, 246
    de-skilling and, 22, 41
    German unions and, 239–241
    Nazi rise to power, 57
    US unions and, 234–238
United Auto Workers (UAW), 40,
    233–238
US Bureau of Labor Statistics (BLS), 3
    automation and, 233, 238
    common brick manufacturing, exam-
    ple, 24–26
    1898 technological change study, 17–
    19 (see also *Hand and Machine Labor*
    [BLS])
    labor productivity index of eleven US
    industries, 29–32, 33t, 34
    Marshall Plan Productivity Program,
    84–89
    Productivity and Technological Devel-
    opment Division, 85
    on productivity and wages, 35–36
    productivity field studies, early 1920s,
    21–24
    quantitative productivity measure-
    ments, late 1920s, productivity
    index, 27–35
    red brick manufacturing, example,
    19–21
US Chamber of Commerce, 29, 30, 92
US Department of Labor, 28, 238
US Steel, 40

Varieties of capitalism, 7–8, 172,
251–252
Vogel, Herbert, 240–241
Volksgemeinschaft, 70–73
Volkswagen, 71–74
von Gottl-Ottlilienfeld, Friedrich, 56, 57
von Moellendorf, Wichard, 58
Vonnegut, Kurt, 229
von Siemens, Carl Friedrich, 57, 58

Wagner Act. *See* National Labor Relations Act (Wagner Act)
Wall Street Journal, 169
Watson, Arthur K., 200, 212
Watson, Jeanette K., 201–204, 207
Watson, Thomas J., 50–51, 121, 195,
198–199, 202–207
Watson, Thomas J., Jr., 213, 244
Weimar RKW. *See* Reichskuratorium für Wirtschaftlichkeit (Weimar RKW)
Welfare capitalism, 250
definition, 44–45
Ford and, 51, 54
German views of, in United States, 60,
61, 65–66
in Germany, 99, 164
at IBM, in Germany, 196–197, 199,
203, 205–206, 208, 210, 214
US union views of, 129–130
Wessel, Hubertus (Bert), 138, 158–159,
165
Westermann, Franz, 46, 55–56, 59, 68
Wiener, Norbert, 224, 229, 234
Woitschach, Max, 244
Work-Study Training for Productivity
(WSTP), 135–140, 148–149, 151–158,
166, 167. *See also* German transatlantic visitors; Marshall Plan Productivity Program
World Federation of Trade Unions
(WFTU), 123, 126
"World Peace through World Trade"
(slogan), 50–51, 199, 201

Wright, Carroll, 17
WSTP. *See* Work-Study Training for Productivity (WSTP)

YMCA, 48

Zellerbach, James D., 113, 131
Zentralarbeitsgemeinschaft.
See Central Working Group
(Zentralarbeitsgemeinschaft)
Zulauf, Harold C., 146
Zuse, Konrad, 9

History of Computing
William Aspray and Thomas J. Misa, editors

Janet Abbate, *Recoding Gender: Women's Changing Participation in Computing*

John Agar, *The Government Machine: A Revolutionary History of the Computer*

William Aspray and Paul E. Ceruzzi, *The Internet and American Business*

William Aspray, *John von Neumann and the Origins of Modern Computing*

Charles J. Bashe, Lyle R. Johnson, John H. Palmer, and Emerson W. Pugh, *IBM's Early Computers*

Martin Campbell-Kelly, *From Airline Reservations to Sonic the Hedgehog: A History of the Software Industry*

Paul E. Ceruzzi, *A History of Modern Computing*

I. Bernard Cohen, *Howard Aiken: Portrait of a Computer Pioneer*

I. Bernard Cohen and Gregory W. Welch, editors, *Makin' Numbers: Howard Aiken and the Computer*

James Cortada, *IBM: The Rise and Fall and Reinvention of a Global Icon*

Nathan Ensmenger, *The Computer Boys Take Over: Computers, Programmers, and the Politics of Technical Expertise*

Thomas Haigh, Mark Priestley, and Crispin Rope, *ENIAC in Action: Making and Remaking the Modern Computer*

John Hendry, *Innovating for Failure: Government Policy and the Early British Computer Industry*

Mar Hicks, *Programmed Inequality: How Britain Discarded Women Technologists and Lost Its Edge in Computing*

Michael Lindgren, *Glory and Failure: The Difference Engines of Johann Müller, Charles Babbage, and Georg and Edvard Scheutz*

David E. Lundstrom, *A Few Good Men from Univac*

René Moreau, *The Computer Comes of Age: The People, the Hardware, and the Software*

Arthur L. Norberg, *Computers and Commerce: A Study of Technology and Management at Eckert-Mauchly Computer Company, Engineering Research Associates, and Remington Rand, 1946–1957*

Emerson W. Pugh, *Building IBM: Shaping an Industry and Its Technology*

Emerson W. Pugh, *Memories That Shaped an Industry*

Emerson W. Pugh, Lyle R. Johnson, and John H. Palmer, *IBM's Early Computers: A Technical History*

Kent C. Redmond and Thomas M. Smith, *From Whirlwind to MITRE: The R&D Story of the SAGE Air Defense Computer*

Alex Roland with Philip Shiman, *Strategic Computing: DARPA and the Quest for Machine Intelligence, 1983–1993*

Raúl Rojas and Ulf Hashagen, editors, *The First Computers—History and Architectures*

Corinna Schlombs, *Productivity Machines: German Appropriations of American Technology from Mass Production to Computer Automation*

Dinesh C. Sharma, *The Outsourcer: A Comprehensive History of India's IT Revolution*

Dorothy Stein, *Ada: A Life and a Legacy*

Christopher Tozzi, *For Fun and Profit: A History of the Free and Open Source Software Revolution*

John Vardalas, *The Computer Revolution in Canada: Building National Technological Competence, 1945–1980*

Maurice V. Wilkes, *Memoirs of a Computer Pioneer*

Jeffrey R. Yost, *Making IT Work: A History of the Computer Services Industry*

Printed in the United States
by Baker & Taylor Publisher Services